Satellite Communications

Satellite Communications: Principles and Applications

David Calcutt
Department of Electrical and Electronic Engineering, University of Portsmouth, Portsmouth, UK

Laurie Tetley
Principal Lecturer in Communications Electronic Engineering

Edward Arnold
A member of the Hodder Headline Group
LONDON MELBOURNE AUCKLAND

© 1994 David Calcutt and Laurie Tetley

First published in Great Britain 1994

British Library Cataloguing in Publication Data
Available on request

0340 61448 X

All rights reserved. No part of this publication may be reproduced or transmitted in any form or by any means, electronically or mechanically, including photocopying, recording or any information storage or retrieval system, without either prior permission in writing from the publisher or a licence permitting restricted copying. In the United Kingdom such licences are issued by the Copyright Licensing Agency: 90 Tottenham Court Road, London W1P 9HE.

Typeset in Great Britain by Computape (Pickering) Ltd. Printed in Great Britain for Edward Arnold, a division of Hodder Headline PLC, 338 Euston Road, London NW1 3BH by St. Edmundsbury Press, Bury St. Edmunds, Suffolk.

Preface

The advent of the satellite has revolutionized telecommunications by allowing, as it does, links to be established within the area covered by the satellite footprint (which for global coverage is typically over 40% of the surface of the earth). The satellite is particularly useful for establishing a link between areas which would be inaccessible, or at least difficult to access, by other means. The use of satellites however has brought with it new problems, such as: designing the communications sub-system to meet the requirements of the system; designing a structure which contains the communications and support sub-systems, which can be launched into the geo-stationary orbit; providing a control system which, among other things, allows the maintenance of station-keeping within fine limits for satisfactory operation.

There are constraints on the ability of the satellite to handle the traffic required by the system operators and the capacity of the satellite may be limited by the available power on the satellite or the bandwidth available for the service. Notwithstanding the problems, the communications satellite is providing the user with an effective means of transmission of telephony, video and data and is reaching a wider market than ever before. One of the factors which has brought satellite communications to the wider market has been the dramatic technological developments in the receiver noise figure at the frequencies used and this, together with high gain, means a ground station can be made available at a fraction of the cost of earlier systems. This is particularly true of stations designed to receive television signals by satellite, either directly in an individual home or by a cable network, making the system affordable by a single household.

In designing this book the authors have tried to set out the principles involved in satellite communications and to show, with reference to particular systems, how the principles are effected in practice. Because our aim is to show the communications aspect of the use of satellites, information on the launching of satellites, and the design problems associated with ensuring the structure can be accommodated within the launch vehicle, has been kept to a minimum.

The book is divided into three sections. Section one provides the principles of satellite communications and includes link parameters, multiple access techniques, modulation/demodulation methods, coding etc, and looks at earth stations and satellites in some detail. Section two looks at communications with mobile earth stations and deals with the Inmarsat systems, because of their important marine applications.

The various Inmarsat systems are explained together with details of the ground and space segments required to support the service. Other mobile earth station systems are mentioned briefly in Section three, although this section is designed to deal predominantly with the fixed satellite service and gives details of the Intelsat and Eutelsat systems, including the ground and space segment provision and information on services. Finally, a list of abbreviations is provided which gives details of all those curious acronyms and buzzwords which abound in the jargon of satellite communications.

There are a multitude of satellite networks currently in orbit operated by various nations of the world. Because the principles are common to all communication satellites, Section one of this book could be applicable to many of the networks not mentioned by name. Omission of a particular network is not an indication of the relative merit of that network but simply an indication of the constraints on the size of this book. Named networks in Sections two and three have been arbitrarily chosen to give an indication of the application, for those systems, of the principles outlined in Section one.

The intention of the book is to explain in clear and simple terms the principles and practice of satellite communications at a level that makes it suitable for students undertaking courses at undergraduate level in this specialization, for those who simply have an interest in the topic, and for practising engineers who could use the book as a reference text.

D. Calcutt. M.Sc., C.Eng., M.I.E.E.
L. Tetley. I.Eng., F.I.E.I.E.

Acknowledgements

A book of this complexity owes much to the co-operation of various individuals, equipment manufacturers and organizations. In many cases the authors have had no personal contact with individuals but despite this they gave freely of their time when information was requested. To mention individuals by name would require a lengthy acknowledgement list, so although no names are included we hope the individuals will take the credit that is due to them.

We are extremely grateful for the assistance given by the following companies and organizations during the writing of this book. We are particularly indebted to those organizations who permitted us to reproduce copyright material.

Our thanks go to the following:
Asea Brown Boveri (ABB).
ABB NERA AS.
British Telecommunications plc.
COMSAT Technical Review.
COSPAS-SARSAT Secretariat.
EB Communications (Great Britain) Ltd.
Japan Radio Company (JRC)
The Eutelsat Organization.
The Inmarsat Organization.
The Inmarsat quarterly magazine *Ocean Voice*.
The Intelsat Organization.

Contents

Preface		*page* v
List of abbreviations used in the text		xii

SECTION ONE   COMMUNICATIONS PRINCIPLES APPLICABLE TO SATELLITE OPERATION

Introduction		1
1	**Basic Principles**	3
	1.1 Introduction	3
	1.2 Why use satellites?	4
	1.3 Basic concepts	6
2	**Satellite Link Parameters**	16
	2.1 Introduction	16
	2.2 Basic transmission concepts	17
	2.3 Link budgets	36
	2.4 Summary	52
3	**Multiple access**	53
	3.1 Introduction	53
	3.2 FDMA	55
	3.3 TDMA	74
	3.4 CDMA	76
	3.5 SDMA	79
	3.6 Packet access	81
	3.7 Random access	82
4	**Modulation and Demodulation**	86
	4.1 Introduction	86
	4.2 Amplitude modulation	89
	4.3 Frequency modulation	93
5	**Digital Transmission**	108
	5.1 Introduction	108
	5.2 Digital Speech Interpolation (DSI)	110
	5.3 Some commercial coding systems	110

x *Contents*

6 Digital Modulation   123
   6.1 Introduction   123
   6.2 Phase shift keying (PSK)   125
   6.3 Quadrature phase shift keying   128
   6.4 Probability of bit error rate   131

7 Coding   136
   7.1 Introduction   136
   7.2 Block codes   138
   7.3 Cyclic codes   141
   7.4 Convolutional codes   145
   7.5 Decoding   149
   7.6 Error correction   157
   7.7 Pseudo-noise   158

8 Earth Stations   162
   8.1 Introduction   162
   8.2 Earth station operating FDM/FM/FDMA   162
   8.3 Earth station operating TDM/QPSK/TDMA   165
   8.4 High-power amplifiers (HPAs)   165
   8.5 Low-noise amplifiers (LNAs)   169
   8.6 Antennae   169
   8.7 Monitoring and control   177

9 Communications Satellites   179
   9.1 Introduction   179
   9.2 The support sub-systems   179
   9.3 Communications sub-systems   189
   9.4 Satellite switching   196
   9.5 Satellite antennae   197

SECTION TWO   COMMUNICATIONS WITH MOBILE EARTH STATIONS

Introduction   203

10 The Inmarsat Organization   205
   10.1 Introduction.   205
   10.2 The Inmarsat Organization   205

11 The Inmarsat-A System   222
   11.1 Introduction   222
   11.2 Details of the Inmarsat-A system   228
   11.3 Inmarsat-A SES equipment   250
   11.4 Inmarsat-A transportable equipment   278
   11.5 Inmarsat-A aeronautical service   282

12 The Inmarsat-B System   285
   12.1 Introduction   285
   12.2 Outline system specifications   285

Contents xi

13	**The Inmarsat-C System**	295
	13.1 Introduction	295
	13.2 Outline system specifications	296
	13.3 Mobile earth stations	308
14	**The Inmarsat-M System**	314
	14.1 Introduction	314
	14.2 Outline system technical specifications	315
15	**Satellite Mobile Frequency Bands**	319

SECTION THREE COMMUNICATIONS WITH FIXED SATELLITE SERVICE

Introduction		321
16	**The Intelsat Organization**	323
	16.1 Introduction	323
	16.2 Development of the Intelsat space segment provision	325
	16.3 Ground network	355
17	**Intelsat Services**	362
	17.1 Introduction	362
	17.2 Public switched services	363
	17.3 Private network services	364
	17.4 Other services	366
18	**The Eutelsat Organization**	370
	18.1 Introduction	370
	18.2 Eutelsat space segment	372
	18.3 Ground segment	380
	18.4 Eutelsat services	382

Appendix: The decibel (dB)	388
Index	391

Abbreviations used in the text

A/D	Analogue-to-digital signal conversion
ADE	Above decks equipment
ADM	Adaptive delta modulation
ADPCM	Adaptive differential pulse code modulation
AFC	Automatic frequency control
AFTN	Aeronautical fixed telecommunications network
AM	Amplitude modulation
AOR	Atlantic Ocean Region
AORE	Atlantic Ocean Region East
AORW	Atlantic Ocean Region West
APC	Adaptive predictive coding
APK	Amplitude phase keying
ARQ	Automatic request repeat
ASK	Amplitude shift keying
BDE	Below decks equipment
BER	Bit error rate
BPSK	Bipolar phase shift keying
C/N	Carrier-to-noise ratio
C/N_o	Carrier-to-noise density ratio
C/IM_o	Carrier-to-intermodulation noise density ratio
C/I	Carrier-to-interference power ratio
CBT	Carrier and bit timing
CCC	Intelsat Control Co-ordination Centre
CCIR	International Radio Consultative Committee
CCITT	International Telegraph and Telephone Consultative Committee
CCS	Command Co-ordination System
CDMA	Code division multiple access
CEPT	European Conference of Postal and Telecommunications Administrations
CES	Coast earth station
Codec	Coder/decoder
COMSAT	Communications Satellite Corporation
COSPAS	Cosmicheskaya Sistyeme Poiska Avariynich Sudov
COSPAS-SARSAT	International satellite based emergency alerting and locating system
CSC	Communications system control
CSC	Common Signalling Channel
CSM	Communication system monitoring
CW	Continuous wave
D/A	Digital-to-analogue signal conversion

DAMA	Demand-assigned multiple access
dB	Decibel
DCE	Data circuit terminating equipment
DCME	Digital circuit multiplication equipment
DM	Delta modulation
DNI	Digitally non-interpolated
DPCM	Differential pulse code modulation
DS	Direct sequence
DSBSC	Double sideband, suppressed carrier
DSI	Digital speech interpolation
DTE	Data terminal equipment
E_b/N_o	Energy per bit/noise density ratio
EBU	European Broadcasting Union
E_c/N_o	Energy per symbol/noise density ratio
EDI	Electronic data interchange
EGC	Enhanced group call. Group calling on Inmarsat-C
EIRP	Effective isotropic radiated power
EME	Externally mounted equipment
EPC	Electronic power conditioner
EPIRB	Emergency position indicating radio beacon
ESA	European Space Agency
Eutelsat	European telecommunications satellite organization
FANS	Future air navigation systems
FCC	Federal Communication Commission
FDM	Frequency division multiplex
FDMA	Frequency division multiple access
FEC	Forward error correction
FET	Field effect transistor
FM	Frequency modulation
FSK	Frequency shift keying
FSS	Fixed Satellite Service
G/T	Receive gain/system noise temperature ratio
GaAs	Gallium arsenide
GEO	Equatorial geostationary orbit
GES	Ground earth station
GMDSS	Global Maritime Distress and Safety System
GNS	Global Network Service
HF	High frequency
HPA	High-power amplifier
HSD	High-speed data
IBS	Intelsat Business Service
ICAO	International civil aviation organization
IDR	Intermediate digital rate
IESS	Intelsat Earth Station Standards
IF	Intermediate frequency
IM	Intermodulation
IMBE	Improved multi-band excitation
IMN	Inmarsat MES identification number
IMO	International Maritime Organization
Inmarsat	International maritime satellite organization
Intelsat	International telecommunications satellite consortium

IOR	Indian Ocean Region
ISDN	Integrated services digital network
ISL	Interstation signalling link
ISO	The international organization for standardization
ITA	International telegraph alphabet code
ITU	International Telecommunications Union
LAN	Local area network
LES	Land earth station
LHCP	Left-hand circular polarization
LNA	Low-noise amplifier
LO	Local oscillator
LRE	Low-rate encoding
MAC	Multiplexed analogue components
MCPC	Multiple channels per carrier
MCS	Maritime communications sub-system
MES	Mobile earth station
MIC	Microwave integrated circuit
MSK	Minimum shift keying
MSS	Mobile Satellite Service
NBDP	Narrow band direct printing
NCC	Network Control Centre
NCS	Network co-ordination station
O-QPSK	Offset quadrature phase shift keying
OCC	Operations Control Centre. Inmarsat
PC	Personal computer
PCM	Pulse code modulation
PFD	Power flux density
PLL	Phase-lock loop
PM	Phase modulation
PN	Pseudo-random noise
POR	Pacific Ocean Region
PSDPN	Packet switched public data network
PSK	Phase shift keying
PSTN	Public switched telephone network
QPSK	Quadrature phase shift keying
RF	Radio frequency
RHCP	Right-hand circular polarization
S/N	Signal-to-noise ratio
SACE	Signalling and access control equipment
SAW	Surface acoustic wave
SCC	Satellite Control Centre
SCPC	Single channel per carrier
SCPT	Single carrier per transponder
SDMA	Space domain multiple access
SES	Societé Europeanne des Satellites
SES	Ship earth station
SIS	Sound in synchronization
SMATV	Satellite master antenna TV
SMS	Satellite multiservice system
SNG	Satellite news gathering
SOLAS	Safety of Life at Sea Convention

SS	Switched satellite
SSAM	Single sideband amplitude modulation
SSB	Single sideband
SSBSC	Single sideband, suppressed carrier
SSMA	Spread-spectrum multiple access
SSPA	Solid-state power amplifier
TDM	Time division multiplex
TDMA	Time division multiple access
TT&C	Telemetry, tracking and control
TTC&M	Telemetry, tracking command and monitoring
TVRO	TV receive-only
TWTA	Travelling-wave tube amplifier
UTC	Co-ordinated universal time
UW	Unique word
VCO	Voltage controlled oscillator
VDU	Video display unit
VSAT	Very small aperture terminal
WAN	Wide area network
WARC	World Administrative Radio Conference

Section One
COMMUNICATIONS PRINCIPLES APPLICABLE TO SATELLITE OPERATION

Introduction

Man, by nature, is gregarious and inventive. The ability to communicate has developed with time from the sign language of civilization's dawn to the spoken and written word developed using a multitude of languages across the world. The urge for man to communicate has progressed to the point where communication from one site to another can be achieved using either land links, a terrestrial radio link, a satellite link or a combination of these. The information to be transmitted could be speech, data or video; it is even feasible for machines to 'talk' to each other across vast distances for the benefit of man who utilizes the data.

This section of the book looks in detail at the principles that are applicable to satellite communications. Sections two and three show how the principles are applied in practice to satellite communications by international and regional system operators.

1
Basic principles

1.1 Introduction

Consider the situation where two persons are the only occupants of a large room. They are conversing quietly at one end of the room. A communication link has been established. Suppose one of the occupants moved to the far end of the room leaving the second occupant at the near end; communication would still be possible although both persons would probably need to raise their voices to be heard. There is a limit to the separation distance the two talkers could endure while still being able to make sense of what they heard. Speech causes compression and rarefaction in the air and these perturbations spread out from the speaker, with a very limited range. The range could be increased by converting the speech signal to an electrical signal, by means of a microphone, and transmitting the electrical signal by way of, say, a telephone link utilizing a cable or radio connection between the speakers. Speech is reconstructed at the receiving end by means of a loudspeaker.

Returning to the two conversationalists in the large room, what would be the effect of adding an extra hundred or so people in the room, all of them striking up conversations with near neighbours? Our two conversationalists could probably still converse provided they were close together. If each were constrained to be at opposite ends of the room the likelihood is that neither would hear the other because their speech would be ineffective in the presence of the noise produced by others in the room. The twin requirements of communication over long distance have been established—there is a need to modify the information from its original form to a form that will travel long distances and to do so in the presence of interference of any kind.

The block diagram of Fig. 1.1 is frequently used to represent a communications link.

The function of the transducer is to convert the input data to a form suitable for transmission. An example of a transducer is the telephone, which contains a microphone to convert the compression and rarefaction associated with the air close to a speaker's mouth to an electrical signal containing frequencies in the audio band which

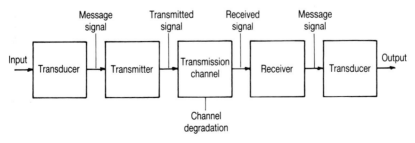

Fig. 1.1 Block diagram representation of a communications link

ranges from dc up to about 20 kHz. Not all of this frequency band need be used for an audio baseband however, since the range contains many harmonics which could be eliminated without seriously affecting the intelligibility of the speech. Typically, the speech range is from 300 – 3000 Hz. Other transducer inputs may contain other information such as telex, computer data, facsimile, television signals, telemetry, navigational information etc. The message signal could be analogue or digital in form.

Speech is a form of analogue signal containing wide variations in amplitude with time. The output of a teleprinter on the other hand has characters represented by a series of binary digits (bits) which can have a fixed level to represent binary 1 and another to represent binary 0. Speech too can be digitized for transmission and this is becoming the rule for latter-day transmission purposes.

The transmitter converts the input message signal to a form suitable for feeding to the transmission channel. A carrier frequency of a value suitable for transmission is modulated by the transducer output signal. Modulation is a systematic variation of some parameter of the carrier, i.e. its amplitude, frequency or phase, according to a function of the input message signal.

Why is a carrier used and why is it modulated? Among the reasons are:

- the requirement for transmission over long distances;
- to minimize interference over the channel;
- to enable several transmissions to exist in a single channel (multiplexed signals);
- for channel assignment;
- to overcome equipment limitations.

The transmitter may perform other functions such as filtering, amplification and connection of the output signal to the channel input, which could be a radiating element (antenna).

The channel could be a line link such as that between two telephone subscribers, a radio link such as that between a commercial radio transmitter and a domestic radio receiver or a communications satellite link. In all cases the channel will affect the signal and cause it to be degraded as it continues along its path. The degradation could be noise but there are other forms of interference which may help to distort the required signal.

The function of the receiver is to extract the required signal at the output of the channel and reconvert it, using demodulation, to the original baseband signal. Some amplification may be necessary to restore what could be a very weak signal to a level suitable for demodulation. Since the receiver itself is noisy it is often a delicate matter to restore the signal to a value sufficiently above the noise level. The signal-to-noise (S/N) ratio at the receiver output is often quoted as a means of establishing how well the receiver achieves its purpose.

The output transducer completes the link. Its function is to restore the form of the original input data. This could be a speech signal via a loudspeaker, the printed output of a teleprinter, data input to a computer etc.

This book is mainly concerned with the satellite communications link and all of the chapters and sections that follow will enlarge on the above with particular reference to the requirements of satellite links to fixed and mobile stations.

1.2 Why use satellites?

Long distance communications have been established in the past using terrestrial connections such as HF radio links or submarine telephone links. Indeed, long distance telephony cables still operate successfully on transoceanic routes with increased capacity over the years. However, it was due to congestion on the long distance cable routes and the difficulties encountered in attempting to relay television signals that satellites first became a viable option. The main objection to the use of satellites at that time had been the transmission delay that occurs on a satellite link. Transmission delay occurs because a signal travelling at the speed of light would require about 270 ms to complete the up-link and down-link path via a satellite in a geosynchronous orbit with a range of approximately 40,000 km. A telephone subscriber using the satellite link for two-way telephony would therefore need to wait in excess of 540 ms before receiving a response.

The effect of time delay is compounded by the echo effect whereby a speaker hears an echo of his/her own voice. This effect is liable to occur on terrestrial links but its effect on purely terrestrial links is limited because terrestrial transmission delay is small (about 30 ms).

A transmission path via a satellite link and a terrestrial link would be likely to enhance the echo effect because of the increased time delays involved. However, the echo effect has been successfully minimized in satellite links by the use of echo control circuitry. Time delay has also been found to be important for digital communications and some form of error detection and/or correction has been found to be necessary.

The Telstar 1 satellite first provided television signals across the Atlantic in 1962 and Intelsat 1 in 1965 provided a telephony/television service between Europe and North America. Telstar utilized a single transponder (a transponder is the electronics package on the satellite that receives signals, processes them and frequency translates them for retransmission to earth. Transponders are discussed in detail in Sections two and three). The Telstar orbit was elliptical with its plane inclined to the equator at about 45°. The three earth stations involved (one in North America and two in Europe) were only able to access the satellite for about 30 minutes at a time, three or four times a day.

Intelsat 1 was the first commercial geostationary communications satellite allowing continuous access. Other satellite systems followed while the major Intelsat system also developed into its present day structure of 19 satellites covering different regions of the earth. Satellite services can be identified by the type of link provided, i.e. point-to-point, point-to-multipoint and variations thereof. This is one advantage of satellite communication compared with a cable link which is generally point-to-point. Other advantages include:

- the length of a satellite link does not affect the cost of a service whereas cable link cost increases with distance;
- satellites provide a service in difficult terrain which would be inaccessible to cables;
- satellites provide a service in remote areas which would be prohibitive in terms of the cost of providing the cable link;
- satellites uniquely provide a service to mobile terminals such as ships, aircraft etc;
- satellites offer a large channel capacity in each transponder channel.

Although the early satellites were used as alternative links to the terrestrial routes, improvements in the capacity of terrestrial links, especially by the use of fibre optics, has meant that true competition now exists for many services. It may well be that a communications link will consist of several stages of which the satellite link is only a part. None the less, the satellite communication link will increasingly play an important part in the development of modern communications and will develop into new spheres of influence. Already satellite television plays an important part in everyday life bringing major events, or specialist programmes, directly into most peoples' homes, either directly or via a cable network. This application is expected to expand further. Another developing area of utilization is the provision of high-speed data services; the

6 *Basic principles*

use of very small aperture terminals (VSATs) already enables low data rate services to be integrated into an overall system (Intelsat's Intelnet system is a case in point). Such services should experience great expansion with time, especially if the operating costs can be kept low. Another specialist application is the part satellites play in the Global Maritime Distress and Safety System (GMDSS) which allows a rapid response in the case of mobile emergencies. The use of the Inmarsat satellites for GMDSS is dealt with fully in the companion volume *Understanding GMDSS*.

1.3 Basic concepts

Communications satellite systems may be broadly divided into space and ground segments.

Space segment

The satellites to be considered in this book operate in a geostationary orbit. Such satellites have an equatorial orbit and travel at a speed and direction such that, relative to the earth, the satellite appears stationary. On-board controls adjust for any slight deviation in position and inclination that may occur with time. Such a satellite may be accessed by an earth terminal and provide a link to another terminal. The satellite acts as a repeater station allowing the up-link signal to be amplified and translated in frequency for the down-link. The space segment refers to the satellites and all their on-board equipment and the telemetry, tracking and control (TT & C) system. Other features of the space segment include:

- the use of solar cells which provide power to the satellite sub-systems. Batteries allow for standby power when solar energy is not available during an eclipse;
- provision of antenna systems to provide the required coverage area for reception and transmission of signals;
- a thermal control system which maintains a required temperature within the satellite despite the extremes of temperature suffered on the outside, and also provides for heat dissipation from the sub-systems.

The earth segment

The earth segment consists of the earth terminals together with their associated equipment, antennae, electronic circuits etc and the links to terrestrial networks. Another aspect is the ownership of the earth station responsible for organizing the use of the services. Earth stations may be fixed or mobile with the added complexity for the latter being the requirement to provide tracking and alignment facilities.

Satellite orbital parameters

Whilst the subject of satellite orbital parameters is extremely complex and may appear to the reader to be somewhat academic when considering the operation of satellite systems, it will aid system understanding if the rudimentary principles of satellite orbits are considered. As an example, the space segment of the Global Maritime Distress and Safety System (GMDSS) currently includes two types of satellite each with a very different type of orbit.

- GMDSS distress alerting may be achieved using the international emergency alerting and locating system (COSPAS/SARSAT) polar orbiting satellites; the system is

Fig. 1.2 A USAF DELTA rocket carries an Inmarsat-2 F2 satellite into orbit (courtesy *Ocean Voice*)

activated as the satellite passes above a maritime beacon known as an EPIRB (Emergency position indicating radio beacon)
- The geostationary orbiting (GEO) satellites which, as part of the Inmarsat organization, provide constant earth coverage for radio communications.

The above example illustrates that different orbits are useful to perform particular functions. However, the type of orbit of major interest for communications purposes is the geostationary orbit which causes the satellite to appear to remain in a fixed position relative to the earth. Further information on the GMDSS can be found in a companion volume *Understanding GMDSS*.

8 *Basic principles*

Although details of the launch of a satellite into its orbit are not covered in this book, Fig. 1.2 is included out of interest to show the launch of an Inmarsat-2 F2 satellite into orbit using a USAF Delta rocket.

Polar orbit As the term suggests, a satellite in this orbit will travel its course over the geographical north and south poles and will effectively follow a line of longitude. However, it must be remembered that the earth is revolving below the orbit and consequently the satellite, over a number of orbits determined by its specific orbit line, will pass over any given point on the surface of the earth.

The orbit may be virtually circular or elliptical depending upon requirements. If two such satellite orbits are spaced at 90° to each other, the time between satellite passes over any given point will be halved. More satellites in orbit will reduce the time even further. The parameters concerning this type of orbit are fully described in the chapter entitled 'Navigation by Satellite' in the companion volume *Electronic Aids to Navigation (Position Fixing)*. A polar orbiting satellite is rarely used for communication purposes because it is in view of a specific point on the earth's surface for only a short period of time. Also, complex steerable antenna systems would be needed to follow the satellite as it passed overhead. Such a satellite is used in conjunction with omnidirectional antennae for navigation purposes and for distress alerting in the GMDSS, as fully described in the companion volume *Understanding GMDSS*.

Elliptical inclined orbit The inclination of a satellite orbit is the angle which exists between that orbit and the earth's equator. Thus, the geostationary orbit is 0° inclined whilst the polar orbit is 90° inclined. In practice, a satellite can follow an orbit with any angle of inclination. Actually, many satellite orbits are inclined by accident because it is not easy to launch satellites into their pre-determined orbits.

In future, Inmarsat is likely to use a combination of satellites in geostationary and other orbits to provide communications for aircraft. Figure 1.3 shows the arrangement.

Circular equatorial orbit (GEO) A satellite placed into circular equatorial orbit may be made to appear stationary when viewed from the surface of the earth. Such an orbit may be called a geostationary orbit, and satellites in that unique circular orbit become geosynchronous satellites. These are satellites whose orbital period is synchronized to the period of rotation of the earth's surface relative to their distance from it. By the use of three satellites equally spaced in the geostationary orbit it is possible to provide continuous communications from one point on the earth to another. Satellites in this orbit handle the bulk of military and commercial international communications on the earth.

In October 1945 the science fiction writer and noted engineer Arthur C. Clarke, published an article in the magazine *Wireless World* in which he calculated that a geosynchronous orbit would exist directly above the earth's equator at a distance of some 35,855 km. To maintain a position in this orbit a satellite must travel at a velocity of 3.073 km/s and possess an orbital period of 23 hours, 56 minutes, 4 seconds, called one sidereal day. Clarke's 1945 article effectively formed the basis of the international communications network which is enjoyed today, and in his honour the geosynchronous orbit is often called the Clarke Orbit.

The advantages of using satellites in the geostationary orbit are numerous and include the following.

- Fixed ground station antennae need not be fully steerable and consequently could be made to a simpler design. Mobile earth station (MES) antenna systems may be

1.3 Basic concepts   9

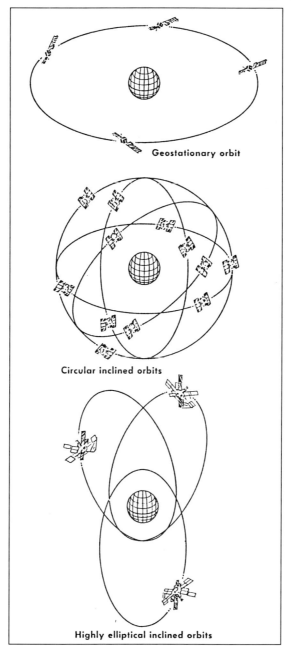

Fig. 1.3 Satellite constellations for aeronautical mobile services (courtesy *Ocean Voice*)

smaller and simpler. Omni-directional antennae are practicable and eliminate the need for complex gyroscopics. These are used, for example, in the Inmarsat-C Ship Earth Station (SES).
- There is effectively no relative movement between the satellite and a fixed earth station and, consequently, no effective Doppler-shift of the communications

10 Basic principles

frequency is introduced which, with polar orbiting satellites, requires more complex processes to eliminate errors.
- Continuous, instantaneous communications, between a mobile and a fixed station are possible because the satellite is always in view.

A major disadvantage of the GEO orbit is that its altitude is very large and a considerable amount of power is necessary to lift a payload into orbit. In practice, an easterly rocket launch is used in order to take advantage of the earth's velocity and a point as close to the equator as possible is used as the launch site. Earth velocity is greatest at the equator. The satellite is first launched into a highly elliptical orbit whose point of apogee (the furthest point from the earth surface) is equal to the altitude required for a geostationary orbit. When the satellite reaches its apogee point an 'apogee motor' is fired to cause the satellite to follow the new orbit. This is shown diagrammatically in Fig. 1.4.

Basically, a satellite remains in orbit when two forces, one caused by the gravitational pull of the earth and the other by the centripetal acceleration due to its angular velocity, are in balance.

The velocity (v) of a geostationary satellite should ideally be zero relative to the earth's surface, although small variations do occur. Satellite orbital velocity is 3.073 km/s to enable the satellite to maintain geosynchronism. Small orbital variations do occur due to the influence of other heavenly bodies but they are of no consequence to the reader as their effects are counteracted in the ground control station.

The altitude of the geostationary orbit may be readily calculated.

For a circular orbit at altitude (h) above the earth's equator, the circumferal path is given as:

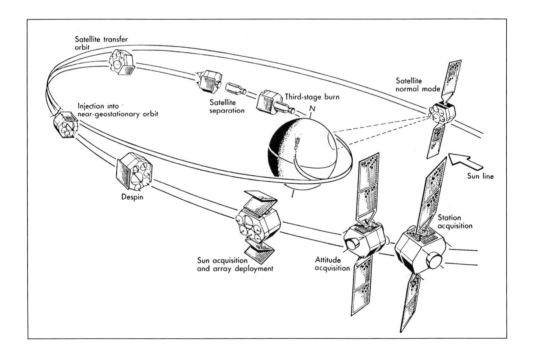

Fig. 1.4 Typical launch to geostationary orbit by expendable vehicle (courtesy *Ocean Voice*)

$$2\pi(a + h)$$

where a = the average earth radius (6371 km)
h = altitude above earth surface.

The circumferential velocity (v) is constant, therefore the period of one orbit is
$$T = 2\pi(a + h)/v$$

All satellites maintain their orbits with reference to velocity, mass and earth gravity. The centripetal force on a satellite with a mass (m) is

$$mv^2/(a + h)$$

and the earth's gravitational pull is the product of mass and gravity (mg'), where the gravitational force $g = 9.81\ m\ s^{-2}$. The gravitational acceleration g' is therefore

$$g' = g[a/(a + h)]^2$$

Balancing centripetal force against gravitational force in order to maintain orbit:
$$mg[a/(a + h)]^2 = mv^2/(a + h)$$
therefore
$$v = a[g/(a + h)]^{1/2}$$

Substituting v above into the orbital period T formula and including numerical values gives
$h = (5075T - 6371)$ km
$h = 35{,}855$ km (for a 24 hour period)

Angle of elevation

This is the angle between a tangent drawn to the visual horizon from a given point on the earth's surface and the direct line-of-sight path to a satellite from that same point. See Fig. 1.5.

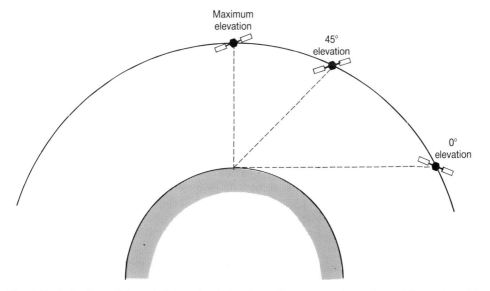

Fig. 1.5 Indications of the satellite angle-of-elevation with respect to the surface of the earth and the corresponding range change

12 *Basic principles*

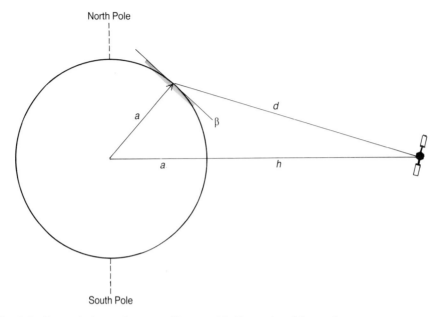

Fig. 1.6 Transmission path range with respect to the centre of the earth

If the observer is standing on the equator directly beneath a geostationary satellite the angle of elevation will be a maximum of 90°. If the observer now moves in any direction away from that point to the outer edge of the satellite footprint coverage area the angle of elevation decreases until it will ultimately be zero. However, in practice, the minimum angle of elevation for a commercial signal path is approximately 5°. The amplitude of the received signal will progressively decrease as the observer moves away from the 90° elevation point.

Transmission path range variation

Consider the angle of elevation made by a satellite with respect to the surface of the earth as shown in Fig. 1.6.
The angle of elevation is $\beta°$.
By applying the cosine rule

$$(a + h)^2 = a^2 + d^2 - 2ad\,(\cos 90 + \beta)$$
$$= a^2 + d^2 - 2ad\,\sin\beta$$
$$d = [(a + h)^2 - (a\cos\beta)^2]^{1/2} - a\sin\beta$$

When the observer is beneath the satellite $\beta = 90°$ so that $\cos\beta = 0$ and $\sin\beta = 1$.

and $d = [(a + h)^2]^{1/2} - a$
$= a + h - a$
therefore $d = h$

When the observer is in a position where $\beta = 0°$

$$d = [(a + h)^2 - a^2]^{1/2} - 0$$
$$= 41{,}745 \text{ km}$$

Because signal attenuation increases with distance it follows that the received signal will become progressively weaker as the angle of elevation decreases.

1.3 Basic concepts 13

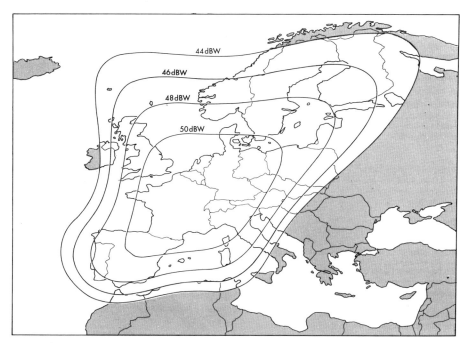

Fig. 1.7 Gain contours for one of the transmit beams of the Eutelsat II regional fixed sat-coms spacecraft. The beam is shaped to concentrate maximum power on western Europe (courtesy Eutelsat/*Ocean Voice*)

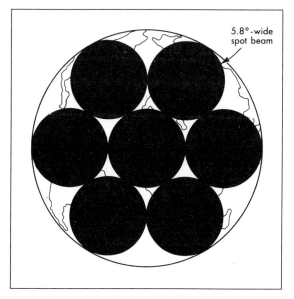

Fig. 1.8 Earth coverage by seven spot beams (courtesy *Ocean Voice*)

Satellite footprint

As the diagram of Fig. 1.7 illustrates, the footprint describes the area of usable signal strength for a given beam width. The beam of a satellite down-link transmission may be wide and produce an almost circular footprint of effectively one third earth coverage. Alternatively, the transponder beam may be designed to be narrow and shaped to cover a specific region as used by television broadcasting and some communication services. The gain figures provide an indication of the drop in signal strength, evident as the receiving antenna is positioned away from the centre of the beam.

In future, narrow circular beams from geostationary satellites will be used to produce footprints as shown in Fig. 1.8. The main advantage with this system is that specific earth areas can be covered more accurately than with a wide beam. Also, greater power density per unit area for a given input power can be achieved, when compared with that produced by a global circular beam, leading to the use of smaller receiving antennae.

Transmission path loss

Almost all of the transmission path loss of signal amplitude occurs because of spreading of the transponder beam over a large area. For a given amplitude of signal at the satellite transponder, the larger the footprint created on the earth's surface the greater will be the transmission path loss for the worst case (longest) path length. The transmission path loss in decibels may be approximated as:

$$L_p = 32.5 + 20 \log d + 20 \log f \ \text{dB} \tag{1.1}$$

where, d = distance in km between transponder and receiver,
f = transmission frequency in MHz.

The effects of path loss on the link budget are discussed in more detail in Chapter 2, Section 2.3.

All signals from a satellite must obviously travel through the earth's ionosphere and troposphere, both of which will absorb and scatter signal energy leading to signal loss. Transmission signal path losses caused by each of these natural phenomena is proportional to the path length within the medium. The path length will be much greater at low elevations rising to a maximum at 0° elevation, leading to the maximum signal loss due to this factor. In practice, when the angle of elevation subtended by a receiver is 90° the radio wave will travel directly through the stratosphere and ionosphere with a path length through the medium of approximately 120 km, whereas when the angle of elevation is 0° the path length increases to approximately 720 km. This sixfold increase in path length within the attenuation medium leads to a corresponding decrease in signal strength.

Signal attenuation due to atmospheric effects varies inversely with frequency. In addition, the atmosphere possesses two natural absorption peaks, one at 60 GHz due to the oxygen molecule effect, and the other at 22.2 GHz caused by excessive water vapour. Both of these frequencies are above the bands currently used for commercial satellite communications.

Modulation techniques

Early satellite systems used techniques already established for terrestrial networks. Telstar 1 for example used 6 GHz for the up-link transmissions and 4 GHz for the down-link—frequencies already in use by microwave terrestrial links. The high gain needed for satellite transponders to amplify the received weak signals from earth could cause instability due to feedback from the output to the input of the amplifiers. This effect is eliminated due to the use of the different frequencies for reception and transmission. The modulation method utilized was frequency modulation (FM) which gave an acceptable compromise between signal power levels and RF bandwidth. Intelsat 1 also used FM, the carriers being modulated by a baseband signal that was either television or multi-channel telephony (240 voice circuits). Intelsat 1 used a hard limited non-linear transponder, which can cause intermodulation products (signals that are a combination of all sum and difference frequencies). The energy required to support these intermodulation (IM) products causes a loss of useful signal energy and, at the same time, the IM products appear in the bandwidth of other signals and cause interference. For Intelsat 1 this effect kept the number of simultaneous accesses to just two; one station was in North America and the other in Europe. Later satellites provided multiple carriers per transponder using the transponders in the linear region and reduced the effect of IM distortion. This improved multiple access capability but at the cost of reduced amplifier power efficiency. Frequency modulation is still widely used in satellite systems although digital techniques are now coming to the fore. Digital modulation occurs when a binary signal alters the parameters of the carrier in some way. Most digital transmissions using satellite links are modulated on to the carrier using phase-shift keying (PSK) techniques, which alter the phase of the carrier by 0° or 90° according to whether the binary signal is a logic 0 or logic 1. Other phase shift keying techniques are also used, such as offset PSK which gives phase shifts of 0°, 90°, 180° or 270° according to a dual binary state. Further details of modulation techniques are to be found in the relevant chapters that follow.

Multiple access

Multiple access is that facility whereby several earth stations can access the same satellite or transponder. In a satellite with many transponders it is possible that, say, one transponder carries a single carrier (television or trunk telephony) accessed by a single earth station while other transponders carry multi-carrier signals from a multiplicity of, say, small earth terminals. Multiple access can comprise frequency, time, code and space domains as well as access by contention using systems such as ALOHA. All systems and their applications will be discussed in greater depth in Chapter 3.

2
Satellite link parameters

2.1 Introduction

The design of a satellite communication system using geostationary satellites must take into account the total cost of providing the service; this will include the long-term cost of positioning and controlling the satellite and maintaining the ground stations. The designer should take care to ensure the payload is of adequate size to support sufficient capacity to enable the financial return to provide a viable commercial service. Dedicated frequency bands are allocated for satellite communication systems and satellites using similar frequency bands must be sufficiently separated in space to avoid possible interference problems.

It is generally cheaper to use the lower frequency bands but this has the disadvantage of being more likely to incur interference since many terrestrial microwave links use frequencies within these bands. Also, the bandwidth of the system will be less than at higher frequencies. Although the use of higher frequencies offers a wider bandwidth there is the disadvantage that propagation difficulties are more severe above about 10 GHz. A major factor in the selection of a frequency band for communication purposes is attenuation. This is not so significant, however, for a satellite communication path with an elevation angle of 5°, or higher, at frequencies below about 10 GHz; this is because of the short path length through the atmosphere of the earth and the low attenuation value per kilometre for frequencies under 10 GHz. Above 10 GHz the effect of rain on attenuation increases, the effect being variable according to the intensity of the rainfall.

Methods have been devised for estimating the attenuation due to rain using statistical data for the region under consideration. The total attenuation suffered by a link is due to the cumulative effect of losses due to the slant range, atmosphere (due to the effect of gases in the atmosphere causing attenuation, especially at low elevation angles), precipitation, scintillation (rapid fluctuations in attenuation) and Faraday rotation (caused by the effect of the ionosphere on a wave which has a different polarization on leaving compared with that on entry. The effect is worse at lower frequencies). The effect of sky noise is reduced at low frequencies so that systems with low noise temperatures are possible at those frequencies. Many geostationary satellites use 6/4 GHz, which is at the lower end of the allocated frequency spectrum, for the reasons outlined above. Other satellite systems operate with a frequency of 14/11 GHz and others are proposed at 30/20 GHz in order to increase the traffic capacity. The use of microwave frequencies for satellite communications has the advantage of allowing the radiated signal to be concentrated into a narrow beam for transmission to the satellite. In turn, the satellite can retransmit the signal to earth in a beam shaped to provide a desired footprint for a particular requirement. See the section in Chapter 1 entitled 'satellite footprint'.

Where many earth stations share a satellite there are many interconnecting paths. Since some of the paths may be through a single transponder the capacity of the transponder must be shared between the earth stations. This sharing is known as multiple access and may be achieved in practice in several ways, see Chapter 3.

The communication system should be designed to meet a certain minimum performance specification, within the limitations of transmitter power and signal bandwidth. One important performance criterion is the signal-to-noise ratio (S/N) at the baseband frequency. Because the strength of a baseband signal received by the satellite, or the earth station, is small and comparable in size with noise levels, it is essential that the signal can be recovered in the presence of the noise and this will determine the minimum signal-to-noise ratio in a receiver baseband channel. The carrier-to-noise ratio (C/N) of the RF or IF signal in the receiver will affect the signal-to-noise ratio in a baseband channel. Other factors which must be taken into account include the type of modulation used to impress the baseband signal on to the carrier and the IF band baseband channel bandwidths in the receiver. For the design of a satellite communication link it is necessary to calculate the carrier-to-noise ratio; this is illustrated in the following section.

2.2 Basic transmission concepts

System power levels

If an isotropic transmitter, in free space, radiates a total power P_T watts uniformly in all directions, then at a distance d metres from the source, the power is spread over the surface of a sphere of radius d metres. The power flux density across the surface of the sphere of radius d metres is:

$$PFD = \frac{P_T}{4\pi d^2} \text{ W/m}^2$$

Practical satellite systems use directive antennae which concentrate the radiated power in a given direction. The gain $G(\theta)$ of the antenna in a direction θ is defined as the ratio of power per unit solid angle radiated in a given direction, to the average power radiated in a unit solid angle:

$$G(\theta) = \frac{P(\theta)}{P_0/4\pi}$$

where $P(\theta)$ is the power radiated per unit solid angle by the test antenna,
P_0 is the total power radiated by the test antenna,
$G(\theta)$ is the gain of the antenna at an angle θ.

The reference for the angle θ is usually the direction in which maximum power is radiated, called the boresight. The antenna gain is thus the value of $G(\theta)$ where $\theta = 0°$ and is an indication of how much the power radiated from the antenna is increased compared with the value in that direction from an isotropic source emitting the same value of total power.

If the gain of the directive antenna is G_T compared with the isotropic radiation level, then for a source radiating power P_T watts, the power flux density in the direction of the antenna boresight at a distance d metres is:

$$PFD = \frac{P_T G_T}{4\pi d^2} \text{ W/m}^2 \tag{2.1}$$

Since the product of $P_T G_T$ is known as the effective isotropic radiated power (EIRP), equation (2.1) can be rewritten as:

18 Satellite link parameters

$$\text{PFD} = \frac{\text{EIRP}}{4\pi d^2} \text{ W/m}^2 \qquad (2.2)$$

For aperture-type antennae with aperture area A_T large compared with λ^2 (where λ is the wavelength of the transmitted signal) then maximum gain of the antenna is given by:

$$G_T = \frac{4\pi A_T}{\lambda^2}$$

Power P_R intercepted by a receiver antenna is the product of the receiver aperture area A_R and the power flux density at the aperture:

$$P_R = \left(\frac{P_T G_T}{4\pi d^2}\right) A_R \qquad (2.3)$$

Relating the receiver aperture antenna to its maximum gain gives:

$$G_R = \frac{4\pi A_R}{\lambda^2}$$

and received power becomes:

$$P_R = \frac{P_T G_T G_R \lambda^2}{(4\pi d)^2} \qquad (2.4)$$

This expression for received power relative to transmitted power is known as the Friis transmission equation. The equation for received power includes loss of power from isotropic spreading of the transmitted signal and the term $(4\pi d/\lambda)^2$ is known as the path loss (L_p). L_p was given by equation (1.1). Since L_p can be written as:

$$L_p = (4\pi d f/c)^2$$

where $\lambda = c/f$
and c is the speed of light (2.998×10^8 m/s)

then:

$$L_p = k + 20\log d + 20\log f$$

where k is a constant $= 32.5$ when d is measured in km and f is measured in MHz.

Other losses such as absorption and scattering of the signal can be accounted for by an additional loss factor L_a so that:

$$P_R = \frac{P_T G_T G_R \lambda^2}{(4\pi d)^2 L_a}$$

It should be noted at this point that a practical receiver antenna with a physical aperture A_R m^2 will not deliver the power suggested by equation (2.3). There is an effective aperture A_{eff} where:

$$A_{\text{eff}} = \eta A_R$$

η is the aperture efficiency of the antenna and accounts for all losses between the incident wavefront and the antenna output port. Losses include the illumination efficiency of the antenna, phase errors, mismatch losses etc. The value of η varies, being in the range 65 – 75% for large parabolic reflector earth station antennae; lower for smaller antennae and higher for large Cassegrain antennae. For horn antennae η can have values of about 90%.

This means that the Friis equation will use a value of G_R determined by $4\pi A_{\text{eff}}/\lambda^2$ which will be less than the value $4\pi A_R/\lambda^2$ used above.

2.2 Basic transmission concepts

Using logarithmic terms the transmission equation becomes:

$$10\log P_R = 10\log P_T + 10\log G_T + 10\log G_R - 20\log[4\pi d/\lambda]$$

$10\log P_R$ can be taken as the received power in decibels referenced to 1 W, commonly referred to as dBW. Since $P_T G_T$ is EIRP then:

$$10\log P_T + 10\log G_T = \text{EIRP in dBW, and:}$$
$$P_R = [\text{EIRP} + G_R - L_p] \text{ dBW}$$

It should be observed from the above that P_T and P_R can be measured in dBW to give absolute values of power referred to 1 W. G_T and G_R give antenna gains in decibels, above the isotropic value, and L_p is a loss factor in decibels. P_R is thus measured in dBW relative to P_T dBW, which itself is altered according to the value, in dBs, of G_T, G_R and L_p. An example should help to make this clear.

Example 2.1

A satellite at a distance of 40,000 km from an earth station radiates a power of 3 W from an antenna with a gain of 15 dB in the direction of the earth station. If the receiving antenna has an effective area A_{eff} of 10 m², find the flux density and the power received by the antenna.

Solution

Using equation (2.1):

$$\begin{align}\text{PFD} &= 10\log P_T + 10\log G_T - 10\log 4\pi - 20\log d \\ &= 10\log 3 + 15 - 11 - 20\log(4 \times 10^7) \\ &= 4.7 + 15 - 11 - 152 \\ &= -143.3 \text{ dBW/m}^2\end{align}$$

and
$$\begin{align}P_R &= -143.3 + 10\log 10 \\ &= -133.3 \text{ dBW}\end{align}$$

The transmission equation is easily modified if it is necessary to include other system losses not mentioned. For example, if there are losses associated with the transmitting antenna (L_{tx}) and receiving antenna (L_{rx}) then P_R becomes:

$$P_R = (\text{EIRP} + G_R - L_p - L_a - L_{\text{tx}} - L_{\text{rx}}) \text{ dBW}$$

This condition is shown diagrammatically in Fig. 2.1.

The Friis equation is commonly used for the calculation of received power in a transmission link where the parameters have been set out as a link budget using decibels. A system designer is thus able to vary link parameters in order to achieve a target value of received power.

Most satellite links use a modulation method, for analogue and digital systems, in which the carrier amplitude remains unchanged by the modulation process. Because of this, for such links the carrier power C is the same value as received power P_R.

System noise

In a conducting medium there will be a random movement of charge carriers. Nyquist has shown that a noise voltage will appear across a resistor R ohms at a temperature T Kelvin in a bandwidth B hertz such that the mean-square value of noise voltage is given by:

20 Satellite link parameters

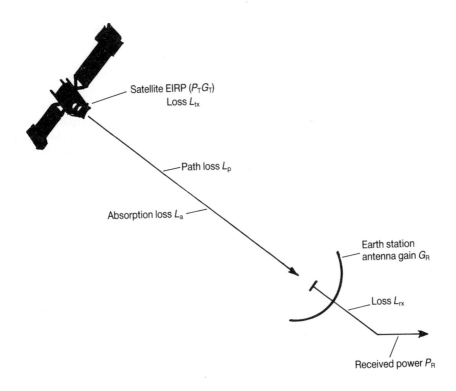

Fig. 2.1 Power gains and losses in a satellite link

$$\overline{e_n^2} = 4kTBR \text{ volts}$$

where e_n is noise voltage
  k is Boltzmann's constant 1.38×10^{-23} Joules/Kelvin
  R is resistance in ohms
  B is noise bandwidth in hertz
  T is absolute temperature in Kelvin

The available noise power P_n into a matched resistive load is thus given as:

$$P_n = \frac{\overline{e_n^2}}{4R} = kTB \text{ watts} \qquad (2.5)$$

P_n is independent of frequency and is known as 'white' noise.

Since P_n represents the power transferred to a matched resistive load then the noise power is independent of the value of resistance. The equation may be used to define the equivalent noise temperature, T_n, of any noise source:

$$T_n = \frac{P_{n(max)}}{kB}$$

where $P_{n(max)}$ is the maximum noise power the source can deliver in a bandwidth B hertz.

2.2 Basic transmission concepts

Noise temperature is a useful means of determining the amount of thermal noise generated by devices in a system. At microwave frequencies all elements with a physical temperature T_p, greater than zero Kelvin, will generate noise at the system frequency within the system bandwidth.

The term kT_n is a noise spectral density, in watts/hertz. The density is constant for all radio frequencies up to 300 GHz.

The communications signals carried over a satellite system inherently have to traverse large distances, thus suffering considerable attenuation resulting in very low signal strength at the receive end of the link. Reduction of noise level is important in order to achieve good carrier-to-noise ratios, and one contribution towards this is by designing the receiver bandwidth, usually in the IF stages, such that only the signal and immediate sidebands are accommodated.

Besides this, systems are designed to keep the noise temperature as low as possible. The use of GaAs FET amplifiers, or uncooled parametric amplifiers, allow noise temperatures better than 200 K to be achieved. Even better values are obtainable if the front-end amplifier is cooled to keep its physical temperature low; this is possible in large earth stations but is costly and not feasible for a mobile end of the link.

In order to measure the overall performance of a receiver it is necessary to know the total thermal noise against which the signal must be demodulated. Consider a typical receiver layout shown in Fig. 2.2.

Let the noise temperature at the input of the RF stage be T_s, where T_s is that noise value which, applied at the input of a noiseless receiver, produces the same noise power as the actual receiver, measured at the output. Thus, T_s may be defined as a system noise temperature of the complete receiver, measured at the input, which when multiplied by kB gives the same noise power as the actual receiver. If the overall gain of the RF and IF stages is G then the noise power at the demodulator input is:

$$P_n = kT_s BG$$

If the signal power delivered by the antenna to the receiver is P_R watts, then the signal power at the demodulator input is $P_R G$ and this represents the power in the carrier and sidebands after RF amplification and frequency down-conversion. Thus, the carrier-to-noise ratio at the input to the demodulator is:

$$\frac{C}{N} = \frac{P_R G}{kT_s BG} = \frac{P_R}{kT_s B} \qquad (2.6)$$

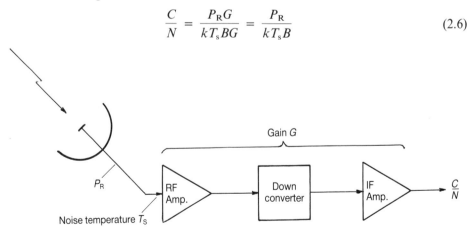

Fig. 2.2 Block diagram of earth station receiver

22 Satellite link parameters

It has already been established that, using logarithmic terms:

$$P_R = P_T + G_T + G_R - 20\log[4\pi d/\lambda]$$
or $$P_R = \text{EIRP} + G_R - 20\log[4\pi d/\lambda]$$
since $(P_T + G_T)\text{dBW} = \text{EIRP dBW}$

So that combining this with the C/N value given by equation (2.6):

$$C/N \text{ dB} = \text{EIRP dBW} - 20\log[4\pi d/\lambda]\text{dB} + G_R \text{ dB} - 10\log(kT_sB)\text{dB}$$

It is usual to express the signal-to-noise ratio C/N (where $C/N = C/kTB$) as the ratio at the output of the IF filter of the receiver. The ratio prior to the filter is usually expressed as C/kT (C/N_o) which is related to C/N by the expression $C/N = C/N_o - 10\log B$ i.e.:

$$C/N_o = (\text{EIRP} - 20\log[4\pi d/\lambda] + G_R - 10\log T_s) \text{ dBHz}$$

Noise figure and noise temperature

Noise figure is a convenient means of defining the noise produced within a system. One definition gives the noise figure, F, as the ratio of signal-to-noise at the system input to the signal-to-noise at the output i.e.:

$$F = \frac{(S/N)_{i/p}}{(S/N)_{o/p}}$$

thus
$$F = \frac{S_{i/p}N_{o/p}}{N_{i/p}S_{o/p}} \qquad (2.7)$$

and since $S_{o/p} = GS_{i/p}$, where G is the gain of the system:

$$F = \frac{N_{o/p}}{GN_{i/p}} \qquad (2.8)$$

If the system is noiseless then $N_{o/p} = GN_{i/p}$ and $F = 1$. This would be an ideal system and, in practice, F will be greater than unity indicating that the system will introduce some noise. The ratio for F involving noise powers only is another way of defining noise figure, i.e. F is the ratio of the noise output power of a system ($N_{o/p}$) to the noise output power of a noiseless system ($GN_{i/p}$). If the system is assumed to introduce a noise power N_{sys}, then $N_{o/p} = GN_{i/p} + N_{sys}$ and equation (2.8) can be rewritten as:

$$F = \frac{GN_{i/p} + N_{sys}}{GN_{i/p}}$$

thus
$$F = 1 + \frac{N_{sys}}{GN_{i/p}} \qquad (2.9)$$

If $N_{i/p}$ is a source power of value $N_{i/p} = kT_sB$ then F becomes:

$$F = 1 + \frac{N_{sys}}{GkT_sB}$$

If T_s is made equal to T_0 (290 K), for reference purposes, in order to standardize a noise figure then:

$$F = 1 + \frac{N_{sys}}{GkT_0 B} \qquad (2.10)$$

Equation (2.10) shows that for systems with a high value for gain ($G \gg 1$), the value of F approaches the ideal condition.

For extremely low-noise devices it is more convenient to work in terms of noise temperature. Since N_{sys}/GkB has the dimensions of temperature it is possible to specify F in terms of temperature so that:

$$F = 1 + \frac{T_{eff}}{T_0} \qquad (2.11)$$

T_{eff} is the effective noise temperature of the system, depending only on the parameters of the system. It is a measure of how noisy the system is, referred to the input, since it is the temperature required of a resistance placed at the input of a noiseless system in order to produce the same value of noise power at the output as that produced within the system itself.

Example 2.2

A system has a quoted noise figure of 2.7 dB. What is its equivalent noise temperature?

Solution

Converting the decibel value to a ratio, then:

$$\text{Since } F = 1 + (T_{eff}/T_0)$$
$$\text{then } T_{eff} = T_0(F - 1)$$
$$T_{eff} = 290(1.86 - 1) = 250 \text{ K}$$

Figure 2.3 shows a typical communications receiver in equivalent circuit form. In

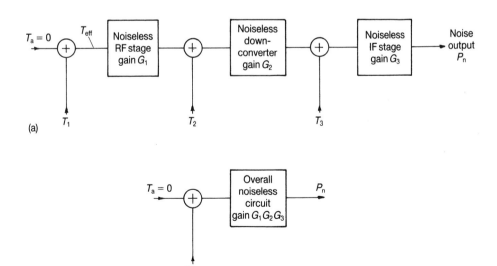

Fig. 2.3 (a) Equivalent circuit of receiver showing effect of noise temperature. (b) Simplified equivalent circuit showing effect of a single noise source.

Fig. 2.3(a), all noisy elements in the receiver are represented as noiseless circuits with their associated noise temperatures at the element inputs. The noise source at the input is assumed to be at a temperature T_{eff}.

Ignoring for the moment the effect of antenna noise, the total noise power at the output is given by P_n where:

$$P_n = G_3 k T_3 B + G_3 G_2 k T_2 B + G_3 G_2 G_1 k T_1 B$$

where G_1, G_2 and G_3 are the gains of each element of the receiver stage and T_1, T_2 and T_3 are the equivalent noise temperatures respectively of each element of the stage.

The equation for P_n could be rewritten as:

$$P_n = G_1 G_2 G_3 kB[T_1 + (T_2/G_1) + (T_3/G_1 G_2)]$$

A single source of noise, with noise temperature T_{eff}, at the input to the receiver would generate the same noise power P_n at the output provided:

$$P_n = G_1 G_2 G_3 k T_{\text{eff}} B$$

It follows then that:

$$T_{\text{eff}} = T_1 + \frac{T_2}{G_1} + \frac{T_3}{G_1 G_2} \qquad (2.12)$$

This is shown in the equivalent circuit of Fig. 2.3(b).

From equation (2.12) it can be seen that the first stage of the receiver contributes the most noise and the effect of the noise of the following stages is reduced as shown. If the first stage is a low-noise, high-gain stage the overall value of noise could be reduced to that of little more than that of the first stage only.

Example 2.3

A 4 GHz receiver consists of an RF stage with $G_1 = 25$ dB, $T_1 = 30$ K; a down-converter with $G_2 = 10$ dB, $T_2 = 350$ K; an IF stage with $G_3 = 15$ dB, $T_3 = 1200$ K. Calculate the value of the system noise temperature.

Solution

Converting decibel values to ratios,

$$T_{\text{eff}} = T_1 + (T_2/G_1) + (T_3/G_1 G_2)$$

hence, $T_{\text{eff}} = 30 + (350/316) + (1200/3160)$
$= 30 + 1.11 + 0.38 = 31.49$ K

Calculations can also be made in terms of the noise figure. Since it has already been established from equation (2.11) that $F = 1 + (T_{\text{eff}}/T_0)$, then:

$$F = 1 + \frac{T_1}{T_0} + \frac{T_2}{G_1 T_0} + \frac{T_3}{G_1 G_2 T_0}$$

This equation then reduces to:

$$F = F_1 + \frac{F_2 - 1}{G_1} + \frac{F_3 - 1}{G_1 G_2} \qquad (2.13)$$

2.2 Basic transmission concepts

Example 2.4

The receiver in the previous example has an RF stage gain $G_1 = 25$ dB, noise figure $F_1 = 0.42$ dB; a down-converter with gain $G_2 = 10$ dB, noise figure $F_2 = 3.44$ dB, and an IF stage with gain $G_3 = 15$ dB, noise figure $F_3 = 7.14$ dB. Calculate the overall noise figure and the value of the system noise temperature.

Solution

Converting decibel values to ratios and using equation (2.13) gives:

$$F = 1.1 + (3.83 \times 10^{-3}) + (1.32 \times 10^{-3})$$
$$F = 1.105$$

and since
$$T_{\text{eff}} = T_0(F - 1)$$
$$T_{\text{eff}} = 290(1.105 - 1) = 30.45 \text{ K}$$

Compare this value of T_{eff} with the value obtained in example 2.3.

Attenuator noise figure and noise temperature

If a purely resistive attenuator is inserted into a system such that a loss of factor L occurs in the power available between input and output, then:

$$P_{\text{o/p}} = P_{\text{i/p}}/L = GP_{\text{i/p}}$$
where $G = 1/L$

If the attenuator is resistive and at the same temperature, T_s, as the source resistance at its input, then:

$$P_{\text{o/p}} = kT_s B$$

Using $T_{\text{eff}} = N_{\text{sys}}/GkB$ for the attenuator, the total noise power at the attenuator output is given by:

$$P_{\text{o/p}} = GkT_s B + GkT_{\text{eff}} B = GkB(T_s + T_{\text{eff}})$$

Substituting for $G = 1/L$ gives:

$$P_{\text{o/p}} = \frac{kB(T_s + T_{\text{eff}})}{L}$$

Hence:

$$\frac{kB(T_s + T_{\text{eff}})}{L} = kT_s B$$

from which:

$$T_{\text{eff}} = T_s(L - 1)$$

This equation gives the effective temperature of a noise source, at temperature T_s, followed by an attenuator of loss factor L.

Example 2.5

A receiver system similar to that of Fig. 2.1 has an antenna with a feeder system having a loss factor $L = 2$ dB. The receiver stages are the same as in example 2.4 with the values

Satellite link parameters

for gain and noise figure as given in that example. Calculate the overall noise figure and system noise temperature.

Solution

Converting decibel values to ratios,
for the feeder system, $F = 1 + (T_{\text{eff}}/T_0)$ or:
$$F = 1 + [(L - 1)T_s/T_0]$$
If the attenuator is at the reference temperature, T_0, then:
$$F = 1 + [(L - 1)T_0/T_0] = L$$

The overall noise figure is thus:
$$F_{o/a} = F + \frac{F_1 - 1}{G} + \frac{F_2 - 1}{GG_1} + \frac{F_3 - 1}{GG_1G_2}$$

where $G = 1/L$

$$F_{o/a} = L + L(F_1 - 1) + L\frac{(F_2 - 1)}{G_1} + L\frac{(F_3 - 1)}{G_1G_2}$$

$$F_{o/a} = L\left[1 + (F_1 - 1) + \frac{(F_2 - 1)}{G_1} + \frac{(F_3 - 1)}{G_1G_2}\right]$$

$$F_{o/a} = L\left[F_1 + \frac{F_2 - 1}{G_1} + \frac{F_3 - 1}{G_1G_2}\right]$$

$F_{o/a} = (L + F)$ dB $= (2 + 0.43)$ dB $= 1.75$ dB.
and T_{eff} for the system $= T_0(F - 1) = 290(1.75 - 1)$
$= 217.5$ K

Example 2.5 has shown that the effect of the attenuation of the feeder system is to increase the overall system noise. The attenuation of the feeder system also reduces the amplitude of the signal so that the carrier-to-noise ratio will deteriorate.

The antenna itself is a source of noise. For a directive antenna pointed skywards at a particular elevation angle, thermal noise sources such as the sun and stars will have an effect. Side-lobes may also pick up terrestrial noise sources. An effective noise temperature (T_a) for an antenna may be defined as that temperature that would produce the same thermal noise as actually measured for the antenna i.e.:

$$T_a = (P_a/kB) \text{ Kelvin}$$

Taking antenna noise and feeder noise into account, the overall system noise temperature is given by:

$$T_s = \frac{T_a}{L} + \frac{(L-1)}{L}T_f + T_1 + \frac{T_2}{G_1} + \frac{T_3}{G_1G_2} \text{ Kelvin} \qquad (2.14)$$

where T_s is the system noise temperature in Kelvin
T_a is the antenna noise temperature in Kelvin
T_f is the feeder noise temperature in Kelvin
L is the attenuation loss factor
T_1 and G_1 are noise temperature and gain respectively of the RF stage of the receiver

2.2 Basic transmission concepts

T_2 and G_2 are noise temperature and gain respectively of the down-converter stage of the receiver

T_3 is the noise temperature of the IF stage of the receiver.

Equation 2.14 specifies the system temperature at the receiver input terminals. To find the system temperature at the antenna terminals simply multiply each term in equation (2.14) by L.

Example 2.6

(a) The receiver system of example 2.5 is fed from an antenna with a value of $T_a = 50$ K. Calculate the system noise temperature under these conditions with reference to receiver input terminals.

(b) What is the effect on the system noise temperature of referring to the antenna output terminals?

Solution

Converting decibel values to ratios,

(a) $$T_s = \frac{T_a}{L} + \frac{(L-1)}{L}T_f + T_1 + \frac{T_2}{G_1} + \frac{T_3}{G_1 G_2} \quad \text{Kelvin}$$

Assuming $T_f = T_0 = 290$ Kelvin,
$L = 2$ dB $= a\log 0.2 = 1.58$
$T_s = 50/1.58 + (0.58/1.58)290 + 30 + 1.11 + 0.38 = 169.6$ K

(b) T_s in this case is the value found in (a) above multiplied by L, i.e.:

$$T_s = 1.58 \times 169.6 = 268 \text{ Kelvin}$$

Example 2.6 shows the effect of the attenuation prior to the RF stage of the receiver. As attenuation increases so does the system noise temperature. For low-noise systems it is essential to keep the losses prior to the receiver RF stage to a minimum. As an exercise, consider the effect on the system noise temperature in example 2.6 if the value of L were to increase to 3 dB.

G/T ratio for earth stations

The received power equation (equation (2.4)) can be rewritten, in terms of C/N at the earth station, by adding the effect of system noise:

$$\frac{C}{N} = \frac{P_T G_T G_R}{k T_s B}\left(\frac{\lambda}{4\pi d}\right)^2$$

thus
$$\frac{C}{N} = \frac{P_T G_T}{kB}\left(\frac{\lambda}{4\pi d}\right)^2 \frac{G_R}{T_s} \quad (2.15)$$

$P_T G_T$ has already been specified as the EIRP (equivalent isotropic radiated power) and $(4\pi d/\lambda)^2$ is the free space loss (L_p) where λ is the signal wavelength.

Thus, equation (2.15) could be written as:

$$\frac{C}{N} = \frac{\text{EIRP}}{L_p}\left(\frac{G_R}{T_s}\right)\frac{1}{kB}$$

The term in brackets, usually shortened to G/T, is a factor, sometimes called a figure of merit, widely quoted for satellite receiving systems. The units for G/T are dB/K but since T is often quoted in dB relative to 1 K, or dBK, the ratio of G/T may be referred to as dBK. The Inmarsat-B specification gives minimum receive system G/T for a coast earth station (CES) at C-band (6/4 GHz) as 32 dBK or 30.7 dBK depending on which operational satellite is to be used. The full specification would include the elevation angle (5° for the Inmarsat-B system) since T_s depends on sky-noise temperature, which increases as the elevation angle is reduced below 10°.

The Intelsat system specification, using C-band (6/4 GHz) gives minimum receive system G/T for earth stations between the range 35.0 dB/K (for Earth Station Standard A) and 22.7 dB/K (for Earth Station Standard F1); using Ku-band (14/11/12 GHz) the value of system G/T ranges from 37.0 dB/K (for Earth Station Standard C) to 25.0 dB/K (for Earth Station Standard E1).

G/T values may be low or even negative in value. The Inmarsat-B specification, for example, quotes a minimum receive antenna system G/T of +2 dB for CES with an L-band (1.5/1.6 GHz) capability where reception of inter-station signalling channels and the C-to-L band AFC pilot is required. The Inmarsat-B system ship earth station (SES) has a value of G/T equal to or greater than -4 dBK in the direction of the satellite. A negative value of G/T simply means that the numerical value of T_s is greater than the numerical value of G_R.

It is worth mentioning at this point that the ratio G/T is better for earth stations than for satellites. The earth station antenna is 'pointing' at the sky and, assuming a 'perfect' antenna with no pick-up from sidelobes, the noise temperature is typically better than 100 K. The satellite antenna 'sees' the earth, which is assumed to be at 290 K.

Example 2.7

(a) An earth station antenna has a diameter of 18 metres with an overall efficiency of 70%. The received signal is at 4.15 GHz and at this frequency system noise temperature is 85 K when the antenna points at an elevation of 10°. Calculate the earth station G/T under these conditions.

(b) If heavy rain causes the sky temperature to increase, causing the system noise temperature to rise to 90 K, what is the new G/T value?

Solution

Assuming a circular aperture for the antenna,

$$GR = \eta(4\pi A/\lambda^2) = \eta(\pi d/\lambda)^2$$

At 4.15 GHz, $\lambda = 0.0723$ m and:

$G_R = 0.7 \times [(\pi \times 30)/0.0723]^2 = 428219.3 = 56.32$ dB
T_s in dBs relative to 1 Kelvin $= 10\log 85 = 19.3$ dBK
$G/T = 56.32 - 19.3 = 37$ dBK
If $T_s = 90$ K because of rain,
$G/T = 56.2 - 19.3 = 36.8$ dBK

Intermodulation distortion

Any amplifier operating close to its saturation level will exhibit non-linearity between the input and output powers. This can be seen by reference to the travelling wave tube

2.2 Basic transmission concepts

power transfer curve shown in Chapter 8 (Fig. 8.5). A mathematical approach to determining the effect of the non-linearity can be established using a series of the form:

$$V_0 = aV_1 + bV_1^3 + cV_1^5 \ldots \quad (2.16)$$

where V_0 is the output voltage

V_1 is the input voltage, given by:

$$V_1 = \sum_{i=1}^{n} V\cos\omega_i t \quad (2.17)$$

where n is the number of equal carriers and a, b etc are constants.

The Taylor series used above, represents the expected intermodulation effects since only the odd order terms of the transfer characteristic will generate products which fall into the frequency band of other signals and cause interference with those signals. Assuming three carrier frequencies are used, namely f_1, f_2 and f_3, the intermodulation products can be established as follows.

Assuming a total input power P_I of $nV^2/2$ watts then substituting equation (2.17) into equation (2.16) and expanding to determine the amplitudes of all intermodulation products gives, for each of the n equal carriers:

$$V = a\sqrt{\frac{2P_I}{n}}\left(1 + 3\frac{b}{a}P_I + 15\frac{c}{a}P_I^2 + \ldots\right) \quad (2.18)$$

where the term of equation (2.18) not in brackets represents the linear component of the transfer characteristic and the term in brackets represents the deviation from linearity around saturation.

Amplitudes of intermodulation products of frequency $(2f_1 - f_2)$ are given by:

$$I_{n1} = \frac{3b}{4}\left(\frac{2P_I}{n}\right)^{3/2}\left[1 + \frac{2c}{3b}\frac{P_I}{n}(12.5 + 15(n-2)) + \ldots\right]$$

If terms in the series of third-order and above are neglected this reduces to:

$$I_{n1} = (3b/4)(2P_I/n)^{3/2} \quad (2.19)$$

Amplitudes of intermodulation products of frequency $(f_1 + f_2 - f_3)$ are given by:

$$I_{n2} = \frac{3b}{2}\left(\frac{2P_I}{n}\right)^{3/2}\left[1 + \frac{10cP_I}{bn}(1.5 + (n-3)) + \ldots\right]$$

Again if third-order and above terms are neglected this reduces to:

$$I_{n2} = (3b/2)(2P_I/n)^{3/2} \quad (2.20)$$

It can be seen from equations (2.19) and (2.20) that the products given by I_{n2} are 6 dBs higher in amplitude than those given by I_{n1}.

Within the transponder bandwidth, intermodulation products fall on the same frequencies as carriers. For n carriers the maximum number of products falling on the centre carriers is given by:

$N = (n-2)/2$ for $(2f_1 - f_2)$ products $\quad (2.21)$
$N' = (n-2)(3n-4)/8$ for $(f_1 + f_2 - f_3)$ products $\quad (2.22)$

If the intermodulation products are assumed incoherent then the ratio of carrier to intermodulation power can be written as:

$$(C/I)_n = V^2/(NI_{n1}^2 + N'I_{n2}^2)$$

and from equations (2.18), (2.19) and (2.20):

$$\left(\frac{C}{I}\right)_n = \frac{4n^2\left(\frac{a}{b}\right)^2\left(1 + 3\frac{b}{a}P_I + 15\frac{c}{a}P_I^2\ldots\right)^2}{9P_I^2\,(N + 4N')}$$

Substituting for $N + 4N'$ (which reduces to $(3/2)(n-1)(n-2)$ from equations (2.21) and (2.22):

$$\left(\frac{C}{I}\right)_n = \frac{8n^2\left(\frac{a}{b}\right)^2\left(1 + 3\frac{b}{a}P_I + 15\frac{c}{a}P_I^2\ldots\right)^2}{27P_I^2(n-1)(n-2)} \qquad (2.23)$$

Putting $n = 2$ and solving equation (2.23) to find the ratio of a/b gives:

$$\frac{a}{b} = 0.75P_I\left[\sqrt{\left(\frac{C}{I}\right)_2} - 3\right] \qquad (2.24)$$

The ratio $(C/I)_2$ is used by amplifier manufacturers and this information should readily be available.

Substituting the result of equation (2.24) into equation (2.23) gives:

$$\left(\frac{C}{I}\right)_n = \frac{n^2}{6(n-1)(n-2)}\left[\sqrt{\left(\frac{C}{I}\right)_2} + \left(\frac{n-2}{n}\right)\right]^2$$

which, as $n \to \infty$, reduces to:

$$\left(\frac{C}{I}\right)_\infty = \frac{1}{6}\left[\sqrt{\left(\frac{C}{I}\right)_2} + 1\right]^2$$

The level of interference due to intermodulation products may be reduced by a suitable choice of carrier frequencies. Assignment of frequencies may be arranged so that most of the intermodulation products fall outside the transponder band. The expressions above have ignored the higher terms in the series and hence will not be accurate close to saturation where the higher order terms will come into effect.

The effect of intermodulation distortion can be minimized by operating the power amplifier device at a point below saturation. The operating point would be several decibels below the saturation power level giving an input backoff BO_i dB which, in turn, causes an output power backoff BO_o dB. This is discussed in more detail in Chapter 8, Section 4.

Intermodulation distortion caused by AM to PM conversion

Many amplifiers, including the travelling-wave tube amplifier (TWTA), produce phase modulation of the output signal, which is a function of the square of the envelope of the input signal, i.e. amplitude modulation (AM) converted to phase modulation (PM). Amplitude variations can exist even when a constant amplitude modulation technique, such as FM or PSK, is used and will occur for single carrier and multiple carrier signals. For single carrier operation, band limiting of the signal causes a ripple to be produced in the signal envelope while for multiple carrier operation the input signal is made up of many different incoherent carrier frequencies, which causes the carrier envelope amplitude to fluctuate. The AM to PM conversion will produce a spectrum of unwanted output frequency components in much the same way as discussed in the previous section. The effect of AM to PM conversion is usually specified in terms of a transfer coefficient k_o which has units of degrees of phase shift per decibel of amplitude change (°/dB). Figure 2.4 shows a typical amplifier curve for k_o plotted against input power.

2.2 Basic transmission concepts

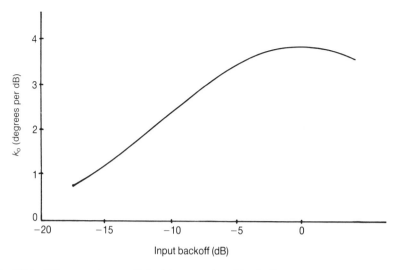

Fig. 2.4 AM to PM conversion coefficient k_o versus input backoff

The phase modulation $\theta(t)$ produced at the amplifier output for all power levels, including saturation, can be given by a polynomial of the square of the input signal envelope:

$$\theta = aV^2(t) + b(V^2(t))^2 + c(V^2(t))^3 + \ldots \quad (2.25)$$

If the input $V_1(t)$ consists of n frequency modulated carriers then:

$$V_1(t) = \sum_{i=1}^{n} V_i \cos(\omega_i t + \phi_i(t)) = \sum_{i=1}^{n} V_i \cos W_i(t) = V(t)\cos\omega_o(t) \quad (2.26)$$

where $V(t)$ is the envelope of $V_1(t)$ and is the term which allows for the phase modulation shown in equation (2.25), and ω_o is an arbitrary value of angular frequency.

Assuming the amount of modulation is small compared to the input power:

$$V(t) = \left[\sum_{i=1}^{n} V_i^2\right]^{1/2} \left[1 + \frac{\sum_{j=1}^{n}\sum_{k=1}^{n} V_j V_k \cos(W_j(t) - W_k(t))}{\sum_{i=1}^{n} V_i^2}\right]^{\frac{1}{2}} \quad (2.27)$$

This expression uses only two values and assumes that $j \neq k$. The first term in equation (2.27) is constant while the second term is the AM contribution.

With reference to equation (2.25), the coefficient a can be found in terms of the transfer coefficient k_o for small values of k_o:

$$a = \frac{20\log(1 + k_o)}{57.3} = 0.1516 k_o \quad (2.28)$$

Thus, the output, together with the added phase modulation $aV^2(t)$ to each carrier, is given by:

$$\sum_{i=1}^{n} V_i \cos\left[W_i(t) + 0.1516 k_o \frac{\sum_{j=1}^{n}\sum_{k=1}^{n} V_j V_k \cos(W_j(t) - W_k(t))}{\sum_{i=1}^{n} V_i^2}\right] \quad (2.29)$$

where $j \neq k$.

The output signal can be found from equation (2.29) and for small values of k_o gives, for each of the n carriers:

$$V_i\cos(W_i(t)) - V_i 0.1516k_o \left(\frac{\sum_{j=1}^{n}\sum_{k=1}^{n} \frac{V_j V_k}{2} \sin[W_i(t) \pm (W_j(t) - W_k(t)]}{\sum_{i=1}^{n} V_i^2} \right) \quad (2.30)$$

where $j \neq k$.

The second term in equation (2.30) is the distortion contribution.

If $W_i \pm (W_j - W_k)$ is equal to the angular frequency of another carrier then the distribution effect on the required signal will be apparent. In the special case where $V_i = V_j = V_k$ for n carriers, equation (2.30) reduces to:

$$\left(\frac{C}{I_{\text{AM-PM}}} \right)_n = \left(\frac{n}{0.1516k_o} \right)^2 \quad (2.31)$$

Equation (2.31) gives C/I noise for a product of the $i \neq j \neq k$ type. All other combinations (such as $i=j$, $i=k$) are fewer in number and, since they contribute less to output noise power, they can safely be ignored.

For the Intelsat V and VI satellites the specifications for the TWTAs, in terms of output phase shift, intermodulation product values and AM–PM transfer coefficient are as shown in Tables 2.1 to 2.3. For Intelsat VI the opportunity for manufacturers to use solid-state power amplifiers (SSPAs) was incorporated and the specification for these devices is also shown in Tables 2.1 to 2.3. Because the performance with SSPAs was expected to be better than with TWTAs the specification was more stringent.

Interference

The effect of noise and intermodulation distortion on system performance has been established earlier in this chapter. Noise and intermodulation products are, however, not the only unwanted signals to affect the reception of a wanted signal; there are other unwanted signal effects which may be classified broadly as interference. Possible sources of interference include:

- interference from a satellite link into a terrestrial link where the same frequency bands are in use
- interference between satellites sharing the same frequency bands
- interference from a terrestrial link into a satellite link where the same frequency bands are in use.

The ITU Radio Regulations list the frequency spectrum allocation for particular services such as Fixed Satellite Service (FSS), Mobile Satellite Service (MSS), Fixed (terrestrial) Service, Mobile (terrestrial) Service etc. Frequency allocation is accomplished on a regional basis and the regulations specify whether a particular satellite service, for example, can use any, or all, of a frequency band in specified regions. The restriction is aimed at the prevention of interference with users of the same band in the specified regions.

The Fixed Satellite Service shares frequency bands with terrestrial networks in the 6/4 GHz and 14/12 GHz bands and thus it is possible that a terrestrial network could affect a satellite on the up-link, or that a terrestrial network may be affected by the down-link from a satellite, or even directly by an earth station. Additionally, the satellite up-link may be affected by radiation from earth stations and, equally, earth stations may be affected by reception of signals from adjacent satellites. Figure 2.5 shows the possible interference pattern between adjacent satellites and adjacent earth stations.

Table 2.1 Output transmitter phase shift (reproduced by permission of *COMSAT Technical Review*, CTR, Vol. 20, No. 2, Fall 1990, p 300)

Relative flux density* (dB)	Maximum phase shift (deg)	
	TWTA	SSPA
0	46	20
3	38	14
6	28	7
9	18	3
12	12	1
14	9	1

Table 2.2 AM–PM transfer coefficient (reproduced by permission of *COMSAT Technical Review*, CTR, Vol. 20, No. 2, Fall 1990, p 301)

Relative flux density* (dB)	AM–PM Transfer coefficient (°/dB)	
	TWTA	SSPA
0	8	2
3	9	2
6	9	2
9	8	2
12	5	1
14	3	1
>14	3	1

* Measured below the flux density that produces single-carrier saturation.

Table 2.3 Intermodulation product values (reproduced by permission of *COMSAT Technical Review*, CTR, Vol. 20, No. 2, Fall 1990, p 301)

Relative flux density per carrier* (dB)	Maximum C/I** (dB)	
	TWTA	SSPA
3	−10	−12
10	−15	−25
17	−26	−38

* Flux density illuminating the spacecraft for each of two equal-amplitude carriers, below the flux density which produces single-carrier saturation.
** Maximum level of third-order intermodulation product relative to the level of each RF carrier, measured at the output of each transmission channel.

34 Satellite link parameters

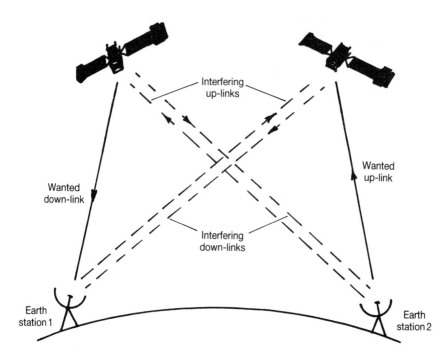

Fig. 2.5 Interference between systems using adjacent satellites

Since one of the interference problems is caused by the power contained in the sidelobes radiated by an antenna, this effect can be minimized by careful antenna design and by siting earth stations and terrestrial network stations such that they are unlikely to receive unwanted main beam or sidelobe radiation. If such radiation is present then the interference level should be such that it is an acceptable level compared with the required signal level and its effect can be minimized. The EIRP of an interfering transmitter depends on the transmitter power and the antenna gain in the direction of the interference signal. To avoid interference from satellites into terrestrial networks a limit is placed on the power flux density (PFD) that is permitted to arrive at the surface of the earth at low angles of elevation. Similarly, earth station radiation must have restricted values of EIRP allowed at elevation angles of less than 5°.

Because of the increasing number of satellite systems there are more satellites in the geostationary orbit and these may have the same frequency bands and footprints that overlap. The ability of the space segment to cope with the increased traffic has been aided by the use of new frequency bands and reuse of existing frequency bands. Frequency reuse involves spatially separating the antenna footprint or by employing orthogonal polarization. Types of orthogonal polarization include linear polarization (vertical and horizontal) and circular (left-hand and right-hand). The use of orthogonal polarization can result in cross-polarization isolation levels of 30 dB or more and enables a doubling of the usable bandwidth, although at the expense of some co-channel interference. Because co-channel interference occurs on both the up-link and the down-link then, even if the separate links had 30 dB isolation, the composite value of the complete link is only 27 dB.

The effect of interference from an adjacent satellite can be evaluated in terms of signals received at the earth station from the required satellite and those received from the interfering satellite. Interference from a single satellite is often referred to as single entry interference.

The wanted carrier power C is given by:

$$C = E_W - L_{DWW} + G_W \text{ dB}$$

where:

E_W is required EIRP in dBW
L_{DWW} is the down-link path loss in the direction of the desired satellite, dB.
G_W is the earth station antenna gain in the direction of the desired satellite, dB.

The interfering carrier power I is given by:

$$I = E_I - L_{DWI} + G_I \text{ dB}$$

where:

E_I is the interfering EIRP in dBW
L_{DWI} is the down-link path loss in the direction of the interfering satellite, dB.
G_I is the earth station antenna gain in the direction of the interfering satellite, dB.

Then C/I is simply:

$$C/I = (E_W - E_I) - (L_{DWW} - L_{DWI}) + (G_W - G_I) \text{ dB}$$

The receive station antenna discrimination $DA(\theta)$ is defined as the antenna gain towards the desired satellite minus the gain in the direction of the interfering satellite.

If G_I is the antenna gain in a direction θ to the main beam then, according to CCIR requirements for antennae where the ratio of antenna diameter (D) to operating wavelength (λ) is greater than 100, the gain in that direction must be $32 - 25\log_{10}\theta$ dBi (the FCC require that some antennae have an antenna radiation pattern given by $29 - 25\log_{10}\theta$ dBi). θ is known as the off-axis angle and is measured in degrees. The antenna pattern envelope for an earth station conforming to the CCIR requirements is shown in Fig. 2.6. The FCC requirements are shown on the same diagram.

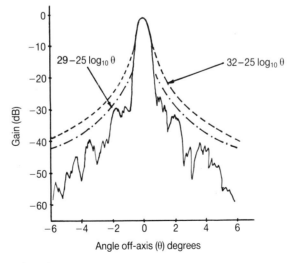

Fig. 2.6 Typical earth station antenna pattern

36 Satellite link parameters

Figure 2.6 represents antenna gain against the off-axis angle in degrees for a small diameter antenna and is, in effect, a cross-sectional 'view' across the beam. The dashed lines represent the sidelobe limits. The pattern is typically symmetrical with the two sides not necessarily being identical.

For ratios of D/λ of less than 100 the allowable gain is $52 - 10\log_{10}(D/\lambda) - 25\log_{10}\theta$.

Thus, $DA(\theta) = G_W - 32 + 25\log_{10}\theta$

and, assuming path losses are identical:

$$C/I = (E_W - E_I) + G_W - 32 + 25\log_{10}\theta$$

or:

$$C/I = (E_W - E_I) + DA(\theta) \text{ dB}$$

Similar calculations can be made for the up-link where a satellite may receive unwanted signals from an interfering earth station. C/I values can also be obtained for cross-polar interference in up-link and down-link paths and, where applicable, adjacent transponder interference caused by intermodulation products when several carriers sharing the same station high power amplifier (HPA) accesses one or more transponders on the same satellite.

Where interference is noise-like it is possible to combine C/I with C/N to give an overall C/N figure for a link. For example if N_T is made up of N and I_T then:

$$C/N_T = C/(N + I_T)$$

or:

$$\frac{1}{\left(\dfrac{C}{N_T}\right)} = \frac{N + I_T}{C} = \frac{N}{C} + \frac{I_T}{C}$$

or:

$$\frac{1}{\left(\dfrac{C}{N_T}\right)} = \frac{1}{\left(\dfrac{C}{N}\right)} + \frac{1}{\left(\dfrac{C}{I_T}\right)}$$

2.3 Link budgets

Introduction

A satellite system needs to be planned in order to ensure that the transmission link can be satisfactorily established, having regard to the conditions under which the link is operated. Previous sections have shown how a link will experience various gains and losses; collecting the gains and losses together produces what is known as the link budget. Such a budget may be defined in terms of an up-link from, say, a coast earth station (CES) or ship earth station (SES) to a satellite and a down-link from the satellite to the CES or SES. Alternatively, the link budget may be defined in terms of a forward link, which may be from a land earth station (LES), say, to a mobile earth station (MES), and a return link which sends the signal from the MES, via the satellite, to the LES.

As an example, details of the Inmarsat-B system are given below for voice (V), data (D) and the Network Co-ordination Station (NCS), Time Division Multiplex (TDM) channels. A typical arrangement for the system is shown in Fig. 2.7.

2.3 Link budgets

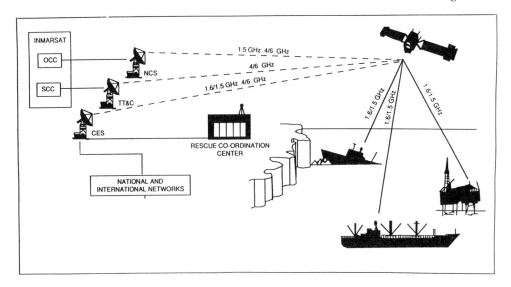

Fig. 2.7 Inmarsat-B links between satellite and earth stations

Table 2.4 Inmarsat-B shore-to-ship link budgets (courtesy Inmarsat)

Channel type		CESV	CESD	NCS/CES TDM*
CES elevation angle		5°	5°	5°
CES EIRP	(dBW)	56.0	56.0	51.0
Path loss	(dB)	200.9	200.9	200.9
Absorption loss	(dB)	0.4	0.4	0.4
Satellite G/T	(dBK)	−14.0	−14.0	−14.0
Mean up-path C/N_o	(dBHz)	69.3	69.3	64.3
Mean satellite C/IM_o	(dBHz)	60.7	60.7	55.7
SES elevation angle		5°	5°	5°
Satellite EIRP	(dBW)	16.0	16.0	11.0
Path loss	(dB)	188.5	188.5	188.5
Absorption loss	(dB)	0.4	0.4	0.4
SES G/T	(dBK)	−4.0	−4.0	−4.0
Down-path C/N_o	(dBHz)	51.7	51.7	46.7
Nominal unfaded C/N_o	(dBHz)	51.2	1.2	46.2
Fading margin	(dB)	4.0	4.0	4.0
Overall C/N_o	(dBHz)	47.2	47.2	42.2
Theoretical C/N_o	(dBHz)	45.9	45.2	39.4
Available margin	(dB)	1.3	2.0	2.8

Note: the above link budgets apply to Inmarsat first and second-generation satellites assuming worst-case transponder conditions.

* This link budget is for the NCS TDM channel used for combined NCSA, NCSC and NCSI functions and also when used for separate (or combined) NCSA and NCSC functions. However, when it is used only for NCSI function (i.e. inter-station signalling) the SES G/T of −4.0 dB becomes CES G/T of 2.0 dB and the fading margin 0 dB.
The case for CES TDM channels is similar.

38 Satellite link parameters

Table 2.5 Inmarsat-B ship-to-shore link budgets (courtesy Inmarsat)

Channel type		SESV	SESD	SES TDMA and SESRQ/SESRP
SES elevation angle		5°	5°	5°
SES EIRP	(dBW)	33.0	33.0	33.0
Path loss	(dB)	189.0	189.0	189.0
Absorption loss	(dB)	0.4	0.4	0.4
Satellite G/T	(dBK)	−11.2	−11.2	−11.2
Mean up-path C/N_o	(dBHz)	61.0	61.0	61.0
Mean satellite C/IM_o	(dBHz)	63.7	63.7	63.7
CES elevation angle		5°	5°	5°
Satellite EIRP	(dBW)	−5.5	−5.5	−5.5
Path loss	(dB)	197.2	197.2	197.2
Absorption loss	(dB)	0.4	0.4	0.4
CES G/T	(dBK)	32.0	32.0	32.0
Down-path C/N_o	(dBHz)	57.5	57.5	57.5
Nominal unfaded C/N_o	(dBHz)	55.2	55.2	55.2
Fading margin	(dB)	4.0	4.0	4.0
Miscellaneous loss*	(dB)	1.5	1.5	1.5
Overall C/N_o	(dBHz)	49.7	49.7	49.7
Theoretical C/N_o	(dBHz)	45.9	45.2	45.2
Available margin	(dB)	3.8	4.5	4.5

* **Note:** including loss due to SES HPA impairment and adjacent channel interference.

The above link budgets represent the worst case for Inmarsat first-generation satellites. Second-generation satellite link budgets have the following differences resulting in improved overall link performance:
- satellite G/T = −12.5 dBK at edge of coverage;
- satellite C/IM_o = 65.1 dBHz;
- satellite EIRP = −3.1 dBW (resulting from improved transponder gain);
- path loss = 195.9 dB on C-band down-link; and
- CES G/T = 30.7 dBK.

Frequencies used for the CES-satellite links are in C-band with 6 GHz for the up-link and 4 GHz for the down-link. Frequencies used for SES-satellite are in L-band with 1.6 GHz for the up-link and 1.5 GHz for the down-link.

Table 2.4 shows a shore-to-ship link budget for CESV channels (voice telephony), CESD channels (9.6 kbit/s SCPC data/facsimile) and NCS/CES TDM channels (telex, low-speed data and signalling).

Table 2.5 shows a ship-to-shore link budget for SESV channels (voice telephony), SESD channels (9.6 kbit/s SCPC data/facsimile) and SES TDMA/SESRQ/SESRP channels (telex, low-speed data and signalling).

The link budgets shown in Tables 2.4 and 2.5 represent channel performance under generally worst case satellite transponder conditions for Inmarsat first and second generation satellites (MARECS, INTELSAT-V MCS and Inmarsat-2) and are shown only as examples.

The values shown for 'theoretical C/N_o' for each channel type relate to the corresponding values given in Table 3.1 of Chapter 3 (see page 60). The values shown for 'available margin' relate to the margin above the theoretical C/N_o requirements after

2.3 Link budgets

taking into account fading and other link losses. The margin therefore includes equipment implementation degradations compared with the assumed theoretical performance. This aspect will be discussed further, later in this section.

Inmarsat-B link budget

To consider a link budget in greater detail, take the example of the Inmarsat-B link budget for a telephony channel for a satellite–ship down-link and a ship–satellite up-link, using the INTELSAT-V MCS satellite:

(a) Satellite–ship down-link
Frequency   1535 – 1542.5 MHz
Satellite EIRP   16 dBW
Path loss at 1542.5 MHz at 40000 km range
$\qquad\qquad\qquad\qquad L_p = 20\log[4\pi d/\lambda]$ 188.5 dB
Absorption loss (L_a)   0.4 dB
SES G/T −4 dBK
Boltzmann's constant (k)   −228.6 dBW/Hz/K
$C/N_o = (\text{EIRP} - L_p - L_a + G/T - 10\log k)$ 51.7 dBHz

Table 2.4 gives the down path C/N_o as 51.7 dBHz. A nominal value of 51.2 dBHz together with a fading level of 4 dB gives an overall value of C/N_o of 47.2 dBHz. By comparison with the theoretical value of 45.9 dBHz required to give the BER value under white Gaussian noise (see Table 3.1 of Chapter 3) there is a margin of 1.3 dBHz.

(b) Ship–satellite up-link
Frequency   1636.5 – 1644 MHz
Ship terminal EIRP   33 dBW
Satellite G/T_s −11.2 dBK
Path loss to satellite at 1636.5 MHz
$\qquad\qquad\qquad\qquad L_p = 20\log[4\pi d/\lambda]$ 189 dB
Absorption loss L_a 0.4 dB
Boltzmann's constant (k)   −228.6 dBW/Hz/K
$C/N_o = (\text{EIRP} - L_p - L_a + G/T_s - 10\log k)$ 61 dBHz

Table 2.5 gives the up path C/N_o as 61 dBHz. The effect of intermodulation in the transponder band is given as C/IM_o with a value of 63.7 dBHz. If the effect of intermodulation noise is to be allowed for, the ratio of carrier power to noise power should be $C/(N_o + IM_o)$, so that overall value of $(C/N_o)_{o/a} = C/(N_o + IM_o)$. This can be written as:

$$(N_o/C)_{o/a} = (N_o + IM_o)/C = N_o/C + IM_o/C$$

and $1/(C/N_o) = 1/(C/N_o)_{o/a} - 1/(C/IM_o)$
for $(C/N_o)_{o/a}$ to have a mean value of 61 dBHz (1258925.4) with a value of (C/IM_o) of 63.7 dBHz (2344228.8) then C/N_o must be:

$\quad 1/(C/N_o) \quad = 7.943 \times 10^{-7} - 4.266 \times 10^{-7}$
$\qquad\qquad\qquad = 3.677 \times 10^{-7}$
$\quad$ and $(C/N_o) \quad = 2719608.3 = 64.3$ dBHz.

For the satellite/CES link the satellite EIRP is given as −5.5 dBW. The down-link frequency is between 4192.5 MHz and 4200 MHz. The flux density at the surface of the earth can be calculated using the data given and, assuming a path length of 40,000 km, using equation (2.2):

$$\text{PFD} = (\text{EIRP} - 10\log 4\pi - 20\log(4 \times 10^7)) \text{ dBW/m}^2$$
$$\text{PFD} = (-5.5 - 11 - 152) \text{ dBW/m}^2$$
$$= -168.5 \text{ dBW/m}^2 \tag{2.32}$$

The calculated value of PFD for a given system is often called the 'illumination factor' W and it can be shown that:

$$W = \text{EIRP} - 163.3 \text{ dBW/m}^2 \tag{2.33}$$

where:

W is expressed in dBW/m² and is the total power in the *full* transmission bandwidth.

The figure 163.3 dBW/m² includes slant range attenuation factor and the gain of a 1 m² 'ideal' antenna, at 5° elevation angle and assumes edge of earth coverage.

Typically, for a geostationary satellite, the range to the edge of coverage is 41,680 km while the range to the point directly below the satellite on the equator (sub-satellite point) is 35,786 km. Using the sub-satellite point range, equation (2.17) becomes:

$$W = \text{EIRP} - 162 \text{ dBW/m}^2$$

The difference in range gives a maximum difference of 1.3 dB. It follows that where the satellite footprint is small the value of W will be larger than the value suggested by equation (2.33) but that the maximum value can only be 1.3 dB above that given by equation (2.33).

Using equation (2.33) with an EIRP value of -5.5 gives an illumination factor W of -168.8 dBW/m².

The slight difference between this value and the value of -168.5 dBW/m² used earlier is due to the approximation for the slant range d in equation (2.16).

For a signal having a rectangular spectrum the power flux density is the illumination level W divided by the ratio of the actual transponder bandwidth to the International Radio Consultative Committee (CCIR) specified bandwidth i.e.:

$$\text{PFD}_B = W - 10\log(B_t/B_{\text{CCIR}}) \text{ dBW/m}^2 \text{ per } B_{\text{CCIR}}$$

where:

PFD_B = power flux density at edge of earth coverage
B_t = transponder bandwidth actually occupied in Hertz
B_{CCIR} = CCIR specified bandwidth in Hertz.

Hence:

$$\text{PFD}_B = \text{EIRP} - 163.3 - 10\log(B_t/B_{\text{CCIR}}) \text{ dBW/m}^2 \text{ per } B_{\text{CCIR}}$$

Often, PFD is expressed as a value in dBW/m² without reference to a particular bandwidth. The bandwidth should at least be indicated even if not specified.

There is a relationship between the maximum EIRP value and PFD as a function of occupied transponder bandwidth. In certain types of systems only a small part of the transponder bandwidth may be occupied so as to enable the available EIRP to be concentrated into the narrower bandwidth, thus raising the EIRP density (dBW/Hz) and therefore raising the PFD (dBW/m² Hz). If the PFD value is specified, the available EIRP will set the upper limit to the amount of bandwidth that can actually be used.

The CCIR has set a limit to the maximum value of permissible flux density at the surface of the earth to prevent interference with terrestrial links due to a satellite transmitting at 4 GHz. For shared frequency channels in the band 3.4 GHz to 7.75 GHz, in any 4 kHz bandwidth slot, for a wave arriving at an angle $\theta°$ (where θ is between 0° and 5°), the maximum permitted flux density $\text{PFD}_{max} = -152$ dBW/m².

Where θ is between 5° and 25° above the horizon, the maximum permitted flux density PFD_{max} is given by:

$$\text{PFD}_{max} = \left[-152 + \left(\frac{\theta - 5}{2}\right)\right] \text{ dBW/m}^2$$

Thus, for example, if the elevation angle θ is 25° the value of PFD_{max} is -142 dBW/m²

At 4 GHz, 0° elevation, the flux density is limited to -152 dBW/m² for a 4 kHz band, so if the value of PFD is greater than this level, energy dispersal may be required to prevent all the transponder power being radiated at one frequency if no modulation were applied to the carrier.

Assuming a constant power satellite transmitter, a single narrowband (cw) carrier has the highest PFD because of the concentration of energy. Modulation of the carrier, or the addition of other carriers, causes the power to be spread over a wider bandwidth. PFD per unit bandwidth is thus decreased while the illumination level remains constant (assuming no output 'backoff'). Often, a deliberate spectrum-spreading waveform is used to produce energy dispersion therefore reducing PFD per unit bandwidth to acceptable levels.

For the system under consideration the value of PFD is -168.5 dBW/m², which is well below the limit set as the maximum, so that no energy dispersal is necessary.

Noise at the CES input is given by equation (2.5). Low noise temperatures are possible with helium-cooled parametric amplifiers where a value of $T_n = 30$ K over a 500 GHz band is possible. To this must be added antenna noise temperature which may be, typically, $T_a = 50$ K for a large antenna at 4 GHz. The effects of other noise sources within the receiver could lead to a system temperature of, say, $T_s = 180$ K. Assuming a channel bandwidth of 20 kHz, the system noise can be calculated:

$$\begin{aligned} N = P_n &= kT_sB \\ &= 1.38 \times 10^{-23} \times 180 \times 20 \times 10^3 \text{ watts} \\ &= 4.968 \times 10^{-17} \\ &= 10\log(4.968 \times 10^{-17}) \text{dBW} \\ &= -163 \text{ dBW} \end{aligned}$$

To maintain satisfactory communications, the value of C/N must remain above a threshold value under all conditions. For a frequency modulation (FM) system, the threshold is in the range 4 to 15 dB depending on the type of demodulator used. For phase-shift-keying (PSK) systems, the threshold is, typically, 8 to 15 dB. Assuming that, for the system considered, using PSK with a channel bandwidth of 20 kHz, a threshold of 10 dB is used. Additionally, there is a system margin which allows for propagation and equipment degradation. For example, at 4 GHz, if a fading of 1 dB occurs which degrades C by 1 dB, and sky noise increases to a value which increases N by, say, 1 dB, then there is a reduction in C/N of 2 dB due to propagation factors. A further allowance of, say, 1 dB could allow for equipment degradation, giving a system margin of 3 dB.

When using FM and frequency division multiple access (FDMA), the transponder cannot be operated at maximum output power because of non-linearity of the output power device (often a travelling-wave tube amplifier (TWTA)) at high values of output power. Backoff is usually employed, usually between 3 to 7 dB at the output of the device, to keep intermodulation products down to a level that is acceptable. Thus, for a system with many channels, the backoff will need to increase as the number of multiple

42 Satellite link parameters

accesses, and hence intermodulation level, increases. Intermodulation is caused by the non-linearity of the transponder output device when operating with a multiplicity of signals. The generation of a large number of intermodulation products under these conditions can cause some interference within the transponder passband. The effect may be minor but it does cause an overall increase in noise levels which would reduce the C/N value of the required signal at the earth station. For details of the effects of intermodulation see the section on intermodulation distortion.

Antenna calculations

The size of antenna required to support the parameters of a system can be calculated as shown below. The calculations are included here because they are based on the Inmarsat-B system but they are applicable to any system.
Using equation (2.17):

$$\text{PFD} = (\text{EIRP} - 163.3) \text{ dBW/m}^2 = -168.8 \text{ dBW/m}^2$$
$$\text{But } C/N = P_R/N = (\text{PFD} \times \eta A_R)/N$$

and since the signal into the earth station receiver must be above the C/N threshold level for 99% of the time, and allowing for the system margin, a C/N of 13 dB is needed under clear sky conditions:

$$N = -163 \text{ dBW}$$
hence $\quad C = -150 \text{ dBW}$
but $\quad C = F + \eta A_R \quad$ using decibels
so that $\quad \eta A_R = C \text{ dBW} - \text{PFD dBW/m}^2$
$\quad \eta A_R = -150 \text{ dBW} + 168.8 \text{ dBW/m}^2$
$\quad \eta A_R = 18.8 \text{ dBm}^2$
$\quad = 75.9 \text{ m}^2$

Assuming an antenna aperture efficiency of 70%, the actual area of a dish with an effective area of 75.9 m² is 108.4 m², giving a diameter of 11.75 m. This value for the diameter of the earth station antenna represents the minimum diameter, at an efficiency of 70%, that could be used in the system to provide a C/N of 13 dB with a system noise temperature, T_s, of 180 K. A coast earth station (CES) could use an antenna of diameter 13 – 15 m.

Considering the down-link from the satellite to the SES, and using equation (2.17):

$$\text{PFD} = 16 - 163.3 \text{ dBW/m}^2$$
$$= -147.3 \text{ dBW/m}^2$$

With an earth station antenna of gain approximately 20 dB and G/T of -4 dBK, it follows that T_s for the SES receiver is about 24 dBK, or $T_s = 250$ K.

$$N = kT_sB$$
$$= 1.38 \times 10^{-23} \times 250 \times 20 \times 10^3 \text{ watts}$$
$$= 6.9 \times 10^{-17}$$
$$= -161.6 \text{ dBW}$$

alternatively, using dBs directly:

$$N = -228.6 + 10\log250 + 10\log(20 \times 10^3)$$
$$= -228.6 + 24 + 43$$
$$= -161.6 \text{ dBW}$$

2.3 Link budgets 43

With a 9 dB threshold level for C/N at the receiver:

$$C = -161.6 + 9 \text{ dBW}$$
$$= -152.6 \text{ dBW}$$

and
$$C = \text{PFD} + \eta A_R$$

so
$$\eta A_R = C - \text{PFD}$$
$$= -152.6 + 147.3 \text{ dBW/m}^2$$
$$= -5.3 \text{ dBW/m}^2$$
$$= 0.3 \text{ m}^2$$

With an antenna efficiency of 66%, the actual area of the dish, with an effective area of 0.3 m², is 0.45 m², giving a diameter of 0.76 m. This would correspond well with the expected SES antenna diameter of between 0.9 m and 1.2 m while still receiving the desired signal at a C/N ratio of 9 dB and system noise temperature T_s of 250 K.

Intelsat V link budget

The link budgets to be described are for a 6 GHz/4 GHz link using different beam configurations as indicated, and a 14 GHz/11 GHz link using spot beams.

(1) 6 GHZ/4 GHz link.
(a) Earth station–satellite up-link

	B	A	
Earth station type			
Beam configuration	Hemi	Zone	
EIRP	93.9	93.9	dBW
Satellite G/T_S	−11.6	−8.6	dB/K
Path loss to satellite (L_p) 40,000 km at 6 GHz	200.1	200.1	dB
Satellite illumination level (W)*	−69.4	−69.4	dBW/m²
Boltzmann's constant (k)	−228.6	−228.6	dBW/Hz/K
$C/T =$ ($W + G/T_S − 21.5 − 20\log_{10}f$)†	−121.1	−118.8	dBW/K
$C/N_o =$ (EIRP $− L_p + G/T_s − 10\log_{10}k$)**	110.8	113.8	dBHz

---

Notes:
* From equation (2.17), $W = \text{EIRP} - 163.3 \text{ dBW/m}^2$.
† Includes a 3 dB margin.
** The formulae used for C/T and C/N_o for the up-link give the figures quoted. Since $C/N_o = C/kT$, then:

$$C/N_o = C/T - 10\log_{10}k$$

and for the hemi beam up-link,

$$C/N_o = -121.1 - (-228.6) \text{ dBHz}$$
$$= 107.5 \text{ dBHz}$$

This figure corresponds to the value quoted (110.8 dBHz) when the 3 dB margin is taken into account.

44 Satellite link parameters

(b) Satellite–Earth Station down-link

Earth station type	B	A	
Beam configuration	Hemi	Zone	
Satellite EIRP‡	21.0	22.0	dBW
Earth station G/T_S	31.7	40.7	dB/K
Path loss to Earth station (L_p) 40,000 km at 4 GHz	196.5	196.5	dB
Earth station illumination level (W)§	−142.3	−141.3	dBW/m²
$C/T =$ $(W + G/T_S − 21.5 − 20\log_{10} f)$††	−147.1	−137.1	dBW/K
$C/N_o =$ $(EIRP − L_p + G/T_s − 10\log_{10} k)$‡‡	84.8	94.8	dBHz

The overall C/T ratio for the complete link can be found from the expression:

$$1/(C/T)_{o/a} = [1/(C/T)_u] + [1/(C/T)_d]$$

The figures to be used in this calculation are the ratios for C/T and not the values in dBs.

Hence, the values are, for the hemi beam:

$(C/T)_u = -121.1 \text{ dB} = a\log_{10}(-12.11) = 7.76 \times 10^{-13}$
$(C/T)_d = -147.1 \text{ dB} = a\log_{10}(-14.71) = 1.95 \times 10^{-15}$

so that substituting the figures given for the hemi beam,

$$\begin{aligned} 1/(C/T)_{o/a} &= [1/(7.76 \times 10^{-13})] + [1/(1.95 \times 10^{-15})] \\ &= 1.288 \times 10^{12} + 5.13 \times 10^{14} \\ &= 5.14 \times 10^{14} \end{aligned}$$

Hence $(C/T)_{o/a} = 1.945 \times 10^{-15}$
$= -147.1 \text{ dBW/K}$

It follows that the overall C/N_o is:

$(C/T)_{o/a} - 10\log_{10} k$
$= (C/N_o)_{o/a} = -147.1 - (-228.6) \text{ dBHz}$
$= 81.5 \text{ dBHz}.$

Alternatively,

$$1/(C/N_o)_{o/a} = [1/(C/N_o)_u] + [1/(C/N_o)_d]$$

where the values for C/N are:

$(C/N_o)_u = 107.5 = a\log_{10}(10.75) = 5.6 \times 10^{10}$
$(C/N_o)_d = 81.8 = a\log_{10}(8.18) = 1.51 \times 10^8$

Notes:
‡ Worst case value which includes output backoff.
§ From equation (2.17), $W = EIRP - 163.3 \text{ dBW/m}^2$.
†† Includes a 3 dB margin.
‡‡ The formulae used for C/T and C/N_o for the down-link give the figures quoted. Since $C/N_o = C/kT$, then:
$C/N_o = C/T - 10\log_{10} k$
and for the zone beam down-link,
$C/N_o = -137.1 - (-228.6) \text{ dBHz}$
$= 91.5 \text{ dBHz}$
This figure corresponds to the value quoted (94.8 dBHz) when the 3 dB margin is taken into account.

so that substituting the figures given for the hemi beam,
$$1/(C/N_o)_{o/a} = 1/(5.6 \times 10^{10}) + 1/(1.51 \times 10^8)$$
$$= 1.79 \times 10^{-11} + 6.6 \times 10^{-9}$$
$$= 6.62 \times 10^{-9}$$

so that:
$$(C/N_o)_{o/a} = 150.95 \times 10^6$$
$$= 81.8 \text{ dBHz}$$

The figures quoted above are for the Standard A G/T ratio of 40.7 dB/K for a 30 m antenna. The value has been revised to 35 dB/K allowing a 15 to 18 m antenna.

(2) 14 GHz/11 GHZ link.
(a) Earth station–satellite up-link

Earth station type	C
Beam configuration	Spot
EIRP	90.6 dBW
Satellite G/T_S	−3.3 dB/K
Path loss to satellite (L_p)	
40,000 km at 14 GHz	207.4 dB
Satellite illumination level (W)*	−72.7 dBW/m²
Boltzmann's constant (k)	−228.6 dBW/Hz/K
$C/T = (W + G/T_S - 21.5 - 20\log_{10}f)$†	−128.1 dBW/K
$C/N_o = (\text{EIRP} - L_p + G/T_s - 10\log_{10}k)$**	108.5 dBHz

(b) Satellite–earth station down-link

Earth station type	C
Beam configuration	Spot
Satellite EIRP‡	37.1 dBW
Earth station G/T_S	37.0 dB/K
Path loss to earth station (L_p)	
40,000 km at 11 GHz	205.3 dB
Earth station illumination level (W)§	−126.2 dBW/m²
$C/T = (W + G/T_S - 21.5 - 20\log_{10}f)$††	−140.2 dBW/K
$C/N_o = (\text{EIRP} - L_p + G/T_s - 10\log_{10}k)$‡‡	97.4 dBHz

Notes:
* From equation (2.17), $W = \text{EIRP} - 163.3$ dBW/m²
† Includes a 3 dB margin and a 5 dB loss due to precipitation.
** The formulae used for C/T and C/N_o for the up-link give the figures quoted. Since $C/N_o = C/kT$, then:
$$C/N_o = C/T - 10\log_{10}k$$
and for the spot beam up-link,
$$C/N_o = -128.1 - (-228.6) \text{ dBHz}$$
$$= 100.5 \text{ dBHz}$$
This figure corresponds to the value quoted (108.5 dBHz) when the 8 dB margin is taken into account.
‡ Worst case value which includes output backoff.
§ From equation (2.17), $W = \text{EIRP} - 163.3$ dBW/m².
†† Includes a 3 dB margin, precipitation of 4 dB and an increase in sky noise temperature of 2 dB.
‡‡ The formulae used for C/T and C/N_o for the down-link give the figures quoted. Since $C/N_o = C/kT$, then:
$$C/N_o = C/T - 10\log_{10}k$$
and for the spot beam down-link,
$$C/N_o = -140.2 - (-228.6) \text{ dBHz}$$
$$= 88.4 \text{ dBHz}$$
This figure corresponds to the value quoted (97.4 dBHz) when the 9 dB margin is taken into account.

The overall C/T ratio for the complete link can be found from the expression:
$$1/(C/T)_{o/a} = 1/(C/T)_u + 1/(C/T)_d$$
so that substituting the figures given for the link and using ratios and *not* the dB values,
$$\begin{aligned}1/(C/T)_{o/a} &= 1/(1.55 \times 10^{-13}) + 1/(9.55 \times 10^{-15}) \\ &= 6.46 \times 10^{12} + 1.05 \times 10^{14} \\ &= 1.11 \times 10^{14}\end{aligned}$$
Hence
$$\begin{aligned}(C/T)_{o/a} &= 8.99 \times 10^{-15} \\ &= -140.5 \text{ dBW/K}\end{aligned}$$

It follows that the overall C/N_o is:
$$\begin{aligned}(C/T)_{o/a} - 10\log_{10}k \\ = (C/N_o)_{o/a} = -140.5 - (-228.6) \text{ dBHz} \\ = 88.1 \text{ dBHz}.\end{aligned}$$

Alternatively,
$$1/(C/N_o)_{o/a} = 1/(C/N_o)_u + 1/(C/N_o)_d$$
so that substituting the figures given for the link,
$$\begin{aligned}1/(C/N_o)_{o/a} &= 1/(1.12 \times 10^{10}) + 1/(6.92 \times 10^8) \\ &= 8.93 \times 10^{-11} + 1.45 \times 10^{-9} \\ &= 1.54 \times 10^{-9}\end{aligned}$$
so that:
$$\begin{aligned}(C/N_o)_{o/a} &= 6.5 \times 10^8 \\ &= 88.1 \text{ dBHz}\end{aligned}$$

The figures quoted above are for the Standard C G/T ratio of 37 dB/K for a 12 m antenna.

The above link budgets have ignored the effect of intermodulation and co-channel interference (cci). For specified values of $(C/T)_{IM}$ and $(C/T)_{cci(up)}$ and $(C/T)_{cci(dw)}$ the total available C/T can be calculated using the technique described above i.e.:

$$\frac{1}{\left(\frac{C}{T}\right)_T} = \frac{1}{\left(\frac{C}{T}\right)_{up}} + \frac{1}{\left(\frac{C}{T}\right)_{dw}} + \frac{1}{\left(\frac{C}{T}\right)_{IM}} + \frac{1}{\left(\frac{C}{T}\right)_{cci(up)}} + \frac{1}{\left(\frac{C}{T}\right)_{cci(dw)}}$$

The link budgets for Intelsat VI are similar to those quoted for Intelsat V. A difference with Intelsat VI is that two extra zone beams are provided compared with Intelsat V; the frequency reuse is made possible by spatial isolation of the extra zone beams from other zone beams and polarization isolation from the hemi beam. The effect of the use of the extra zone beams is an increase in usable bandwidth but at the cost of increased co-channel interference. Thus, a higher value of down-link EIRP has been designated for both hemi and zone beams in Intelsat VI to offset the increase in co-channel interference.

Inmarsat-C link budget

In most satellite links the quality of the link is determined by the threshold level of C/N_o at the demodulator of the receiver. Availability is the percentage of time that the level is received above the threshold level. However, in the Inmarsat-C system variations in C/N_o do not affect the quality of the received message and mobile earth stations (MES)

2.3 Link budgets   47

Table 2.6 Inmarsat-C 'worst case' forward link budgets (courtesy Inmarsat)

Forward Link: 80% of time			
		1st GEN	2nd GEN
LES EIRP	(dBW)	61.4	61.0
Path loss	(dB)	200.9	200.9
Absorption loss	(dB)	0.4	0.4
Satellite G/T	(dB/K)	−15.0	−14.0
Mean uplink C/N_o	(dBHz)	73.7	74.3
Mean satellite C/I_o	(dBHz)	55.8	55.8
Satellite mean EIRP	(dBW)	21.4	21.0
Path loss	(dB)	188.5	188.5
Absorption loss	(dB)	0.4	0.4
MES G/T	(dB/K)	−23.0	−23.0
Mean downlink C/N_o	(dBHz)	38.1	37.7
Nominal unfaded C/N_o	(dBHz)	38.0	37.6
Interference loss	(dB)	0.5	0.5
Total RSS random loss (80%)	(dB)	1.2	0.8
Overall C/N_o	(dBHz)	36.3	36.3
Required C/N_o	(dBHz)	35.5	35.5
Margin	(dB)	0.8	0.8

Forward Link: 99% of time			
		1st GEN	2nd GEN
LES EIRP	(dBW)	61.4	61.0
Path loss	(dB)	200.9	200.9
Absorption loss	(dB)	0.4	0.4
Satellite G/T	(dB/K)	−15.0	−14.0
Mean uplink C/N_o	(dBHz)	73.7	74.3
Mean satellite C/I_o	(dBHz)	55.8	55.8
Satellite mean EIRP	(dBW)	21.4	21.0
Path loss	(dB)	188.5	188.5
Absorption loss	(dB)	0.4	0.4
MES G/T	(dB/K)	−23.0	−23.0
Mean downlink C/N_o	(dBHz)	38.1	37.7
Nominal unfaded C/N_o	(dBHz)	38.0	37.6
Interference loss	(dB)	0.5	0.5
Total RSS random loss (99%)	(dB)	2.2	1.6
Overall C/N_o	(dBHz)	35.4	35.5
Required C/N_o	(dBHz)	34.5	34.5
Margin	(dB)	0.9	1.0

can have variations in the received value of C/N_o. Because the system allows for repeats of packets of information, MESs with low values of C/N_o will simply have a higher number of packet repeats. Link budgets for Inmarsat-C are shown in Table 2.6 for the forward link only and show the acceptability levels of 80% and 99%.

The worst-case budgets are taken with MES and LES at 5° elevation angle, minimum value of G/T and EIRP, worst-case transponder loading (i.e. fully loaded transponder and channel with lowest carrier/intermodulation ratio). Additionally variables such as polarization loss, wet radome, noise degradation and precipitation loss are combined by adding mean values and taking rms values of the deviations to produce the 80% and 99% values.

Eutelsat II. Single FM/TV carrier per transponder

For a single FM/TV carrier per transponder the satellite high gain setting is used corresponding to a satellite input power flux density for saturation of -83.0 dBW/m² when the satellite G/T is -0.5 dB/K.

The transmit state nominally assigned EIRP from the -0.5 dB/K satellite receive coverage contour is 80 dBW, a value which leads to saturation of the transponder.

(a) Earth Station–satellite up-link

EIRP	80.0	dBW
Satellite G/T_S	-0.5	dB/K
Path loss to satellite (L_p)		
38,500 km at 14.5 GHz	207.4	dB
Satellite illumination level (W)*	-83.0	dBW/m²
Boltzmann's constant (k)	-228.6	dBW/Hz/K
$C/T = (W + G/T_S - 21.5 - 20\log_{10}f)$	-128.2	dBW/K
$C/N_o = (EIRP - L_p + G/T_s - 10\log_{10}k)$†	100.7	dBHz

(b) Satellite–Earth Station down-link

Satellite EIRP**	47.0	dBW
Earth station G/T_S	16.5	dB/K
Path loss to earth station (L_p)		
38,500 km at 11.2 GHz	205.1	dB
Earth station illumination level (W)‡	-115.7	dBW/m²
$C/T = (W + G/T_S - 21.5 - 20\log_{10}f)$	-141.7	dBW/K
$C/N_o = (EIRP - L_p + G/T_s - 10\log_{10}k)$§	87.0	dBHz

Notes:
* $W = EIRP - 162.7$ dBW/m² but includes 0.3 dB clear sky attenuation.
†   The formulae used for C/T and C/N_o for the up-link give the figures quoted. Since $C/N_o = C/kT$, then:
$$C/N_o = C/T - 10\log_{10}k$$
and for the up-link,
$$C/N_o = -128.2 - (-228.6) \text{ dBHz}$$
$$= 100.4 \text{ dBHz}$$
** Assumes an antenna of diameter 1.0 m with high gain coverage.
‡ $W = EIRP - 162.7$ dBW/m².
§   The formulae used for C/T and C/N_o for the down-link give the figures quoted. Since $C/N_o = C/kT$, then:
$$C/N_o = C/T - 10\log_{10}k$$
and for the down-link,
$$C/N_o = -141.7 - (-228.6) \text{ dBHz}$$
$$= 86.9 \text{ dBHz}$$

Comparison of modulation and access techniques

A source signal may be voice or data and may be processed in analogue or digital form ready for transmission. For analogue transmission the modulation may be amplitude modulation (AM) or frequency modulation (FM). For digital transmission a form of digitizing the source signal where necessary (i.e. voice signals) is needed and may be pulse code modulation (PCM) or some similar system.

The next processing stage is multiplexing. For analogue transmission, channels may be combined using frequency division multiplexing (FDM) which uses carriers, each of which is separated in frequency, supporting one channel. An FDM baseband assembly can be provided by modulating each individual voiceband signal on to a sinusoidal carrier via a balanced AM modulator. Each voiceband signal bandwidth is typically from 300 Hz to 3400 Hz and, using AM single sideband (SSB) techniques, allows AM

Table 2.7 FDM spectrum used by Intelsat

Number of channels	Frequency band (kHz)
12	12 – 60
24	12 – 108
36	12 – 156
48	12 – 204
60	12 – 252
72	12 – 300
96	12 – 408
132	12 – 552
192	12 – 804
252	12 – 1052
312	12 – 1300
372	12 – 1548
432	12 – 1796
492	12 – 2044
552	12 – 2292
612	12 – 2540
792	12 – 3284
972	12 – 4092
1092	12 – 4892

SSB signals each with a nominal bandwidth of 4 kHz to be combined into a 12 channel FDM group with a frequency bandwidth of 48 kHz. A set of conventions established by the CCITT provides an FDM hierarchy which has a *group* as its first level, consisting of 12 channels with a baseband of 12 to 60 kHz. A *supergroup* consists of five groups of 12 channels (i.e. 60 channels), a *mastergroup* (600 channels), etc. Pilot tones within the composite signal achieve the required spectrum centring frequency control and can be used for satellite transmission.

Table 2.7 shows the FDM spectrum schemes used by Intelsat.

Intelsat specify that wherever possible all basebands should be assembled by means of the standard CCITT groups (12 channels) and supergroups (60 channels) with a nominal carrier spacing of 4 kHz.

For digital transmission, the train of pulses from channels can be bit interleaved to produce a composite high level bit-rate signal known as a time division multiplex (TDM) assembly. The assembly may then be directly modulated on to an RF carrier. As well as combining the outputs from digital coders the multiplexing system must also provide framing information for the composite bit stream. The pattern of the frame will establish the start and finish of each frame in the time domain. A frame will consist of samples from each of the system channel inputs together with frame synchronization information.

In a similar manner to the FDM case, digital hierarchies exist consisting of primary multiplexes which can be combined to form second-order multiplexes etc. The basic systems in use consist of the North American system and the European (CEPT) system which is used in Western Europe and South America. Other systems, based on these systems, also exist. Details of the main systems are shown in Table 2.8.

The levels in the North American hierarchy are classified with DS (digital signal) numbers. The first level in the North American hierarchy (DS1) has a bit-rate of 1.544 Mbit/s. 24 voice and/or data channels can be assembled, each with a nominal data rate of 64 kbit/s. The DS1 frame has 24 8-bit time slots each containing a single encoded voice or digital data signal. A frame actually has 193 bits made up of the 8×24 bits of information plus an extra bit for framing. Normally, using PCM telephony, the eighth bit of every time slot is used every sixth frame to transmit signalling information. The time for each frame is 125 μs, corresponding to the standard 8 kHz sampling rate for PCM.

The first level in the European hierarchy has a bit-rate of 2.048 Mbit/s. Figure 2.8 shows the standard Western European frame. There are 32 8-bit slots in each 125 μs frame with only 30 channels used for signal information. One slot (time slot 0) is

Table 2.8 TDM hierarchies

North American system			Western European system		
Number of channels	Bit-rate Mbit/s	Level	Number of channels	Bit-rate Mbit/s	Level
23	1.544	DS1	30	2.048	Primary
96	6.312	DS2	120	8.448	2nd order
672	44.736	DS3	480	34.368	3rd order
4032	274.176	DS4	1920	139.264	4th order
			7680	565.148	5th order

2.3   Link budgets   51

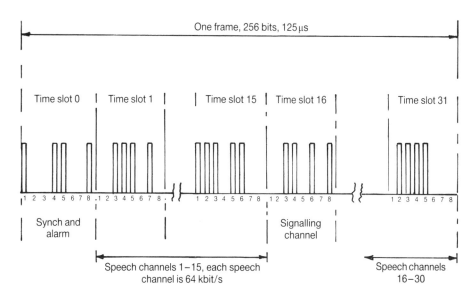

Fig. 2.8 TDM frame of a primary multiplex transmission in the W. European hierarchy

dedicated to carrying synchronization information and another slot (time slot 16) contains signalling information for all 30 channels.

In both systems it can be seen that when multiplexing primary multiplexes to form higher rate groups, the bit-rates are higher than the sum of the individual bit-rates. For example, DS2 is 4 × DS1 (or 4 × 1.544 Mbit/s) which gives 6.176 Mbit/s. The figure quoted, 6.312 Mbit/s, for DS2 allows for extra bits to be accommodated to preserve synchronization for the larger group. The addition of the extra bits is sometimes referred to as 'bit stuffing'.

For FM systems the signal received for any single channel is weak and comparable with the noise value. It is thus essential to use techniques which permit a good signal-to-noise ratio (S/N) to be achieved. It is generally accepted that a value of at least 18 dB for the S/N is required for satisfactory voice operation. However, the type of modulation used can improve the signal-to-noise ratio for a single channel to well above the received value. The use of frequency modulation or phase-shift keying, for instance, could improve the value of S/N to as much as 53 dB in a single channel compared with the input C/N value of 18 dB.

For FM, the increase in S/N results from using a high value for the frequency deviation f_d. As an example, 40 telephone channels could be sent, using FM, with an RF bandwidth of 2.4 MHz giving 60 kHz per channel. The same 40 channels could be sent, using single sideband amplitude modulation (SSAM) in an RF bandwidth of 160 kHz, using 4 kHz per channel, for a similar power per channel. It follows that the improvement in S/N obtained using FM is at the expense of reduced channel capacity.

An improvement in channel capacity can be achieved with digital systems using TDM and PSK modulation. Examples that will illustrate this concept follow in the chapter on systems. (Chapter 3).

The use of Frequency Division Multiple Access (FDMA) is suited to earth stations with only a limited number of channels. This allows the earth station to commence operations with the smallest number of channels needed to give satisfactory quality of service, while allowing the capacity to increase with increasing traffic demands.

52 Satellite link parameters

The use of Time Division Multiple Access (TDMA) enables greater efficiency for both low and high capacity stations but demands great earth station complexity. Also, mixing earth stations with different values of G/T is not possible using this system because each station must receive timing information with the same value of C/N. Multiple Access is dealt with in Chapter 3.

2.4 Summary

The design of a satellite communications system will endeavour to maximize the number of channels used for each transponder. Although system margins are incorporated to ensure adequate quality of received signals, an increase in capacity can be obtained if system margins and signal quality are reduced. Provided the result is not obvious to the user, the increase in channel capacity produced, by such a reduction in system margins and signal quality, can allow a significant difference to the pricing of the system. As an example of one method that has been used to increase channel capacity, consider the use of syllabic companding and time assigned or digital speech interpolation. Since, on average, a one way voice channel only carries speech for about 40% of the time, the channel can be allocated to another customer for part of the remaining 60% of the time. Overloads do occur in such an arrangement but since their effects on quality are brief this usually goes unnoticed by the customer.

3
Multiple access

3.1 Introduction

Access may be defined as the ability to use a facility and multiple access may be defined as the situation where many users obtain entry to the facility. In the case of satellite communications, the facility is the satellite transponder and multiple access occurs when more than one pair of earth stations require the use of the satellite communication channel. Several transponders on the satellite share the frequency band in use, 3.7 to 4.2 GHz say, and each transponder will act independently of the others to filter out its own allocated frequency band and process that signal for retransmission. Access to a transponder may be limited to a single carrier or many carriers may exist simultaneously. The baseband information to be transmitted, whether telephony, data or video, is impressed on the carrier by the process of modulation. The modulation may be by single or multi-channel basebands.

There are many methods by which multiple access may be achieved, but only five will be considered because of their relevance to the satellite systems under consideration. The systems are as follows.

Frequency-division multiple access (FDMA)

This is the separation of signals in the frequency domain. For frequency division multiplexing, each baseband signal is translated to a higher frequency by modulating a carrier frequency which then becomes the centre frequency of the signal. Other user's signals are treated in the same way so that all signals are translated to a new set of frequencies, separated from each other in order to prevent mutual interference. A group of multiplexed signals may then modulate a common carrier for transmission purposes. Such a group of multiplexed signals forms the transponder channel. Figure 3.1 shows the possible arrangement for a typical satellite system with 12 transponders, each with a channel 36 MHz wide and separated from adjacent channels by a guard band of 4 MHz. This would give about 600 voice channels allowing for separation between signals within a channel, assuming analogue transmission. The transmission could be analogue or digital in continuous or burst mode. Analogue transmission suffers badly from the effects of noise and a wider bandwidth is required for good reception. Digital transmission suffers less from the effects of noise and the system described in Fig. 3.1 could probably accommodate about 800 channels.

Advantages of FDMA include: simplicity in earth station equipment; and no complex timing and synchronization techniques required, as is the case for TDMA. Disadvantages include: the likelihood of inter-modulation problems with its adverse effect on the signal-to-noise ratio and the need for fairly complex frequency plans to reduce such effects.

The example illustrated in Fig. 3.1 is shown for the purpose of indicating how the

54 Multiple access

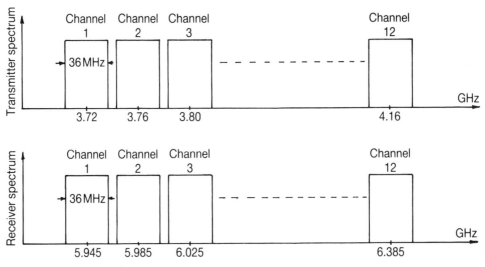

Fig. 3.1 FDMA arrangement

principle of FDMA can be applied. The applications to be discussed in this publication will follow the principle of FDMA but the figures for channel capacity etc will vary according to the system used.

Time-division multiple access (TDMA)

The transponder channel is available to a single carrier for a period of time. Each station transmits a burst of information within its specified time slot; once the burst is complete a second station transmits its information and so on. Each earth station uses the same carrier frequency but since each signal arrives at the satellite at different times there is no danger of interference. When signals are transmitted from the satellite to receiver stations, the receiver station will select only that signal which occurs at a specified time, ignoring all others. The time slots are arranged permanently or assigned as required. Assignment of time slots can be arranged using a common control channel which can also release the station to a common pool once the transmission has been completed. TDMA is suitable only for digital transmissions. An example of TDMA is illustrated in Fig. 3.2.

Advantages of TDMA include: no inter-modulation problems since the transponder only has to deal with one carrier at a time, thus the transponder is able to operate at full power and high system flexibility, allowing channels of differing capacity to be accommodated by altering the number of equally spaced time slots allocated to a user. Disadvantages include the requirement for complex, and expensive, earth stations.

Code-division multiple access (CDMA)

Users share the transponder channel and transmit at the same time as other users. Each transmission spreads its signal over a bandwidth, which is much wider than that required for the information alone, and contention is avoided because each transmission uses a unique code sequence. A receiving station can retrieve the required information by using the same unique code sequence. Advantages of CDMA include privacy and good interference tolerance. Disadvantages include the requirement for a complex earth station and poor frequency utilization.

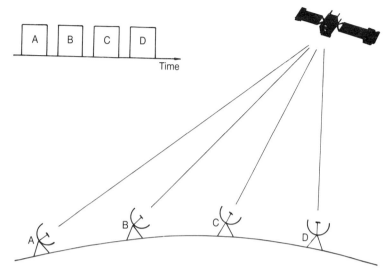

Fig. 3.2 TDMA arrangement

CDMA tends to be a specialist system and has in the past been restricted mainly to military applications.

Packet access

Digital signals may be organized on a packet basis. A packet could comprise traffic from several sources that has been collected and prepared for transmission to the next point in the communications network.

Random access

This is a demand assignment method with little or no central control. Access to the communications network is by contention whereby one earth station transmits a burst of data whenever it desires access to the system, regardless of what other stations may be doing. One form of random access system is the ALOHA system developed by the University of Hawaii for Pacific Island communications.

3.2 FDMA

One possible arrangement of an FDMA system uses single channel per carrier (SCPC) whereby one signal, which could be voice or data, modulates the carrier. The modulation could be frequency modulation (FM), for analogue transmission, or phase-shift keying (PSK) for digital transmission. The carrier is then transmitted using FDMA.

A second arrangement utilizes digital transmission with time division multiplexing (TDM) to combine multiple digital channels. The digital baseband signal then modulates a digital carrier using PSK.

An advantage of using SCPC is to facilitate the use of voice-activated carriers. Such carriers are switched off during the periods between speech activity, thus reducing the power consumption. It has been established that, on average, a speaker will talk for only

56 Multiple access

40% of the time and switching off the carrier for the remaining 60% of the time reduces the satellite power consumption by up to 4 dB. This would enable a corresponding increase in power available and hence channel capacity for the transponder. To minimize inter-modulation between transmissions to acceptable levels, the output of the transponder power device must be backed off; typically this would be about 4 to 6 dB for a travelling wave tube amplifier (TWTA), and about 2 dB for a solid-state power amplifier (SSPA). The SSPA has a reasonably linear characteristic provided it is not over-driven. A small earth station, however, may suffer because of increased down-link noise due to the use of a SSPA and such a station would require a higher value of power per channel than otherwise would be the case. This would require less back-off for the transponder output power device. SCPC has the advantage that the power of individual transmitted carriers can be adjusted to optimize for particular link conditions.

SCPC requires automatic frequency control (AFC) to maintain spectrum centring on a channel by channel basis. This is usually achieved by transmitting a pilot tone in the centre of the transponder bandwidth. A receiving station can use the pilot tone to produce a local AFC system which can control the frequency of the individual carriers by controlling the frequency of the local oscillators.

SCPC/FM/FDMA

The Inmarsat-A system uses SCPC utilizing analogue transmission with frequency modulation for telephone channels.

The Intelsat system utilizes SCPC/FM/FDMA on its Vista telephone service. Using CFM (FM with 2:1 syllabic companding), the service is available at C-band using a wide variety of earth stations.

A basic arrangement for SCPC/FM/FDMA is shown in Fig. 3.3. For the transmit part of the circuit the baseband signal is peak limited to give a desired maximum value of frequency deviation f_d. A filter sets the limit on the baseband signal which is then passed through a compressor circuit, and pre-emphasis, before modulation. The delay circuit

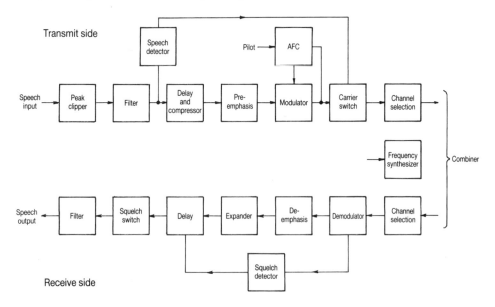

Fig. 3.3 SCPC/FM block diagram of a channel unit

provides time for the speech selector circuit to establish the absence of speech to enable voice-activated carrier operation to be effected. The receive side is basically the opposite of the transmit side. A squelch circuit is required on the receive side because of the effect on the carrier of the voice-activation; the AGC of the receiver is inoperative without a carrier and noise levels could increase significantly unless muting is applied.

In calculating the channel capacity of the SCPC/FM system it is necessary to ensure that the noise level does not exceed the specified levels. The CCIR recommendation for an analogue channel is that the noise power at a point of zero relative level should not exceed 10,000 pWOP with a 50 dB test tone–noise ratio. It is assumed that the minimum required carrier-to-noise ratio per channel is at least 10 dB. The unit pWOP represents picowatts measured at a point of zero relative level and psophometrically weighted. Zero relative level for a telephone channel may be defined as a point in the channel where a test signal would have a power of 1 mW. Psophometric weighting employs a psophometer, which consists of a filter with an amplitude/frequency characteristic similar to that of the human ear, followed by a power meter. Because the psophometrically weighted value of noise in a telephone channel tends to be less than the value measured without the weighting network, it is usual to add a figure, typically 2.5 dB, to the value calculated in order to allow for the difference.

As shown in Chapter 4, the ratio of signal-to-noise for SCPC/FM/FMDA transmissions is:

$$\frac{S}{N} = \frac{C}{N} \times 3B \times \frac{f_d^2}{(f_2^3 - f_1^3)}$$

where S/N = unweighted signal-to-noise power ratio of an FM transmission after demodulation.

C = carrier power at the receiver input in watts.
N = noise power, in watts, in the bandwidth B.
B = RF bandwidth in Hz.
f_d = test-tone frequency deviation in Hz.
f_2 = upper baseband frequency.
f_1 = lower baseband frequency.

It can be seen from this expression that provided the S/N ratio exceeds 50 dB, the number of channels that could be used could be increased by reducing the frequency deviation f_d and hence reducing the FM bandwidth. This could have the effect of increasing the C/N ratio since N is proportional to bandwidth. The Inmarsat-A system has an RF bandwidth of 50 kHz for each voice channel and a test-tone frequency deviation of 12 kHz. Not all the transponder bandwidth is available for voice transmission as some of the bandwidth has to be reserved for TDM/TDMA transmissions. Additionally, special frequency plans are used to minimize intermodulation performance. Typically, some 100 to 125 telephone channels are available on current Inmarsat 2 satellites, occupying a bandwidth of around 7.5 MHz.

The Intelsat Vista service has an allocated satellite bandwidth of 30 kHz per SCPC/CFM carrier providing nominal per channel C/N and C/N_o values of 10.2 dB and 54.2 dB Hz respectively.

SCPC/PSK/FDMA

In this arrangement, each voice or data channel is modulated on to its own radio-frequency carrier. The only multiplexing occurs in the transponder bandwidth where

58 *Multiple access*

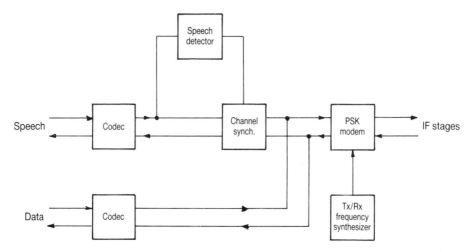

Fig. 3.4 SCPC/PSK block diagram of a channel unit

frequency division produces individual channels within the bandwidth. Figure 3.4 shows a possible arrangement for the SCPC system.

The transmit side has a number of channel units which convert the input voiceband or data digital signal into a PSK-modulated RF carrier for transmission to the satellite. A speech detector determines the presence of active speech so that the RF carrier is only turned on during such periods. The digitally encoded voice-band signal from the CODEC (coder/decoder) is passed to a PSK modem whose function is to produce a four-phase, or quadrature, PSK (QPSK) carrier for transmission and to perform the reverse function for the receive signal. Digital data inputs have a digital interface together with an error-correction channel encoding which improves the error rate without altering the bit-energy-to-noise density ratio (E_b/N_o).

The satellite transponder carrier frequencies may be pre-assigned or demand-assigned. For pre-assigned carriers the frequency is assigned to a channel unit and the PSK modem requires a fixed frequency local oscillator (LO) input. For demand-assignment, channels may be connected according to the availability of a particular carrier frequency within the transponder RF bandwidth. For this arrangement the SCPC channel frequency required is produced by a frequency synthesizer.

Demand-Assigned Multiple Access (DAMA) allows users to share a common link. The channel link is only completed as required and a channel frequency is assigned from a pool of available frequencies within the satellite transponder bandwidth. Such an arrangement is particularly useful for 'small' users who would only pay for the time the link was actually used. A link set-up for pre-assigned channels would be costly to the user and wasteful of system utilization if the link were to be used by the customer for only infrequent periods. The first fully demand-assigned SCPC system was the SCPC, PCM, MA, DA Equipment (SPADE) developed for Intelsat in 1971 by J.G. Puente and A.M. Werth.

In SPADE, the transponder bandwidth is divided into channels each with its own assigned carrier frequency. A two-way link may be established using the channels in pairs. When a user requires a channel, the SPADE terminal at the earth station will select a pair of frequencies for the two-way link from its pool of frequencies and set up the link, via the SPADE terminal at the destination earth station, utilizing the chosen frequencies. When the link is no longer required, the originating earth station notifies all other earth stations that the channel has been freed and the frequencies returned to the pool.

There is a common signalling channel which links all earth stations. The signalling information is obtained by modulating a unique carrier at the bottom end of the transponder bandwidth. The channel is common to all earth stations with each station allocated a time slot of 1 ms allowing a burst of 128 bits with a burst every 50 ms. A SPADE system operates with a transponder bandwidth of 36 MHz with 800 channels, of which all but six are available for pairing.

Developments in the Intelsat systems have evolved from SPADE, utilizing SCPC with pre-assignment and demand-assignment for both analogue and digital transmissions, as mentioned later in this section.

The channel capacity is determined by the value of C/N_o in the RF link. To find the carrier-to-noise ratio necessary to support each carrier, the following equation is used:

$$\left(\frac{C}{N}\right)_{th} = \left(\frac{E_b}{N_o}\right)_{th} - B + R + M \tag{3.1}$$

where $(C/N)_{th}$ = carrier-to-noise ratio at the threshold error rate in dB.
(E_b/N_o) = the bit energy-to-noise density ratio at the threshold error rate in dB.
B = noise bandwidth of the carrier in Hz.
R = data rate of the digital signal in bit/s.
M = system margin to allow for impairments in dB.

From this expression the value of carrier-to-noise density ratio (C/N_o) can be found i.e.:

$$(C/N_o)_t = (C/N)_{th} + 10\log B \text{ dB Hz}$$

where (C/N_o) is the total available carrier-to-noise density for the transponder.

The Inmarsat-B and Inmarsat-M systems use SCPC/FDMA for telephony and data using offset-quadrature phase-shift keying (O-QPSK).

Example 3.1

Considering the Inmarsat-B system, the CESV and SESV (voice) channels specify $(E_b/N_o)_{th}$ of 4.7 dB, for an SES elevation angle of 10°, and 3.3 dB for an SES elevation angle of 5°, $B = 20$ kHz, channel rate, $R = 24$ kbit/s with 3/4 rate FEC encoding. Find the total C/N_o value.

Solution

With 3/4 rate FEC encoding, the rate is effectively 18 kbit/s so that for the elevation angle of 10°:

$(C/N)_{th} = 4.7 - 10\log(2 \times 10^4) + 10\log(18 \times 10^3)$ dB
$= 4.7 - 43 + 42.6$ dB
$= 4.3$ dB
and $(C/N_o)_t = 4.3 + 10\log(2 \times 10^4)$ dB Hz
$= 4.3 + 43$
$= 47.3$ dB Hz

This is the same as the theoretical value quoted for the Inmarsat-B voice channels (which also includes data up to 2400 bit/s).

Table 3.1 Inmarsat-B channel performance objectives (courtesy Inmarsat)

	Channel type	FEC rate	Channel rate (kbit/s)	SES elevation angle	BER (99% time)	E_b/N_o* (dB)	C/N_o* (dBHz)
1	CESV and SESV Telephony (including voice-band data up to 2400 bit/s)	3/4	24	10° 5°	10^{-4} 10^{-2}	4.7 3.3	47.3 45.9
2	CEST and CESDL Forward telex and low-speed data	1/2	6	5°	10^{-5}	4.6	39.4
3	SEST and SESDL Return telex and low-speed data	1/2	24	5°	10^{-5}	4.4	45.2
4	CESD and SESD 9.6 kbit/s SCPC data	1/2	24	5°	10^{-5}	4.4	45.2
5	CESA, NCSC, NCSA and NCSS Forward signalling	1/2	6	5°	10^{-5}	4.6	39.4
6	NCSI and CESI Interstation links	1/2	6	5°	10^{-5}	4.6	39.4
7	SESRQ and SESRP Return signalling	1/2	24	5°	10^{-5}	4.4	45.2

* **Note**: theoretical value required in order to achieve the BER value under additive white Gaussian noise conditions.

Calculations for the Inmarsat-M system would yield different results since the voice channel rate (R) is only 8 kbit/s and the voice channel bandwidth (B) is 10 kHz.

The full Inmarsat-B channel performance objectives are shown in Table 3.1.

The details in Table 3.1 include, in addition to those discussed above, telex and low-speed data; SCPC data at 9.6 kbit/s and signalling/interstation link channel information.

A brief description of the various SCPC/FDMA channels of the Inmarsat-B system, as described in Chapter 12, Section 2.4 is as follows.

Voice channel SCPC digital voice channel using a voice coding rate of 16 bit/s with Adaptive Predictive Coding (APC) used in both the forward and return directions. The channels can support voice-band data, including facsimile, up to 2400 bit/s information rate. Full voice-activation is implemented on forward carriers.

Data channel SCPC digital data channel uses an information rate of 9.6 kbit/s used in forward (CESD) and return (SESD) directions.

For a more complete description of the operation of this system, including signalling information, refer to Chapter 12.

3.2 FDMA

Intelsat utilizes a SCPC digital service for public switched telephony for thin-route traffic (low traffic communications to rural and remote communities) at C-band between Standard A and B earth stations. The modulation is QPSK for both voice, voice-band data and high-speed data services. Transmission parameters quote bandwidth units of 45 kHz with a transmission rate of 64 kbit/s. The service quality is given as 10^{-6} BER for voice and 10^{-9} BER for high-speed data.

MCPC/FM/FDMA

The first multiple access arrangement used in satellite communications utilized analogue MCPC (multiple channels per carrier), which was essentially the same arrangement used in terrestrial multiplex links. Typically in this arrangement, voiceband channels are grouped by an earth station on to a single carrier by frequency division multiplexing to form FDM baseband signals. The FDM basebands are frequency modulated on pre-assigned carriers and transmitted to the satellite. The satellite receives this carrier together with many others, separated in frequency, simultaneously. A receiving station operates the above arrangement in reverse using FDM multiplex equipment to extract the channels assigned to that station. Figure 3.5 shows a possible arrangement.

The advantage of this arrangement is its suitability for limited access use. Because it uses a multi-carrier system the channel capacity falls with an increasing number of carriers. This gives rise to the main system disadvantage since the extra carriers cause more intermodulation products and cause IM prone frequency ranges that cannot be used for traffic.

For channel capacity of MCPC/FM/FDMA the equations used in Chapter 4, Section 3.3 may be utilized. The first step is to determine the carrier-to-noise (C/N) density that is available in the RF channel for each carrier and the bandwidth of each carrier. Next, the number of channels that can be multiplexed on to the carrier must be determined. The signal-to-noise (S/N) ratio after detection can then be used, as determined by equation (4.7). The equation is repeated below for convenience:

$$\frac{S_b}{N_b} = \left(\frac{f_d}{f_m}\right)^2 \frac{B}{b} \frac{C}{N}$$

where:
f_d is the rms test-tone deviation (Hz)
f_m is the highest modulating frequency (Hz)
B is the RF bandwidth of the modulated signal (Hz)
b is the voice signal bandwidth (Hz)
C is the carrier power at receiver input (watts)
N is the noise power, kTB, in bandwidth B (watts).

This equation ignores any improvements produced using pre- and de-emphasis and psophometric weighting.

In a manner similar to that shown earlier for SCPC/FM/FDMA, the CCIR recommendation, for voice transmissions using FDM techniques, gives a value of S/N in the worst-case FDM channel of 50 dB. For each carrier the number of channels can be estimated and the channel numbers adjusted until a value is found that would give a S/N ratio in the worst-case channel of 50 dB. The calculation is repeated for each carrier in the transponder bandwidth to give the transponder capacity, which is the sum total of the individual carriers.

62 Multiple access

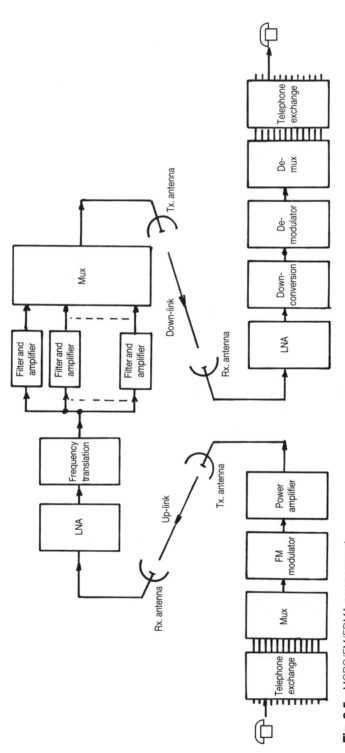

Fig. 3.5 MCPC/FM/FDMA arrangement

MCPC/PSK/FDMA

In this arrangement the incoming baseband signals are digitally encoded. The baseband signal for each carrier could consist of multi-channel PCM-TDM bit streams. In Western Europe the CEPT hierarchy is used whereby at the first level 32 64-kbit/s channels are combined to give a bit-rate of 2.048 Mbit/s. The multiplexed signals are used to phase-shift key (PSK) a carrier which uses FDMA to access the satellite. If the destination for the traffic is a single earth station, digital speech interpolation (DSI) may be used to increase channel capacity. The concept of DSI is discussed in more detail in Chapter 5, Section 5.2.

The earth station size will depend on the traffic it is expected to handle. The receiver must have filters, demodulators and demultiplexers for each incoming link but no network clock timing is needed. The demodulators can have narrow noise bandwidths, which match the carrier, rather than the full transponder width, as would be the case for TDMA or CDMA. Only simple frequency co-ordination is required where transmit and receive local oscillators are stable.

Various methods of digital voice encoding exist and some of these are discussed in Chapter 5.

Channel capacity can be calculated in much the same way as shown in equation (3.1) for SCPC/PSK/FDMA. Using the information that the sum of the carrier to noise densities needed for each individual carrier cannot exceed the total available carrier to noise density, the channel capacity can be iterated in a similar manner to that shown earlier for FDM/FDMA. Once the power-limited capacity of the system is determined, the bandwidth-limited system capacity can be found by summing the bandwidth of all individual carriers.

Intelsat provides an integrated digital communications service, known as intermediate data rate (IDR), using MCPC/PSK/FDMA (also available on SCPC/PSK/FDMA). The system uses QPSK and rate 3/4 FEC (rate 1/2 FEC is also approved for small earth stations). The information bit-rate ranges from 64 kbit/s to 45 Mbit/s and includes the first level rates for the North American hierarchy of 24 64-kbit/s channels at a bit-rate of 1.544 Mbit/s and the Western European hierarchy of 32 64 kbit/s channels at a bit-rate of 2.048 Mbit/s. The allocated satellite bandwidth is 67.5 kHz per 64 kbit/s channel which corresponds to a capacity of approximately 500 channels in a 36 MHz transponder. For voice applications the IDR service allows use of low-rate encoding (LRE) and digital circuit multiplication equipment (DCME) or digital speech interpolation (DSI) equipment. The system specification allows the use of 32 kbit/s LRE with digital speech interpolation allowing up to five derived channels to be obtained from a 64 kbit/s bearer channel, depending on the size of the channel groupings, see Chapter 5.

TDM/FDMA

This arrangement allows the use of a TDM group, or groups, to be assembled at the satellite in FDMA. Phase-shift keying is used as the modulation process at the earth station. Systems such as this are compatible with FDM/FDMA carriers sharing the same transponder and the earth station requirements are simple and easily incorporated.

The channel capacity of this arrangement can be calculated using equation (3.1).

Example 3.2

Inmarsat-B uses TDM/FDMA on the shore–ship link for telex and low-speed data.

64 Multiple access

Details of channel performance are given in Table 3.1 (page 60). Bit rate is quoted as 6 kbit/s but, with 1/2 rate FEC encoding, the effective bit rate is 3 kbit/s. Find the value for the total C/N_o.

Solution

Substituting in equation (3.1) using the figures quoted gives:

$$(C/N)_{th} = 4.6 - 10\log(1 \times 10^4) + 10\log(3 \times 10^3) \text{ dB}$$
$$= 4.6 - 40 + 34.8 \text{ dB}$$
$$= -0.6 \text{ dB}$$
$$\text{and } (C/N_o)_t = -0.6 + 10\log(1 \times 10^4) \text{ dB Hz}$$
$$= -0.6 + 40 \text{ dB Hz}$$
$$= 39.4 \text{ dB Hz}$$

Table 3.1 gives a theoretical value for $(C/N_o)_t$ which agrees with the above.

The Inmarsat-B system for telex/low-speed data uses TDM/FDMA in the shore-to-ship direction only, with the ship-to-shore direction using TDMA/FDMA (see next section). The CES TDM (and SES TDMA) carrier frequency is pre-allocated by Inmarsat. Each CES is allocated at least one forward CES TDM carrier frequency (and a return TDMA frequency). Additional allocations can be made depending on the traffic requirements.

The channel unit associated with the CES TDM channel for transmission consists of a multiplexer, differential encoder, frame (transmit) synchronizer and modulator.

At the SES, the receive part of the channel has the corresponding complementary functions to the transmit end. The full channel unit arrangement for transmit and receive channels is shown in Fig. 3.6.

The CES TDM channels use BPSK with differential coding, which is used for phase ambiguity resolution at the receive end.

The Intelsat IDR system mentioned under the heading MCPC/PSK/FDMA is a form of TDM/FDMA and is often expressed as a TDM/QPSK/FDMA system.

TDMA/FDMA

As mentioned in Section 3.1, TDMA signals could occupy the complete transponder bandwidth. A variation of this is where TDMA signals are transmitted as a sub-band of

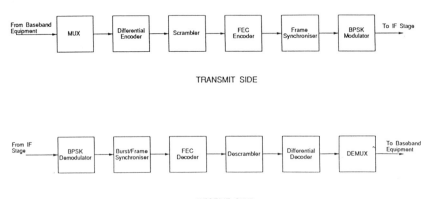

Fig. 3.6 Channel unit configuration for BPSK channels

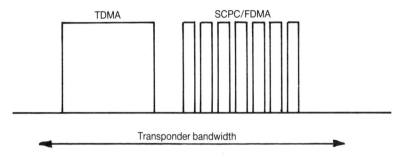

Fig. 3.7 Narrowband TDMA used with SCPC voice channels

the transponder bandwidth, the remainder of the transponder bandwidth being available for, say, SCPC/FDMA signals. The concept is illustrated in Fig. 3.7.

The use of a narrowband TDMA arrangement is well suited for a system requiring only a few channels and has all the advantages of digital transmission. Narrowband TDMA could, however, suffer from intermodulation with the adjacent FDMA channels.

The carrier-to-noise ratio required to achieve the threshold error rate can be calculated using equation (3.1).

Example 3.3

Using the Inmarsat-B return telex/low-speed data information from Table 3.1 for (E_b/N_o) and channel rate R, find the value for total C/N_o.

Solution

From Table 3.1, channel rate = 24 kbit/s but because of 1/2 rate FEC encoding the value for R is 12 kbit/s. Thus:

$$(C/N)_{th} = 4.4 - 10\log(2 \times 10^4) + 10\log(12 \times 10^3) \text{ dB}$$
$$= 4.4 - 43 + 40.8 \text{ dB}$$
$$= 2.2 \text{ dB}$$
$$\text{and } (C/N_o)_t = 2.2 + 10\log(2 \times 10^4) \text{ dB Hz}$$
$$= 2.2 + 43 \text{ dB Hz}$$
$$= 45.2 \text{ dBHz}$$

This agrees with the value quoted in Table 3.1 for the theoretical value of $(C/N_o)_t$ for this system.

The Telex Services of the Inmarsat-B system for shore-to-ship channels has a flexible allocation of capacity for communications and signalling slots depending on traffic requirements. At least one slot (the number of the slot can vary) in each TDM frame is available for signalling messages. Under these conditions the maximum telex capacity in the TDM frame is 56 channels (using seven slots for telex and one slot for signalling). For ship-to-shore channels, each SES TDMA channel provides for a maximum of 32 telex bursts (SEST channel), or a maximum of 16 low-speed data bursts (SESDL channel). The SEST and SESDL functional channels cannot be combined on the same physical TDMA channel. The SES TDMA frame format for telex is shown in Fig. 3.8. Each SES TDMA channel is associated with a forward CES TDM (CEST) channel from which TDMA synchronization timing is derived.

66 *Multiple access*

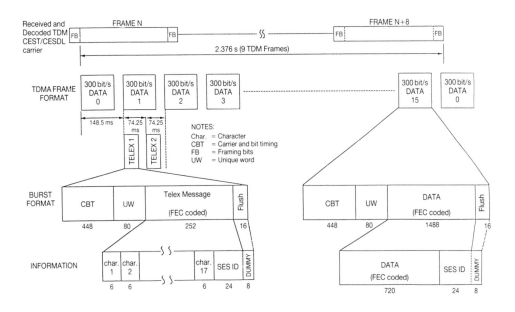

Fig. 3.8 SES TDMA 24 kbit/s channel format

The channel unit associated with the SES TDMA channel for transmission consists of a multiplexer, scrambler, FEC encoder, frame (transmit) synchronizer and modulator.

At the CES, the receive part of the channel has the corresponding complementary functions compared with the transmit end. The SES TDMA channel uses O-QPSK modulation. The full arrangement is shown in Fig. 3.9.

The circuit arrangement shown in Fig. 3.9 is identical to that used by the Inmarsat-B system for the voice channel unit.

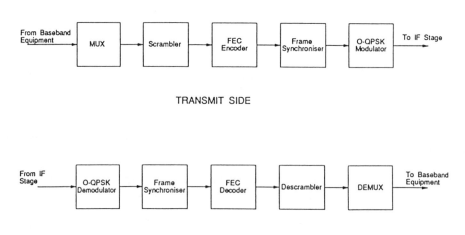

Fig. 3.9 Channel unit configuration for QPSK channels

3.2 FDMA

The Inmarsat-B telex/low-speed data channel is listed here as TDMA/FDMA for the return direction link; similarly, the forward link has been identified as TDM/FDMA. This classification is reasonable since it reflects how the transponder bandwidth is allocated for the Inmarsat-B channels. However, the complete telex/low-speed data link is TDM/TDMA and is referred to as such later.

FM/TV

Television video and audio signals may be transmitted by satellite using one or more carriers per transponder. The usual practice is to provide one or two carriers per transponder producing full transponder TV and half transponder TV respectively. The first use of two television FM carriers in a transponder was by Intelsat for international links in the Atlantic Ocean Region. Domestic services use similar frequency plans and details of the Eutelsat system will be discussed in this section.

The spectrum of a colour television signal is composed of a luminence (brightness) signal, a chrominance (colour) signal and timing (line sync) information. If frequency modulation is used, the associated noise spectrum is triangular increasing linearly from zero (dc) to the edge of the video band. The chrominance sub-carrier, with its sidebands, is at the top of the video band and is thus liable to suffer higher levels of noise. The imbalance between the signal-to-noise ratios in the luminance and chrominance signals can be minimized using pre-emphasis circuits to boost the higher frequencies.

To increase the power/bandwidth efficiency of television transmissions it is usual to combine the video and audio carrier into a single carrier. This is true for the MAC (multiplexed analogue components) system and is achieved with PAL/SECAM/NTSC systems by using an FM sub-carrier combined with the normal video baseband signal and by transmitting the composite signal on an FM modulated single carrier. This could also be achieved using SIS (sound in synchronization) where the audio and video channels are time-multiplexed rather than frequency-multiplexed. In SIS the audio signal is sampled during each line and the samples are PCM encoded, with the sampled bits inserted in the next sync pulse.

Two main types of video transmission are available as follows.

- Systems using a colour sub-carrier such as PAL, SECAM (both of these are used in Europe) or NTSC (used in the USA). The difference between these systems lies in the frequency of the chrominance sub-carrier, the modulation employed and the number of lines/frames. The PAL and SECAM systems use a 625/50 arrangement which indicates a 625 line system with 50 frames transmitted each second. The NTSC system uses a 525/60 arrangement
- Systems using time division multiplexing for the luminance and chrominance components, such as B, D, D2 and C—MAC. The term MAC is derived from Multiplexed Analogue Components. The different MAC systems derive from different sound/data components in the multiplex.

The systems using a chrominance sub-carrier are popular because of the wide availability of low priced equipment needed at the earth stations.

MAC systems are capable of producing better quality pictures and the system is well suited for satellite transmission. However, the variations in the MAC standards, represented by the B, C, D and D2 terms which apply to the sound transmission standard, has meant that no mass-produced receiving equipment is currently available, leading to higher prices.

68 Multiple access

The methods of accommodating the video-associated or additional sound channels currently in use are as follows:

- FM modulated sub-carrier(s) with or without companding,
- digital sound-in-synchronization, SIS,
- digital packets in the MAC systems

The FM sub-carrier is added to the video baseband signal before transmission. At the receiver, the composite signal is demodulated, the video and audio sub-carriers are separated and the individual sub-carriers demodulated. The audio sub-carrier is added to the video baseband signal at the top end of the baseband and endures maximum pre-emphasis in the video network. Over-deviation therefore could cause problems. The FM sub-carrier technique uses readily available, cheap and reliable equipment. Although not as good as the MAC system in terms of quality, it does provide a good compromise between cost and sound quality. It is the system most likely to be used.

To prevent crosstalk between the audio channels of two television transmissions in the same transponder, two sub-carrier frequencies are specified. Intelsat and Eutelsat specify 6.60 MHz and 6.65 MHz as the sub-carrier frequencies. Only the 6.60 MHz sub-carrier is used with full transponder or single half transponder transmissions.

Table 3.2 is an example of the parameters specified for main TV sound.

Table 3.3 shows the improvement factors obtained from using pre/de-emphasis and weighting networks.

Table 3.2 Parameters for the main TV sound (courtesy Eutelsat)

Audio signal	
Compression/expansion	None
Audio bandwidth	15 kHz
Audio low-pass filter	Required
Pre-emphasis	50 µs or CCITT J 17*
Amplitude limiter	Required
Average programme level	0 dBm
Peak programme level	+9 dBm
Sub-carrier	
FM modulator baseband response from 0.04 to 15 kHz	Flat within ± 0.2 dB
FM modulator linearity	Better than 1% in the band 6.1 MHz to 7.1 MHz
Sub-carrier frequency	6.60 MHz or 6.65 MHz†
Sub-carrier frequency tolerance	± 5 kHz
Frequency deviation of the sub-carrier by the audio signal	100 kHz or 300 kHz, peak-to-peak*
Sub-carrier bandwidth (in the composite signal) (5a)	310 kHz or 870 kHz maximum at peak programme level*
Sub-carrier amplitude	100m V peak-to-peak
Sub-carrier amplitude tolerance	± 1m V
Video low-pass filter (before combining the video and audio)	Required
Modulation index (7)	0.26 for a frequency deviation sensitivity of 25 MHz/V

Notes:
 * for TV Distribution services and Temporary TV services via occasional use Transponders respectively.
 † With two TV per transponder operation one carrier has sound sub-carrier frequency of 6.60 MHz while the other one has a sound sub-carrier frequency of 6.65 MHz. This is to eliminate cross-talk effects.

Table 3.3 Combined de-emphasis/weighting networks improvement factors W

pre-emphasis network	CCITT J 17	50 μs
f_o	1.42 kHz, 0 dB or 800 Hz, 3.8 dB	1.42 kHz, 0 dB or 800 Hz, 0.5 dB
W	+ 1.4 dB	+ 0.6 dB

It should be noted that the value of the combined de-emphasis/weighting network improvement factor (W) is for the de-emphasis insertion loss set at 0 dB at the frequency f_o.

The sound-in-synchronization system presently used by the EBU (European Broadcasting Union) has the disadvantages of providing only one mono channel and with a slightly high carrier-to-noise ratio. It is not in popular use, again because of the lack of mass-produced equipment keeping costs high. Stereo SIS is currently under test.

For MAC systems the sound is transmitted in the digital part of the time multiplex. C MAC and D MAC have the ability to accommodate four stereo channels while D2 MAC, having half the bit-rate of D MAC, can accommodate two stereo sound channels.

FM/TV applications Intelsat and Eutelsat both operate TV/FM transmissions with either a complete transponder to a single transmission (full transponder TV) or with two television channels in a single transponder (half transponder TV).

Intelsat V and VI series satellites operating in the 6/4 GHz band have 36 MHz transponders which, when used for full transponder TV, have a transmission with an allocated bandwidth and occupied bandwidth of 30 MHz and a peak frequency deviation of 9 MHz. The over-deviation factor is 0 dB in this case. The 36 MHz transponders used for half transponder TV have an allocated bandwidth of 17.5 MHz, an occupied bandwidth of 15.75 MHz and a peak frequency deviation of 7.5 MHz. The over-deviation factor is 12 dB for the 625/50 transmissions, 6.2 dB for the 525/60 transmissions. Intelsat V and VI satellites also operate with 41 MHz bandwidth transponders which, for half transponder TV use, have an allocated bandwidth of 20 MHz, an occupied bandwidth of 18 MHz and a peak frequency deviation of 10.5 MHz. The over-deviation factor in this case is 10.9 dB for the 625/50 transmissions, 6.8 dB for the 525/60 transmissions.

Eutelsat satellites with 36 MHz transponders operate with a bandwidth in a range from 33 MHz to 36 MHz and a corresponding peak-to-peak frequency deviation of 22.5 MHz to 25 MHz respectively. This is full transponder TV for PAL/SECAM FM TV carriers, which operates without over-deviation. For MAC FM TV carriers, the same transponders would operate with bandwidths and peak-to-peak frequency deviations as follows:

- D and D2 MAC, bandwidth ranging from 33 MHz to 36 MHz with a peak-to-peak frequency deviation of 22.5 MHz;
- C MAC, bandwidth ranging from 33 MHz to 36 MHz with a peak-to-peak frequency deviation of 25 MHz.

Where Eutelsat satellites operate with two TV channels per transponder, a 72 MHz leased transponder is used. The lessee is permitted to set the levels of bandwidth and frequency deviation according to specific needs providing certain limits are not exceeded. The TV channels are specified as the *main* and *additional* TV carrier with limits as follows:

- main TV carrier, maximum bandwidth of 36 MHz and maximum peak-to-peak frequency deviation of 25 MHz;
- additional TV carrier, maximum bandwidth must not exceed 57 MHz—the bandwidth of the main carrier, with a maximum value of 27 MHz. The peak-to-peak frequency deviation must not exceed 40 MHz—the deviation of the main carrier, with a maximum value of 20 MHz.

Interference into FM/TV Video S/N ratio. Interference into FM/TV carriers can be divided into two main categories.

- Noise-like interference which can be treated as thermal noise. This is the case for SCPC carriers.
- Non noise-like interference which cannot be treated as thermal noise, e.g. FM/TV carriers.

For the first category the equation introduced earlier for signal-to-noise ratio, measured in the bandwidth of the modulating signal, can be used. Equation (4.6) of Chapter 4, Section 3.3 is reproduced below:

$$\frac{S}{N} = \frac{3}{2}\left(\frac{f_d}{f_m}\right)^2 \frac{B}{f_m} \frac{C}{N} \qquad (3.2)$$

where: C/N is the carrier-to-noise power ratio in the RF bandwidth B Hz.

f_d is the peak-to-peak frequency deviation in Hertz corresponding to a 1 V peak-to-peak test tone.

f_m is the highest frequency in the baseband modulating signal, in Hertz.

S/N is the ratio of the signal power to the noise power.

The overall value for S/N, $(S/N)_{TV}$, is modified by a combined weighting improvement factor W which is given, in dB, for the 525 line system, as 14.8 dB and for the 625 line system as 13.2 dB. The overall value of $(S/N)_{TV}$ is thus the sum of equation (3.2), in dB, and W dB.

Example 3.4

A 625 line television carrier for TV distribution with a video frequency deviation of 25 MHz is received with a carrier-to-noise ratio of 12.5 dB in a 36 MHz bandwidth. The video bandwidth is 5 MHz. Determine the value of $(S/N)_{TV}$.

Solution

$$(S/N)_{TV} = (3/2)(f_d/f_m)^2(B/f_m)(C/N)W$$

or in dBs:

$$\begin{aligned}(S/N)_{TV} &= 10\log 1.5 + 20\log(f_d/f_m) + 10\log(B/f_m) + C/N + W \text{ dB} \\ &= 10\log 1.5 + 20\log(25/5) + 10\log(36/5) + C/N + W \text{ dB} \\ &= 10\log 1.5 + 20\log 5 + 10\log 7.2 + C/N + W \text{ dB} \\ &= 1.77 + 13.98 + 8.57 + 12.5 + 13.2 \text{ dB} \\ &= 50.02 \text{ dB}\end{aligned}$$

For calculations involving Intelsat TV/FM signals reference is made to the peak frequency deviation that corresponds to a 1 V peak-to-peak test tone at the crossover frequency of the pre-emphasis characteristic. The value of S/N given by equation (3.2) is determined for a sinusoidal test signal with a peak-to-peak deviation of $2f_m$. The test tone power S is proportional to the rms voltage squared, i.e. $(f_m)^2/2$. The television signal produces a peak-to-peak deviation of $2f_m$ using the full range composite signal so that equivalent television signal power S_{TV} is proportional to rms voltage squared, i.e. $(0.7 \times 2f_m)^2$.

The ratio of S_{TV} to S is given by:

$$S_{TV}/S = (0.7 \times 2f_m)^2/(f_m)^2/2$$
$$= [2 \times (0.7)^2 \times (2f_m)^2]/f_m^2$$
$$\approx 4$$

Thus, for television signals, equation (3.2) can be re-written as:

$$\left(\frac{S}{N}\right)_{TV} = 6 \left(\frac{f_d}{f_m}\right)^2 \frac{B}{f_m} \frac{C}{N} \qquad (3.3)$$

Example 3.5

A 625 line television carrier for TV distribution with a video frequency deviation of 10.5 MHz is received with a carrier-to-noise ratio of 17.3 dB in an 18 MHz bandwidth. The video bandwidth is 5 MHz. Determine the value of $(S/N)_{TV}$.

Solution

$(S/N)_{TV} = 10\log 6 + 20\log(f_d/f_m) + 10\log(B/f_m) + C/N$ dB $+ W$ dB
$= 10\log 6 + 20\log(10.5/5) + 10\log(18/5) + C/N + W$ dB
$= 10\log 6 + 20\log 2.1 + 10\log 3.6 + C/N$ dB $+ W$ dB
$= 7.8 + 6.44 + 5.56 + 17.3 + 13.2$ dB
$= 50.3$ dB.

It can be seen (see Chapter 4) that for frequency modulation the spectral distribution of RF power is given by Carson's rule:

$$B = 2(f_m + f_d)$$

where f_m is the maximum baseband frequency, and f_d is the peak value of frequency deviation.

The values of $B = 18$ MHz and $f_d = 10.5$ MHz used in the previous example are taken from Intelsat earth station standards (IESS) module 306. The value of $f_m = 6$ MHz is also a figure specified by Intelsat as the maximum video bandwidth for a 625/50 system. (The corresponding value for the 525/60 system is specified as 4.2 MHz.)

It can be seen that if $B = 18$ MHz and $f_m = 6$ MHz, the value of f_d is given by:

$$f_d = B/2 - f_m$$
$$= (18/2) - 6 \text{ MHz}$$
$$= 3 \text{ MHz}.$$

The figure of 10.5 MHz quoted by Intelsat is considerably greater than 3 MHz, suggesting that the system uses over-deviation by an amount corresponding to:

20log(deviation used/empirical value)
or 20log(10.5/3) dB in this case
or 10.9 dB.

72 Multiple access

The amount of over-deviation for the 525/60 system is:

20log(10.5/4.8) dB
or 6.8 dB.

The use of over-deviation will produce values of instantaneous frequency, at peak deviation, that are well outside the passband of the band-limiting filters. This will cause the suppression of the carrier and the generation of a short burst of noise. The adverse effects of over-deviation on picture quality are usually not too significant providing the amount of over-deviation is not excessive.

Audio sub-carrier S/N ratio The weighted audio sub-carrier signal-to-noise ratio is given by:

$$\frac{S}{N} = \frac{3}{2}\left(\frac{f_d}{b}\right)^2 \frac{B}{b} \left(\frac{C}{N}\right)_{sc} \qquad (3.4)$$

where f_d is the peak frequency deviation of the audio sub-carrier caused by a 9 dBm test tone (peak programme level) at a given frequency, f_o (Hertz):

b is the audio bandwidth (Hertz);
B is the bandwidth of the video signal (Hertz);
$(C/N)_{sc}$ is the carrier-to-noise ratio of the audio sub-carrier at the output of the TV carrier FM demodulator (dBHz).

The carrier-to-noise ratio of the audio sub-carrier, $(C/N)_{sc}$, at the output of the TV carrier FM demodulator is given by:

$$\left(\frac{C}{N}\right)_{sc} = \left(\frac{C}{N}\right)_c \frac{k_f^2}{2} \qquad (3.5)$$

where k_f is the modulation index of the TV carrier by the audio carrier.

$(C/N)_c$ is the carrier-to-noise ratio of the TV carrier (dB).

In turn it can be shown that:

$$k_f = F_d/f_{sc}$$

where F_d is the peak deviation of the TV carrier by the audio sub-carrier and,

f_{sc} is the audio sub-carrier frequency.

Example 3.6

A TV carrier dedicated to TV distribution with a video frequency deviation of 25 MHz is received with a carrier-to-noise ratio of 12.5 dB in a 36 MHz bandwidth. Using the main TV sound parameters given in Table 3.2, find the audio sub-carrier to noise ratio and the audio sub-carrier weighted S/N ratio.

Solution

From Table 3.2 the sound sub-carrier frequency is 6.60 MHz and the peak-to-peak amplitude is 100 mV. Assume 50 μs pre/de-emphasis is used and the peak-to-peak frequency deviation of the sub-carrier for a 0 dBm test tone at 1.42 kHz is 100 kHz. The pre-emphasis insertion loss at this frequency is 0 dB.

The peak frequency deviation (f_d) of the audio sub-carrier for a 0 dBm test tone is thus 50 kHz, and for a 9 dBm test tone, the value of f_d is 140.92 kHz.

3.2 FDMA

From Table 3.2, the value of modulation index k_f is 0.26 and from equation (3.5):

$$(C/N)_{sc} = (C/N)_c (k_f)^2/2$$

or in dB:

$$\begin{aligned}(C/N)_{sc} &= (C/N)_c + 10\log(k_f)^2/2 \text{ dB} \\ &= 12.5 + 10\log(0.26)^2/2 \text{ dB} \\ &= 12.5 - 14.7 \text{ dB} \\ &= -2.2 \text{ dB}.\end{aligned}$$

The audio sub-carrier weighted S/N ratio is given in dB from equation (3.4) as:

$$\begin{aligned}S/N &= + 10\log(3/2) + 20\log(f_d/b) + 10\log(B/b) + (C/N)_{sc} \text{ dB} \\ &= 1.76 + 20\log(140.92/15) + 10\log(36.10^6/15.10^3) - 2.2 \text{ dB} \\ &= 1.76 + 19.46 + 33.8 - 2.2 \text{ dB} \\ &= 52.82 \text{ dB}.\end{aligned}$$

If a weighting improvement factor is included:

$$S/N = 52.82 + W \text{ dB}$$

From Table 3.3, the value of W for a 50 μs emphasis network is 0.6 dB. Thus, the weighted value of S/N is:

$$\begin{aligned}S/N &= 52.82 + 0.6 \text{ dB} \\ &= 53.42 \text{ dB}.\end{aligned}$$

It is possible that the TV audio is digitally modulated on to the audio sub-carrier before being combined with the video baseband signal to form the composite signal. Because the signals form a multichannel baseband signal it is possible to consider the modulated audio carrier as the top channel in an FDM baseband and express the signal-to-noise ratio using an equation similar to that shown in Chapter 4 for FDM/FM systems. Thus, S/N in the bandwidth of the digitally modulated audio sub-carrier after FM transmission of the composite signal is given by:

$$\frac{S}{N} = \left(\frac{f_d}{f}\right)^2 \frac{B}{2b} \frac{C}{N} \qquad (3.6)$$

where f_d is the peak deviation of the main carrier produced by the modulated sub-carrier f in Hz.

B is the video bandwidth in Hz.
b is the bandwidth of the modulated sub-carrier in Hz.
C/N is the carrier-to-noise ratio of the combined baseband signal.

Equation (3.1) (page 59) has established a relationship between the carrier-to-noise ratio and the bit energy-to-noise ratio E_b/N_o, and if it is assumed that S/N is equivalent to C/N in the bandwidth in which the digital carrier is transmitted, then S/N can be expressed as:

$$\frac{S}{N} = \left(\frac{E_b}{N_o}\right) \frac{R}{b} M \qquad (3.7)$$

where b is the bandwidth of the modulated sub-carrier in Hz;

R is the data rate of the digital signal in bit/s;
M is the system margin to allow for impairments (dB).

74 Multiple access

Re-arranging equation (3.7) to make E_b/N_o the subject of the equation gives:

$$\frac{E_b}{N_o} = \left(\frac{f_d}{f}\right)^2 \frac{B}{2RM} \frac{C}{N}$$

A full digital TV service is available for temporary TV transmissions using 34 Mbit/s source encoding equipment as adopted by the EBU. The 34 Mbit/s digital stream is modulated in an Intelsat Intermediate Data Rate (IDR) modem, employing QPSK with rate 3/4 FEC and soft Viterbi detection. CCITT Recommendation V35 scrambling is used to give energy dispersal. The quality of the digital TV picture is superior to that provided by PAL or SECAM. The mode of operation can be full transponder or half transponder. A digital full transponder lease allows the transmission of a single 34 Mbit/s signal in either a 36 MHz or 72 MHz transponder, which is not shared with other carriers. A digital half transponder lease allows the transmission of a single 34 Mbit/s carrier in one half of a 72 MHz transponder, the other half being leased for another application.

3.3 TDMA

The concept of TDMA has been described in Section 3.1 and illustrated in Fig. 3.2. In a single carrier per transponder (SCPT) arrangement, the transponder is fully occupied by a single carrier bandwidth and the carrier is shared in time to allow several stations to transmit information, using digital modulation, in bursts. The satellite receives the earth station bursts sequentially without overlapping interference. The satellite can then retransmit all bursts to all stations. Synchronization is necessary and is achieved using a reference station from which burst position and timing information can be used as a reference by all other stations.

So as correctly to ensure the timing of the bursts from multiple earth stations, TDMA systems use a frame arrangement. Figure 3.10 shows a typical TDMA frame structure.

The start of a frame contains a reference burst from the reference station. This is followed by additional bursts from other stations having been synchronized to the reference burst to fix the timing. Each additional burst has a preamble of fixed length which carries no traffic information, followed by the traffic information.

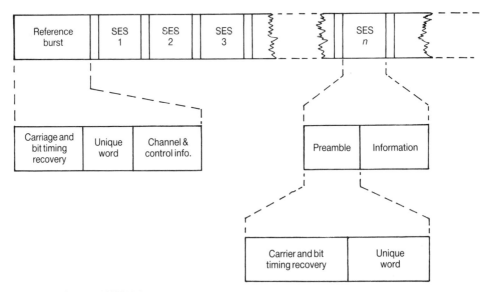

Fig. 3.10 A typical TDMA frame

3.3 TDMA 75

Because different stations have slight differences in frequency and bit rate, the receiving station must be able to establish accurately the frequency and bit rate of each burst. This is achieved using the carrier and bit timing (CBT) recovery sequence. The form of this sequence depends on the modulation method used.

The CBT sequence is followed by a sequence known as the unique word (UW). The function of the UW is to confirm that the burst is present and enables the determination of a timing marker that is used to establish the position of each bit in the remainder of the burst. The timing marker allows the identification of the start and finish of a message in the burst and aids correct decoding. The UW should have a high probability of correct detection.

The CBT and UW form the initial part of the preamble burst. Additional elements in the preamble could contain service information used for exchanging messages (regarding the state of the system), and control and delay channels, used for messages regarding information on acquisition, synchronization and system control.

Figure 3.8 shows the SES TDMA 24 kbit/s channel format of the Inmarsat-B system for ship-to-shore telex and low-speed data. This figure clearly shows the CBT and UW preamble sequence and indicates the number of bits involved.

TDM/TDMA

The Inmarsat-A system also uses the TDM/TDMA arrangement for telex signals. Each CES has at least one TDM carrier and each of the carriers has 22, 50 baud, telex channels and a signalling channel, see Fig. 3.11.

Additionally, there is a common TDM carrier continuously transmitted on the selected idle listening frequency by the NCS for out-of-band signalling. The SES

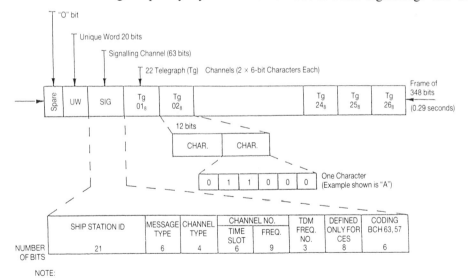

NOTE:

1. The first bit transmitted is written to the left (see the telegraph character "A" above). In the signalling channel this corresponds to the least significant bit.

2. In the telegraph channel the first bit transmitted indicates the type of character field. When the first bit is a "0", the subsequent 5-bit character field represents an ITA No 2 character, when it is a "1", the subsequent 5-bits represent line conditions for signalling.

3. Error detection coding shall be Bose-Chaudhuri-Hocquenghem (BCH) 63, 57.

4. The spare bit preceding the unique word shall be "0". All other spare bits shall be "1's".

Fig. 3.11 Inmarsat-A TDM channel format

remains tuned to the common TDM carrier to receive signalling messages when the ship is idle or engaged in a telephone call.

When a class 1 or class 3 SES is involved in a telex call it is tuned to the TDM/TDMA frequency pair associated with the corresponding CES. Telex transmissions in the return direction (ship-to-shore) form a TDMA assembly at the satellite transponder. Each frame of the return TDMA telex carrier has 22 time slots; each of these slots is paired to a slot on the TDM carrier. The allocation of a pair of time slots to complete the link is received by the SES on receipt of a request for a telex call. The TDMA frame format is shown in Fig. 11.3 of Chapter 11 and a description of how a call is set up, using the TDM/TDMA system can be found in Section 11.2.

Example 3.7

For the Inmarsat-A system, calculate the carrier-to-noise ratio required to achieve the threshold error rate for a bit rate of 1200 bit/s in the forward direction.

Solution

The threshold error rate can be calculated using equation (3.1) (see page 59):

$$(C/N)_{th} = 9.8 - 10\log(25 \times 10^3) + 10\log(1200) \text{ dB}$$
$$= 9.8 - 44 + 30.8 \text{ dB}$$
$$= -3.4 \text{ dB}$$

This figure does not include the necessary fade and implementation margin value which would have the effect of increasing the value.

Value of $(C/N_o)_t$ required is:

$$(C/N_o)_t = -3.4 + 10\log(25 \times 10^3) \text{ dB Hz}$$
$$= -3.4 + 44 \text{ dB Hz}$$
$$= 40.6 \text{ dB Hz}$$

The theoretical value quoted for Inmarsat-A is 43.4 dB Hz, a figure which could thus be met with a margin of 2.8 dB.

Intelsat provides a VSAT (very small aperture terminal) service called Intelnet which allows the design of custom networks for data, voice and video communications. The modulation and access methods available include CDMA, SCPC/MCPC with BPSK/QPSK, TDMA/FDMA and TDM/TDMA.

There are many impairments which could affect the quality of a TDM/TDMA link, including some due to the use of TDMA bursts. These include RF signal leakage between bursts from the earth station, burst delay, or improper burst bit-rate due to jitter or loss (which affects synchronization and hence clock recovery). The terminal equipment has to be able to cope with these impairments. As an example, in the Inmarsat-A system, loss of TDM frame synchronization by the SES could cause TDMA bursts to be transmitted in a time slot not assigned to the SES. A TDMA carrier transmit inhibit signal therefore activates should there be a loss of frame synchronization from the received TDM carrier.

3.4 CDMA

Code Division Multiple Access (CDMA) allows several earth stations to access the same carrier frequency and bandwidth at the same time. It is often referred to as spread-spectrum multiple access (SSMA). Assuming a PCM bit stream is used, each message bit on a link is combined with a predetermined code sequence. The sequence may be a pseudo-random noise (PN) signal which is a type known as an orthogonal code. PN codes are discussed in greater depth in Chapter 7. The bit-rate of the PN sequence must be high enough to spread the signal over the complete bandwidth which means that the bandwidth required for transmission is higher than would be the case for the message signal alone. The PN sequence bits are often referred to as 'chips', to differentiate them from the message bits, and their rate of transmission is known as the 'chip-rate'. The receiver is able to retrieve the message signal by the use of a pseudo-random sequence which is synchronized with the transmitted sequence. At the receiver other transmissions exist which are combined with other (uncorrelated) PN sequences. To the receiver these transmissions simply appear as white noise. CDMA may use direct sequence (DS) techniques where each of N users are allocated their own orthogonal code $a_i(t)$ where $i = 1, 2, 3 \ldots N$. If user codes are orthogonal, the cross-correlation of two codes is zero.

For DS/CDMA a transmitter will assemble a bit stream consisting of message $[m_i(t)]$ and code $[a_i(t)]$ information as a function of time. The product of $a_i(t)$ and $m_i(t)$ will give a signal with a spectrum that is the convolution of the spectrum of $m_i(t)$ with the spectrum of $a_i(t)$. This will be true also for other users ($a_j(t)$ with $m_j(t)$). If the bandwidth of the message signal is small compared with that of the code signal then the product has the bandwidth that approximates to that of the code signal. At the output of the first stage of the receiver the bit stream is of the form:

$$\overline{[a_i^2(t)]}\,[m_i(t)] + \sum_{j=1, j\neq i}^{N} \overline{[a_i(t)][a_j(t)]}\,[m_j(t)] \tag{3.8}$$

where j is an unwanted signal and N is the number of accesses.

Equation (3.8) assumes the receiver is configured to receive the message $m_i(t)$ and that the receiver generates a code $a_i(t)$ synchronized with the received message.

The first term of equation (3.8) is the wanted part of the sequence whereas the remainder is simply noise. The second term is zero for orthogonal codes since:

$$\int_0^T a_i^2(t) = 1$$

and:

$$\int_0^T [a_i(t)\,a_j(t)] = 0 \qquad \text{for } i \neq j$$

In practice, codes are not completely orthogonal, hence cross-correlation between user codes will introduce performance degradation.

Additionally there is a processing gain given by:

$$g = 10\log[\text{number of chips per bit}]$$

For example, if a system has a chip-rate of, say, 2.5 Mbit/s and an information rate of 20 kbit/s, the value of g is given by:

$$10\log[125] = 21 \text{ dB}$$

Since the unwanted component of equation (3.8) is spread across the complete available bandwidth and the receiver only responds to a factor of $1/g$ of this bandwidth, a CDMA system will have the 'noise' component reduced by that factor, i.e. 21 dBs in the example quoted. Additionally, noise is present due to thermal noise and intermodula-

78 *Multiple access*

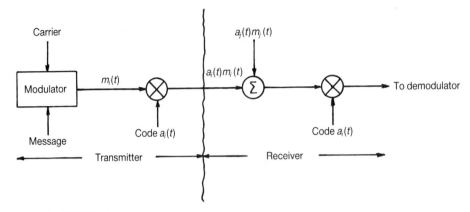

Fig. 3.12 DS/CDMA system

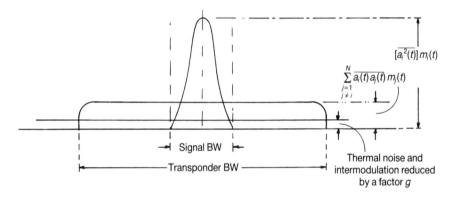

Fig. 3.13 Correlator output spectra for the DS/CDMA system

tion (IM) effects. This noise is also reduced by the gain factor g. The value of g when added to the value C/N produces the useful value of E_b/N_o.

In the frequency domain the input to the DS/CDMA receiver is wideband and contains the wanted and unwanted inputs. If the waveform given by equation (3.8) is applied to the input of a receiver correlator (with the synchronized code input $a_i(t)$), the spectrum after correlation gives the desired message signal over a bandwidth centred at the IF. The unwanted signals remain spread and only that portion of the spread signals within the receiver bandwidth will cause interference. The arrangement for the DS/CDMA system is shown in Fig. 3.12 while the correlator output spectra for the DS/CDMA system is shown in Fig. 3.13.

The transmitting station may be made to frequency-hop within a transponder, again as a function of a PN sequence, to change the frequency of transmission. The result is similar to the basic CDMA, producing a white-noise type signal. Transmitters at the earth station and satellite will operate on a continuous, or quasi-continuous, basis whereas the earth station receiver must use frequency synthesis to track the moving signal using the original PN code for the link.

Advantages of CDMA include the following.

Security With a code allocated to authorized users only, the transmission is secure to unauthorized access.

3.5 SDMA

Resistance to fading When fading occurs in a particular part of the used frequency spectrum the use of FDMA could disadvantage particular users in that part of the spectrum. With frequency-hopping CDMA (FH/CDMA) fading will only affect a user when switched to that part of the spectrum, and it causes the fading effect to be shared between all users.

System flexibility Compared with the use of TDMA, there is no need for precise time co-ordination among the system transmitters. Orthogonality between user transmissions is unaffected by time variations in the transmissions.

3.5 SDMA

Previous systems have used frequency, time and code domains to achieve multiple access. Space Domain Multiple Access (SDMA) uses spatial separation.

A single satellite may achieve spatial separation by using beams with horizontal and vertical polarization or left-hand and right-hand circular polarization. This could allow two beams to cover the same earth surface area being separated by the polarization. Additionally, the satellite could have multiple beams using separate antennae or using a single antenna with multiple feeds.

For multiple satellites spatial separation can be achieved with orbit longitude or latitude and, for intersatellite links, using different planes.

The use of SDMA allows for frequency re-use and on-board switching which, in turn, enhances channel capacity. Additionally, the use of narrow beams from the satellite allows the earth station to operate with smaller antennae and so produces a higher power density per unit area for a given transmitter power.

SDMA is usually achieved in conjunction with other types of multiple access such as FDMA, TDMA and CDMA.

SDMA/FDMA

This arrangement uses filters and fixed links within the satellite to route an incoming up-link frequency to a particular down-link transmitter antenna. A basic arrangement is shown in Fig. 3.14.

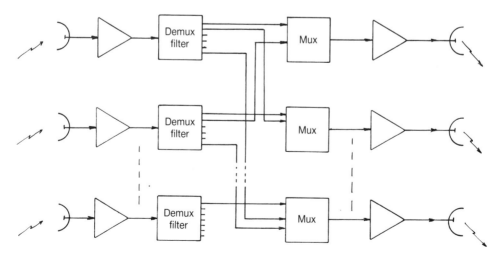

Fig. 3.14 SDMA/FDMA block diagram

80 Multiple access

The fixed links shown in Fig. 3.14 may be set up using a switch that is selected only occasionally. An alternative arrangement allows the filters to be switched using a switch matrix which is controlled by a command link. Such an arrangement would be classed SDMA/SS/FDMA (the term SS standing for switched satellite). The satellite switches are changed only rarely, when it is desired to reconfigure the satellite, to take account of possible traffic changes.

The main disadvantage of the arrangement is the need for filters, which increases the mass of the payload.

SDMA/TDMA

This arrangement is similar to that of SDMA/SS/FDMA in that a switch system allows a TDMA receiver to be connected to a single beam. Switching again is only carried out when it is required to reconfigure the satellite. Under normal conditions a link between beam pairs is maintained and operated under TDMA conditions.

Later systems, such as Intelsat VI, use time division switching to allow TDMA traffic from the up-link beams to be switched to down-link beams during the course of a TDMA frame. The connection exists at a specified time for the burst duration within the frame time before the next connection is made, and so on. For example, beam 1 may be connected to beam 2 for the first 40 μs of a 2 ms frame, beam 1 to beam n for the next 40 μs etc until every connection for the traffic pattern has been completed. The times quoted are indicative only since the values will depend on the traffic conditions. The system is known as SDMA/SS/TDMA. A simplified block diagram of a possible transponder arrangement is shown in Fig. 3.15.

The utilization of the time slots has been described in Section 3.3 and may be arranged on an organized or contention basis. Switching is achieved using the RF signal. On-board processing is likely to be used in the future, allowing switching to take place using the baseband signal. The signal could thus be restored in quality and even stored to allow transmission in a new time slot in the outgoing TDMA frame.

SDMA/CDMA

This arrangement allows access to a common frequency band. CDMA may be used to provide the multiple access to the satellite and each CDMA bit stream is decoded on the

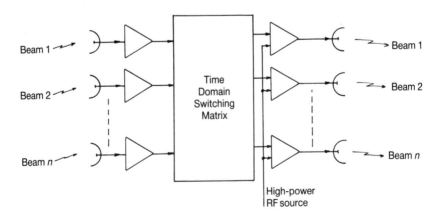

Fig. 3.15 SDMA/SS/FDMA block diagram

satellite in order to obtain the destination address. On-board circuitry must be capable of determining different destination addresses which may arrive simultaneously, while also rejecting invalid users access to the down-link. On-board processors allow the CDMA bit stream to be re-timed, regenerated and stored on the satellite. Because of this the down-link CDMA configurations need not be the same as for the up-link and each link may thus be optimized.

3.6 Packet access

For data transmissions a bit stream may be sent continuously over an established channel without the need to provide addresses, unique words, etc if the channel is not shared. Where sharing is implemented, data are sent in bursts, which thus require unique words, synchronization signals, etc to enable time sharing with other users to be effected. Each burst may consist of one or more packets comprising data from one or, possibly, more sources that have been assembled with time, processed and made ready for transmission. An advantage of packet access, as against the use of dedicated circuits, is the opportunity to store data either simply to transmit at a later time or while waiting for a connection to be established. Packet access can be used in random-access systems, such as ALOHA, where retransmission of blocked packets may be required.

The Inmarsat-C system incorporates automatic repeat request (ARQ) with half-rate convolutional coding with a constraint length of seven. The modulation method is BPSK. The system operates TDM/TDMA with forward channels being operated in continuous mode TDM. Each of the channels incorporates a frame structure within which there may be a packet structure. All frames and packets are an integral number of bytes, and bits within a byte are transmitted in sequence from bit 1 to bit 8. Fields of more than a single byte are transmitted from most significant byte to least significant byte. The TDM channels have fixed length frames of 10368 symbols transmitted at 1200 symbols/s (600 symbols/s for first generation satellites) with a frame length of 8.64 s. Each frame has a 639 byte information field which contains consecutive packets. Any packet overlapping a frame boundary is repackaged as two 'continued' packets, one unfinished in the current frame and the remainder in the next frame. An example of an information field is shown in Chapter 13, Fig. 13.8. As shown in this diagram, the signalling channel descriptor packets are for describing the signalling channels associated with the To-Mobile TDM. This is followed by message and signalling packets as required. If there are insufficient packets to fill the available space the remaining bytes are set to an idle value of all zero.

The bytes of the information field are then scrambled and the scrambled data converted to a serial bit-stream. The bit-stream is then passed to a half-rate convolutional encoder which sends 10240 symbols to the interleave matrix. After assembly the interleave block is transmitted on a row by row basis according to a permuted sequence. Interleaving is undertaken to improve reception of the data in the presence of channel fading. Interleaving is discussed in more detail in Chapter 7.

The MES message channel is similar to the TDM channel except that, since it is quasi-continuous, a preamble is added and frame length is variable between messages. The transmission frame length is $128 + (N + 1)2048$ symbols, where N is a message block size having a value between 0 and 4. Each frame transports $(N + 1)$ message packets which are 127 byte fixed-length packets. Each packet ends with an added zero byte. Block length is thus $128(N + 1)$ bytes. The block arrangement is shown in Fig. 3.16.

The block is scrambled, encoded and interleaved as described for the TDM channel. Rows are transmitted in a permuted order. Empty packets, or the empty part of the final frame, are filled with zero bytes.

82 *Multiple access*

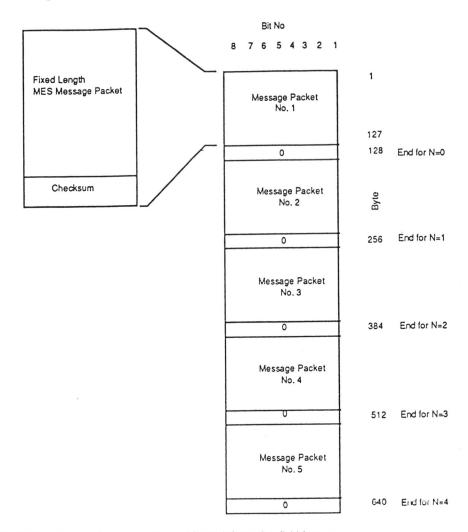

Fig. 3.16 Inmarsat-C message channel frame information field format

3.7 Random access

For random access, access to the communications link is achieved by contention. A user transmits a message irrespective of the fact that there may be other users equally in contention. Because of the random nature of the transmissions there is a possibility that transmissions from other users will collide, causing the data to be blocked from receipt by the earth station. With pure random access if such a collision occurs, then since the destination earth station channel interface equipment monitors the received emission, it would detect the collision and retransmit the message. The retransmissions, which could occur as many times as necessary, are carried out using random time delays. If all stations are entirely independent there is every likelihood that the original two messages that collided will be separated in time on retransmission.

Types of random access systems include ALOHA and Slotted ALOHA. Other forms of ALOHA exist such as Slot reservation ALOHA and Capture ALOHA.

3.7 Random access

ALOHA

This is the basic random access system which was developed in the late 1960s by the University of Hawaii to facilitate Pacific Island communications. Packets of data transmitted by random access may not be received correctly at the satellite at the first attempt and will be delayed by at least the time taken for the data to make the round trip (the originating earth station also receives the transmitter data on retransmission from the satellite). This is about 0.27 s for a geostationary satellite orbiting 40,000 km above the earth. The total delay in seconds is thus 0.27 times the number of retransmissions needed before successful capture of the data. ALOHA has a low saturation capacity, typically $1/2e$ or 18.4%; this is an indication of the utilization of the link, where utilization is defined as the amount of time the channel is earning revenue compared with the total time. The probability of a packet being lost at the initial attempt will depend on the utilization since fewer users attempting to access the link will reduce the probability of collisions. At maximum utilization, over 30% of the packets will fail to reach the satellite in recognizable form owing to the high incidence of collisions.

The advantages of ALOHA are the lack of any centralized control, giving simple, low-cost, stations, and the ability to transmit at any time without having to consider other users.

Slotted ALOHA

Slotted ALOHA, or S-ALOHA, is a form of ALOHA where the time domain is divided into slots equivalent to a single packet burst time. If all users transmit only at the start of a time slot then either the packet will get through or it will be in total contention with another packet. There will be no overlap, as is the case with ALOHA. Figure 3.17 shows a simplified arrangement which illustrates this point.

Because of the reduced risk of collisions, S-ALOHA has a saturation capacity of $1/e$, or 37%, which is twice that of ALOHA. For the same value of utilization as basic ALOHA, the time delay and probability of packet loss are both improved. The disadvantages of S-ALOHA are that more complex earth station equipment is necessary, because of the timing requirement, and that because there are fixed time slots, a

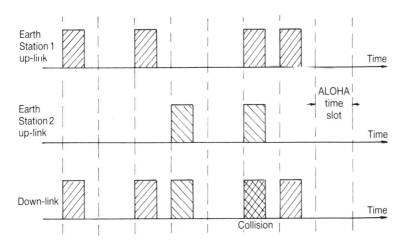

Fig. 3.17 Time slot organization for slotted ALOHA

84 Multiple access

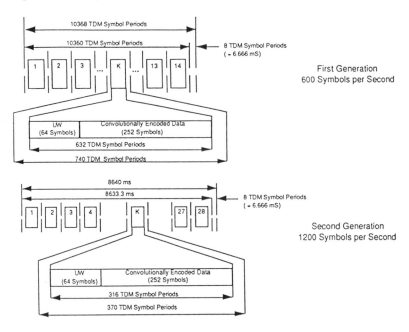

Fig. 3.18 Inmarsat-C signalling channel frame format

user with a small transmission requirement is wasting capacity by not using the time slot to its full availability.

Slot reservation ALOHA

This is simply an extension of the S-ALOHA format whereby time slots may be reserved for a particular earth station. Slot reservation basically takes two forms.

Implicit When a station acquires a slot and successfully transmits, the slot is reserved for that station for as long as it takes the station to complete its transmission. There is a danger that a station with much data to transmit could 'lock-up' the system to the detriment of other users.

Explicit Stations may send reservation requests for a time slot prior to actually sending data. A record of all time slot occupation and reservation requests is kept. A free time slot could be allocated to a requesting station or, if all slots are occupied, the next available slot could be allocated on a priority basis.

Some control for slot reservation is necessary and this could be accomplished by a single station or by all stations being informed of slot occupancy and reservation requests.

All the Inmarsat systems use ALOHA in some form, as part of the Mobile Station Request Channel signalling requirements. As an example, Inmarsat-C uses explicit slot reservation ALOHA. The signalling channel for Inmarsat-C has the same frame length as a TDM channel (8.64 s) with each frame divided into 14 slots for first generation satellites (28 for second generation satellites). The transmission rate for a burst is 600 symbols/s (or 1200 symbols/s for second generation satellites) and the timing of the MES transmission in a slot is taken from the received To-Mobile TDM. Slots are accessed by an MES with slot bursts consisting of a unique word and data as shown in Fig. 3.18.

The unique word is the same as that used for the TDM (after permuting). Details regarding the application of ALOHA to the various Inmarsat systems can be found in Section 2.

4
Modulation and demodulation

4.1 Introduction

The information to be transmitted over any radio frequency communication link consists essentially of voice, data or video signals. Each of these signals may need to be processed so that they are in a form suitable for transmission; this is baseband signal processing. The modified baseband signal is then superimposed on to a higher frequency carrier wave; the signal thus modulates the carrier and in this process undergoes frequency translation to a value suitable for propagation over the transmission link. The process of modulation at the transmit end of the link must be accompanied by demodulation, at the receive end of the link, in order to recover the baseband signal. The circuit that performs the process of modulation is the modulator while the circuit that recovers the baseband information is the demodulator. For digital systems the circuit containing a **mo**dulator for transmission, and a **dem**odulator for reception, of the RF carrier is known as a **modem**.

The process of modulation and demodulation will apply to terrestrial links as well as satellite links. Where satellite links are concerned there could also be a terrestrial link involved between the user terminal and the earth station; this terrestrial link could be by radio link or a land line or both.

For the purposes of this section it is proposed to consider modulation methods for voice, data and video systems for transmission over an RF link using either analogue or digital transmission methods.

Telephone (voice) signals

The range of frequencies that can be received by the human ear is up to about 20 kHz. However, the range of frequencies for speech is less than this and is typically band limited, to a range of 300 to 3400 Hz or even 3000 Hz, by the telephone instrument and the transmission network.

The quality of a received analogue voice signal has been specified by the CCITT to give a worst-case baseband signal-to-noise ratio for a voice signal, for transmission over a long distance, as 50 dB. Here, the signal is considered to be a standard 'test-tone' and the maximum allowable noise in the baseband is 10,000 picowatts.

Speech is characterized by having a large dynamic range of up to 50 dB to accommodate the volume difference between a whisper and a shout. Speakers also tend to pause often while talking, giving bursts of energy of random duration and random separation. It has been found that, on average, a speaker will talk for only about 40% of the time available, the remainder of the time the link is idle.

For digital transmission, the quality of the reconstituted speech at the receive end will depend, among other factors, on the number of bits transmitted per second and the number of bits received in error (bit error rate or BER). In general, the BER necessary to give good quality speech is considered to be about 10^{-4} (1 bit error per 10,000 bits) and this value could be used as a design threshold. Some systems will have values superior to this and 10^{-5}, or better, is common.

Data signals

These signals can be broadly classified into three ranges, namely: narrowband data ($\leq$ 300 bit/s); voice-band data (300 bit/s to 16 kbit/s), and wideband data ($>$ 16 kbit/s). This type of classification by bit rate approximates to the transmission facilities required to support them. As an example of a system application for data services, Inmarsat-B uses the following:

- narrowband data via the SES TDMA 24 kbit/s channel with possibly 16 data bursts, each of 300 bits/s; modems would connect the CES to the PSTN.
- 9.6 kbit/s full duplex data on the 24 kbit/s data channel to permit packet data communications using, for example, the CCITT X.25 recommendation for interface between Data Terminal Equipment (DTE) and Data-Circuit Terminating Equipment (DCE) for terminals operating in packet mode and connected to the PSPDN by dedicated circuits. This channel also supports CCITT Group-3 facsimile services; with this service also being available in the SCPC voice channel using 2.4 kbit/s data rate and APC voice codecs.

Wideband data will be supported as a later system with rates of at least 64 kbit/s.

Inmarsat-A provides wideband data facilities at a rate of 56 kbit/s. This high-speed data transmission uses a voice channel on a dedicated frequency with a special modem. During transmission a Standard-A earth station would need to have its EIRP increased by 2 dB (because of the use of a QPSK modulator instead of the FM voice modulator) to achieve the necessary quality for this service. The data stream is convolutionally encoded (rate 1/2) at the SES terminal and is decoded at the CES using a Viterbi soft decision decoder. Connection to the PSPDN is then possible in the same way as for the Inmarsat-B 9.6 kbit/s service.

Intelsat operates the IBS system which, it is claimed, provides ISDN quality data communications. Other applications include interconnection of computers and Local/Wide Area Networks (LANs/WANs), Electronic Data Interchange (EDI), facsimile transmission (CCITT groups 3 or 4) service etc. The carrier information data rates range from 64 kbit/s to 8448 kbit/s, and higher, in 64 kbit/s increments.

The Intelsat IDR system is also suitable for ISDN and data transmissions, digital television etc using information bit-rates ranging from 64 kbit/s to 45 Mbit/s. Rate 3/4 FEC is used for large earth stations together with convolutional encoding and Viterbi decoding.

Video signals

Television pictures are composed of electrical signals produced when the picture is scanned at a suitable rate. A scanning spot traverses the picture image in a zigzag fashion and determines the luminence of the image at every point along its path. The scanning process may scan only a fraction of the picture elements, with the remaining elements covered in subsequent scans, and a single scan process is called a field. For television signals, scanning of each picture occurs using two fields so the complete picture is scanned twice during the picture period. One field follows another at a rate of 50 or 60 per second giving a picture rate of 25 or 30 pictures per second. (50 fields/second is a European standard while 60 fields/second is used in the USA). The picture rate is

88 Modulation and demodulation

chosen to be sufficiently high to avoid a 'flicker' of the picture. Experiments have shown that the human eye will not detect a flicker at above 20 pictures/second. Since a picture is composed of 625 or 525 lines, (625 lines is a European standard while 525 lines is used in the USA) the number of lines per field are 312.5 and 262.5 respectively. The two fields produced on alternate scans are merged to form a frame. This is known as interlaced scanning. For the first field the scan begins at the top left-hand side and completes at the bottom right-hand side of the picture. A vertical retrace then occurs and the second scan begins at the top centre of the picture which permits the second field scan to fall midway between the positions covered by the first field lines. Thus, two vertical retrace lines occur every picture period and each retrace requires a synchronizing pulse. The vertical synchronization frequency is the same as the picture frequency. The vertical synchronization pulse is superimposed on top of a longer 'blanking' pulse which turns off the electron beam during the retrace period.

The picture definition is defined by the size of the smallest picture element (usually referred to as a pixel). Figure 4.1 shows a number of pixels within a picture of height H and width W.

If N = number of lines in a frame then the line height (H) is H/N.

The number of pixels in the horizontal direction is given by aN, where a is the aspect ratio given by W/H. Typically a is given by the ratio 4/3.

The above ignores the fact that some lines are suppressed during frame synchronization. To allow for this, in practice the value of N would be reduced slightly.

The total number of lines in a field is $N/2$ and since the field frequency is $2P$ per second (where P is the picture repetition rate in pictures/second), the number of lines generated per second is ($N/2 \times 2P$) or NP. A synchronization pulse is required at the end of each line with a frequency given by $f_h = NP$ pulses/second. As an example, the 625 line system has $N = 625$, $P = 25$ frames/second, so that: $f_h = 625 \times 25 = 15625$ pulses/second. The corresponding line time (t_h) is given by $1/f_h$ or, in this example, 1/15625 seconds or 64 μs. The corresponding values for the 525 line system are $f_h = 15750$ pulses/second and $t_h = 63.5$ μs.

The video bandwidth depends on two successive pixels with the 'worst-case' pattern,

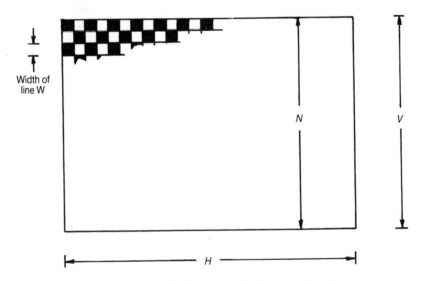

Fig. 4.1 Television picture composed of alternate black and white pixels

as shown in Fig. 4.1 where the pixels alternate from 'black' to 'white', to generate one cycle of video output. The highest video frequency (f) required is given by:

$$f = (aN/2)/t_h$$

For the 625 line system, $f = (4/3) \times 625/2 \times 10^6/64$

or $f = 6.51$ MHz.

This has ignored the number of lines suppressed during frame synchronization and the suppression time required for line synchronization. Experiments have also shown that the figure for bandwidth is high and could be reduced in practice without any detrimental effect on the received picture quality. The actual bandwidth chosen for the 625 line systems is 5.0 MHz and for the 525 line system, 4.2 MHz. For FM/TV the bandwidth chosen for the video low-pass filter is 6 MHz which ensures universal compatibility with all television standards. The television signal will consist of four separate components:

- sound,
- brightness (luminence) scan information,
- synchronization for horizontal and vertical scans,
- colour chrominance.

For the European 625 line systems (PAL and SECAM) and the USA 525 line system (NTSC), the chrominance channel is modulated on to a sub-carrier at a frequency of about 3.5 MHz or 4.5 MHz depending on the system. For the MAC (multiplexed analogue components) systems the luminance and chrominance components are transmitted using time-division multiplexing techniques.

Figure 4.2 shows the waveform for a single scan line of the luminance channel of the composite television signal. The chrominance component is not shown but would occur during the line blanking period. The total amplitude of the luminance (plus synchronization pulse) signal is 1 V peak-to-peak, referred to a point of zero reference. The peak-to-peak amplitude of the luminance signal alone is 0.7 V.

4.2 Amplitude modulation

Amplitude modulation is not a modulation process used in the satellite link. It can be used, however, to modulate individual voice channels before combining such channels

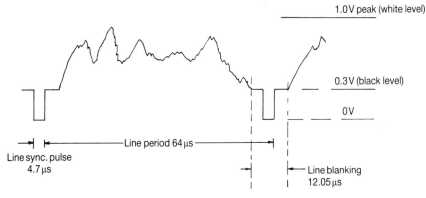

Fig. 4.2 Waveform for a single scan line of the luminance channel of the composite television signal

90 Modulation and demodulation

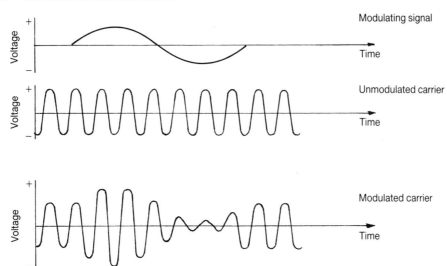

Fig. 4.3 Modulating a carrier wave with a single sinusoidal tone (a) modulating signal; (b) unmodulated carrier; and (c) modulated carrier

using FDM. Thus, amplitude modulation will be discussed since it may form part of the overall link.

For a carrier wave to be amplitude modulated, the amplitude of the carrier is caused to vary according to the instantaneous value of the amplitude of the modulating signal at a rate proportional to the frequency of the modulating signal. If the modulating signal is a pure sine wave (to consider the simplest case) then the amplitude of the carrier varies sinusoidally as shown in Fig. 4.3.

In practice, the modulating signal at any instant could contain frequency elements within the baseband, giving a more complicated variation than shown in Fig. 4.3(c). However, any modulating signal is made up of a fundamental frequency and harmonics so that the concept of Fig. 4.3(c) is valid.

Frequency spectrum for amplitude modulation

Consider a system where a carrier has peak amplitude V_c and angular frequency ω_c radians/s, and the modulating signal is a single sinusoidal tone of peak amplitude V_m and angular frequency ω_m radians/s. Then for amplitude modulation:

$$V(t) = (V_c + V_m \cos\omega_m t)\cos\omega_c t$$
$$= V_c(1 + m\cos\omega_m t)\cos\omega_c t \quad (4.1)$$

where $m = V_m/V_c$ is the modulation index.
The value for m should never exceed unity, i.e. $m \leq 1$.
Thus, expanding equation (4.1):

$$V(t) = V_c\cos\omega_c t + \frac{mV_c}{2}\cos(\omega_c - \omega_m)t + \frac{mV_c}{2}\cos(\omega_c + \omega_m)t$$

This is shown as a frequency spectrum in Fig. 4.4 and can be seen to consist of the carrier (f_c), upper side frequency ($f_c + f_m$) and lower side frequency ($f_c - f_m$).

4.2 Amplitude modulation

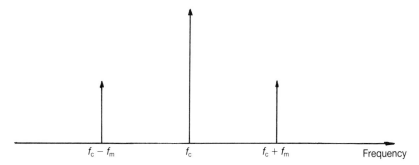

Fig. 4.4 Frequency spectrum of an amplitude modulated carrier, modulated with a single sinusoidal tone

It can be deduced from Fig. 4.4 that the bandwidth required to transmit the 'information' contained in the single frequency modulating tone is:

$$(f_c + f_m) - (f_c - f_m) = 2f_m$$

The power contained in the signal P_T is given by:

$$P_T = \overline{V^2(t)} = \frac{V_c^2}{2} + \frac{m^2 V_c^2}{8} + \frac{m^2 V_c^2}{8} \qquad (4.2)$$

where $V_c^2/2$ is the power, P_c, in the carrier and $m^2 V_c^2/8$ is the power in each side frequency P_{LSF} and P_{USF}.

If $m = 1$ (100% modulation), then P_T becomes:

$$P_T = V_c^2/2 + V_c^2/4 \qquad (4.3)$$

where $V_c^2/4$ is the total side frequency power.

Equation (4.3) shows that only one-third of the total power is used to carry 'information'.

Noise in an AM system

Assume that the transmitted signal is received at the input of the demodulator in the presence of noise. If the bandwidth of the receiver is $B\ (=2f_m)$ and if the noise has a uniform power density N_o in watts/Hz, the total noise power, N, in the receiver bandwidth is given by:

$$N = 2f_m N_o \text{ watts}$$

The noise output of the demodulator can be represented by a value:

$$N_b = AN$$

where N_b = noise power in the baseband, or demodulated, signal.

A = a scaling factor for the demodulator.

Equation (4.2) shows that the power in each side frequency is a quarter of the value of that of the carrier. If S represents signal power in a side frequency and C represents the power in the carrier:

$$S_L = S_U = C/4$$

and after demodulation:

$$S_{bL} = S_{bU} = A(C/4)$$

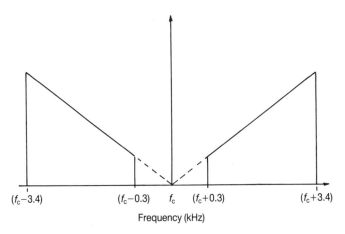

Fig. 4.5 Frequency spectrum of an amplitude modulated carrier, modulated by a band of frequencies

Since both side frequencies are identical before and after demodulation, they will add coherently in the demodulator to produce:

$$S_b = 2A(S_{bL} + S_{bU}) = AC$$

At the demodulator output the signal-to-noise ratio S_b/N_b can be related to the carrier-to-noise ratio C/N at the demodulator input since:

$$S_b/N_b = C/N = C/N_o B$$

This equation represents the case where $m = 1$; for other values of m the equation becomes:

$$S_b/N_b = m^2(C/N)$$

The above has assumed a single frequency modulating tone whereas, in practice, the modulating signal would have a baseband of, say for telephony, 3100 Hz (3400 − 300 Hz). The effect on the frequency spectrum would be as shown in Fig. 4.5.

Fig. 4.5 shows clearly that the process of amplitude modulation produces two sidebands, each of which contains information from the baseband signal, and the carrier which contains no baseband information. If the carrier and one of the sidebands is suppressed then the resulting AM signal still contains the required baseband information. The single sideband, suppressed carrier version of AM (SSBSC) is popular because it reduces the bandwidth necessary for transmission giving more channels in a given RF bandwidth. For telephony, the baseband frequency range is 300 to 3400 Hz requiring an RF bandwidth of just 3100 Hz, although 4 kHz is used in practice in order to provide a guard band between adjacent channels.

One method of eliminating the carrier and a sideband is shown in Fig. 4.6.

The modulating device in Fig. 4.6 is a balanced modulator which eliminates the carrier, while the filter removes one of the sidebands.

In the receiver, envelope detection cannot be used and coherent demodulation of the received signal with a local insertion of the original carrier must be used. The modulation index no longer has any relevance and the sideband power is proportional to the modulating signal power. The value of this power will depend on the power ratings of the transmitter and the transmission channel.

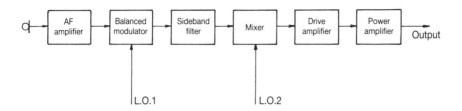

Fig. 4.6 SSB/SC AM block diagram

The signal-to-noise ratio at the output of the demodulator can be established in a similar manner to that of conventional AM. Assume the received signal has a sideband power S with noise in the receiver bandwidth B given by:

$$N = N_o B = N_o f_m$$

where f_m is the highest frequency in the baseband.
The signal power at demodulator output is:

$$S_b = AS$$

The noise power at demodulator output is:

$$N_b = AN = AN_o B = AN_o f_m$$

The baseband signal-to-noise ratio is:

$$S_b/N_b = S/N = S/(N_o f_m)$$

In this case the total power is S and the value for S_b/N_b is independent of the waveform of the modulating signal.

Investigation of the theory of full AM and SSBSC suggests that the transmitted power would need to be three times greater for full AM compared to SSBSC. Also, if the modulation index is less than unity, or if the peak-to-average ratio of the modulating signal is greater than that of a single sinusoidal tone, there is an increase in the ratio of power in full AM to that in SSBSC.

Considering a DSBSC system, it can be shown that the output of a coherent demodulator gives a value of S/N equal to that from the coherent demodulator of an SSBSC system, i.e.:

$$(S/N)_{SSBSC} = (S/N)_{DSBSC}$$

In practice, SSBSC is preferred since only frequency coherence is required whereas for DSBSC, frequency and phase coherence is necessary. Also, SSBSC requires half the bandwidth of DSBSC.

4.3 Frequency modulation

A carrier represented by $V(t) = V_c \sin\omega_c t$ may be modulated by varying $\omega_c t$. Frequency modulation (FM) is defined as that form of modulation whereby the carrier frequency is caused to vary by an amount proportional to the amplitude of the modulating signal, at a rate proportional to the frequency of the modulating signal, with the amplitude of the carrier remaining constant. A graphical representation of FM is shown in Fig. 4.7

It can be seen that the frequency variations of curve (i) of Fig. 4.7(b) are caused by a

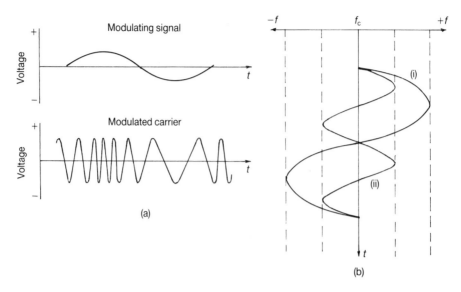

Fig. 4.7 Graphical representation of FM (a) modulation effect on carrier frequency; (b) frequency swing of FM carrier

modulating signal of twice the amplitude and half the frequency of the modulating signal for curve (ii).

Terms used in frequency modulation

Frequency swing The difference between the maximum and minimum values of the instantaneous frequency of the modulated wave.

Frequency deviation (f_d) The peak difference between the instantaneous frequency of the modulated wave and the carrier frequency in a cycle of modulation.

Modulation index (m_f) The ratio of frequency deviation to modulating frequency f_m, i.e. $m_f = f_d/f_m$

The above terms refer to a particular signal while the following terms refer to a particular system.

Rated system deviation ($f_{d.max}$) The maximum allowable frequency deviation in a particular system.

Rated maximum modulating frequency ($f_{m.max}$) The maximum modulating frequency that a particular system can carry.

Deviation ratio (D) Ratio of rated system deviation to the rated maximum modulating frequency of the system. Thus:

$$D = f_{d.max}/f_{m.max}$$

For FM the modulation factor m is the ratio of the frequency deviation to the rated system deviation and is equivalent to the same term in AM i.e.:

4.3 Frequency modulation

$$m = f_d/f_{d.max} = m_f f_m/f_{d.max}$$

The instantaneous frequency of an FM wave is:

$$f_i = f_c + V_m \sin\omega_m t$$

where f_m ($=\omega_m/2\pi$) is the modulating frequency
V_m is the peak value of modulating signal
f_c is the carrier frequency
hence $f_i = f_c + f_d \sin\omega_m t$
since f_d is proportional to the amplitude of the modulating signal.

Thus $\omega_i = \omega_c + \omega_d \sin\omega_m t$
and since $\omega = d\theta/dt$, then:
$\theta_i = \int \omega_i dt$
$= \int (\omega_c + \omega_d \sin\omega_m t) dt$
$= \omega_c t - (2\pi f_d/\omega_m)\cos\omega_m t$ (neglecting the constant of integration)
hence $\theta_i = \omega_c t - (f_d/f_m)\cos\omega_m t$
$\theta_i = \omega_c t - m_f \cos\omega_m t$
and, as $V(t) = V_c \sin\theta$
$V(t) = V_c \sin(\omega_c t - m_f \cos\omega_m t)$

Phasor representation of FM

The way in which the modulating signal causes the carrier frequency to change can be seen by reference to a phasor diagram. Representing FM by a single phasor is complicated by the fact that, due to modulation, the carrier phasor rotates at varying speeds. When the carrier is frequency modulated by, say, a sinusoidal frequency of constant peak amplitude, the phasor will speed up and slow down with changes in carrier frequency. However, by using phasor speed when unmodulated as a reference (see Fig. 4.8(a)), the modulated waveform may be represented by a phasor which oscillates about a mean position at a frequency equal to the modulation frequency. See Fig. 4.8(b).

It can be observed that the phasor is momentarily at rest at the extremes of its travel so the carrier is instantaneously at its unmodulated angular velocity ω_c at these positions (A, C and E in Fig. 4.8(b)).

At the centre of its swing the phasor is moving at its fastest in either direction. At B,

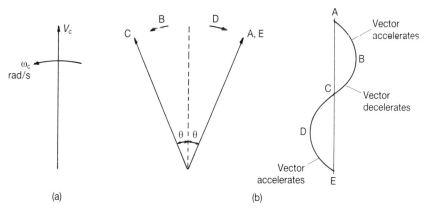

Fig. 4.8 Phasor representation of FM (a) phasor swing; (b) frequency deviation

96 Modulation and demodulation

phasor speed has increased to represent $f_c + f_d$ while at D, phasor speed has fallen to $f_c - f_d$. Hence, the reference in the phasor diagram corresponds to the limits of frequency deviation. For constant amplitude modulating voltage, carrier deviation is fixed, by definition, and for different values of modulating frequencies the speed of the phasor past the reference will be constant.

The length of the phasor sweep, or magnitude of angle θ, will depend on the frequency deviation f_d and the time for one cycle of modulating signal, i.e.:

- the greater the frequency deviation, the greater the angle of sweep.
- the higher the modulating frequency, the shorter the time taken to complete one cycle of sweep.

Hence, length of arc θ is given by:

$$\theta = f_d/f_m = m_f$$

This can be shown by putting values on the waveform and phasor diagram of Fig. 4.8(b) according to the FM equations. Figure 4.8(b) has been redrawn in Fig. 4.9 and labelled, in terms of frequency and sweep angle, over one cycle of modulation.

Using the formula $f_i = f_c + f_d \sin\omega_m t$ for Fig. 4.9(a) it can be seen that:

- at point A, $\omega_m t = 0$ and $f_i = f_c$
- at point B, $\omega_m t = \pi/2$ and $f_i = f_c + f_d$ etc

Using the formula $\theta_i = \theta_c - m_f \cos\omega_m t$ for Fig. 4.9(b) it can be seen that:

- at point A, $\omega_m t = 0$ and $\theta_i = \theta_c - m_f$
- at point B, $\omega_m t = \pi/2$ and $\theta_i = \theta_c$ etc.

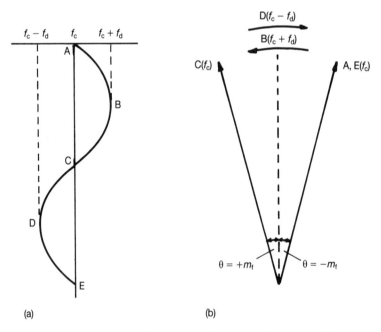

(a)   (b)

Fig. 4.9 Examples of maximum frequency and phase deviation for FM

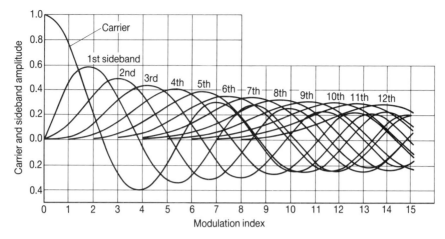

Fig. 4.10 Bessel function curves

Frequency spectrum for FM

It may appear from previous work that information can be carried by an FM wave using a bandwidth of twice the frequency deviation, i.e. $2f_d$. This is true for certain applications but is not generally true since, as the instantaneous frequency of the carrier varies, the carrier waveform is altered and new frequencies generated. To avoid excessive distortion, all those frequencies with an amplitude greater than about 1% of that of the unmodulated carrier should be transmitted. This causes the required bandwidth to exceed $2f_d$.

An angular modulated wave has side frequencies of $f_c \pm f_m, f_c \pm 2f_m \ldots f_c \pm nf_m$. Most of the higher order side frequencies are insignificant in terms of magnitude and can be neglected. The two frequencies of each side pair have equal amplitudes but the amplitudes of the side frequencies compared with the carrier amplitude are related by Bessel functions, see Fig. 4.10.

The graphs drawn from the Bessel functions indicate:

- the number of effective side frequencies increases with m_f;
- at certain values of m_f the amplitude of the carrier is zero;
- where $m_f < 0.5$, the bandwidth is about the same as AM $(2f_m)$. This is *narrowband FM*.
- where $m_f > 0.5$, the system is *wideband FM*.

To determine system bandwidth from the curves:
- find D on the horizontal axis;
- note the highest side frequency of effective amplitude;
- multiply this figure by $f_{m.max}$;
- double this to give the required bandwidth.

As an example, take $D = 5$ and from the Bessel curves of Fig. 4.10 the highest side frequency of effective amplitude is 8. Suppose $f_{m.max}$ is 3 kHz, then the bandwidth required is:

$$2(8 \times 3) \text{ kHz} = 48 \text{ kHz}.$$

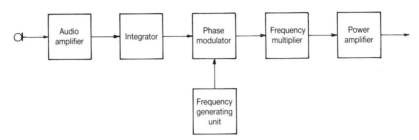

Fig. 4.11 Block diagram of an indirect FM transmitter

An empirical rule for determining system bandwidth was proposed in 1939 by J.R. Carson. His rule states:

$$B = 2(f_d + f_m)$$
or $$B = 2f_m(m_f + 1)$$

and, using the above figures:

$$B = 2 \times 3(5 + 1) \text{ kHz} = 36 \text{ kHz}.$$

This shows a particular value of B assuming particular values for f_d and f_m. For a system the values used would be the rated system values so that:

$$B = 2f_{m.max}(D + 1)$$

The signal power in an angular modulated wave is constant. Whether modulated or not and irrespective of the depth of modulation, the power is the same under all conditions. As the number of side frequencies generated increases, more energy is thrown out to these frequencies.

Modulation of a carrier using FM can be achieved in various ways. A block diagram of one system is illustrated in Fig. 4.11.

The audio amplifier includes a filter to restrict the range of audio frequencies to that required for the system, and a limiter to ensure that excess audio amplitudes do not cause the system deviation figure to be exceeded. The integrator causes the amplitude of the modulating signal to become inversely proportional to frequency. The frequency generating unit may consist of switched single crystals or a frequency synthesizer to give a range of carrier frequencies. Frequency multiplication is necessary to give the required output carrier frequency and frequency deviation. The power amplifier could use class-C bias for greatest efficiency. This method of producing wideband FM is known as Indirect Frequency Modulation.

Demodulation of an FM signal is produced by a circuit that gives an output voltage

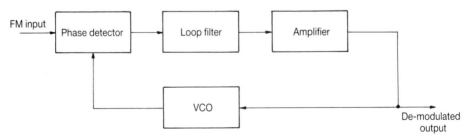

Fig. 4.12 FM demodulator using a PLL

proportional to the frequency deviation of the input. Such a circuit is known as a discriminator. A phase-lock loop (PLL) circuit may be used as a discriminator as the arrangement of Fig. 4.12 shows.

In this PLL circuit the phase detector gives an output proportional to the phase difference between the circuit input and the output of the voltage controlled oscillator (VCO). The phase detector output is filtered, amplified, and applied to the input of the VCO. The VCO is essentially a frequency modulator, the frequency deviation of its output being proportional to its input signal and approximately equal to the circuit input signal deviation. Since the circuit input frequency deviation depends on the modulating signal, it follows that the VCO input is proportional to the circuit input frequency deviation and is thus the demodulated output for an FM input, provided the PLL operates in lock.

Because of its special properties the PLL could also be used as a frequency modulator.

Noise in an FM system

The AM system has a passband, at worst, of twice the highest modulating frequency. The passband filter preceding the FM discriminator must be greater than twice $2f_m$ in order to accommodate all the side frequencies generated by the signal. The input noise to the system will therefore correspond to this bandwidth, which in turn depends on the value of m_f.

The FM signal accompanied by noise can be expressed as:

$$V(t) = V_c \cos[\omega_c t + \phi_c(t)] + V_n(t)\cos[\omega_c t + \phi(t)]$$

where $\phi_c(t)$ is the message modulation and
$V_n(t)\cos[\omega_c t + \phi(t)]$ describes the band-limited noise.

The resultant signal will vary in amplitude and phase. The phasor diagram of Fig. 4.13 shows the effect when $V_c \gg V_n$.

The unwanted amplitude variations can be removed by using an amplitude limiter prior to the discriminator. From Fig. 4.13, the instantaneous phase $\phi_R(t)$ of the resultant signal may be expressed as:

$$\phi_R(t) = \omega_c t + \phi_c(t) + \tan^{-1} \frac{V_n(t) \sin[\phi(t) - \phi_c(t)]}{V_c + V_n(t) \cos[\phi(t) - \phi_c(t)]}$$

and assuming $V_c \gg V_n$:

$$\phi_R(t) = \omega_c t + \phi_c(t) + \frac{V_n(t)}{V_c} \sin[\phi(t) - \phi_c(t)] \qquad (4.4)$$

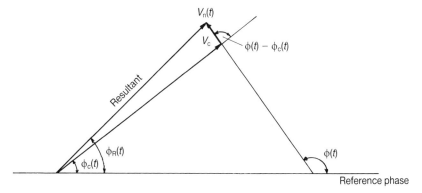

Fig. 4.13 Phasor diagrams of signal plus noise for $V_c \gg V_n$

100 Modulation and demodulation

where $\omega_c t + \phi_c(t)$ represents the desired information and the remainder of equation (4.4) represents the noise.

Since the signal and noise, for the condition $V_c \gg V_n$, are additive, the process of finding the signal-to-noise ratio can be performed in two stages. Considering the signal input first.

The modulating signal can be represented by:

$$V(t) = V_c \cos\left[\omega_c t + 2\pi \int_0^t k_f m(t)\, dt\right]$$

where $m(t)$ is the baseband signal and

k_f is the constant of proportionality between modulation signal amplitude V_m and the peak frequency deviation.

The mean output power from the discriminator is proportional to:

$$k_f^2 \overline{m^2(t)}$$

For the noise alone the output power is found for an input which consists of an unmodulated carrier and many noise sinusoids of random phase. The output contributions of any significance are due to the noise sinusoids beating with the carrier. If each noise sinusoid is of the form:

$$V_{nx}\cos(\omega_n t + \phi_n)$$

where V_{nx} is the peak amplitude of the noise power component, the beating of the noise sinusoid with the carrier results in a carrier–noise component of instantaneous phase:

$$\frac{V_{nx}}{V_c} \sin[(\omega_n - \omega_c)t + \theta_n]$$

Since frequency variations are equal to $d\phi/dt$, then:

$$\frac{d\phi}{dt} = (\omega_n - \omega_c)\frac{V_{nx}}{V_c} \cos[(\omega_n - \omega_c)t + \theta_n]$$

Mean output power from the discriminator is proportional to:

$$\frac{1}{2}(f_n - f_c)^2 \left(\frac{V_{nx}^2}{V_c^2}\right)$$

which can be written as:

$$(f_n - f_c)^2 \left(\frac{N_o}{V_c^2}\right)$$

where N_o is the single-sided noise power spectral density of a non-bandwidth limited system.

The total output power is the result of summing all carrier–noise sinusoids within the baseband b. Thus, total noise output power from the discriminator is proportional to:

$$\frac{N_o}{V_c^2} \int_{f_c - b}^{f_c + b} (f - f_c)^2\, df$$

and if N_o is constant over the frequency range considered, the total output noise power from the discriminator is proportional to:

$$\frac{2b^3 N_o}{3V_c^2}$$

4.3 Frequency modulation

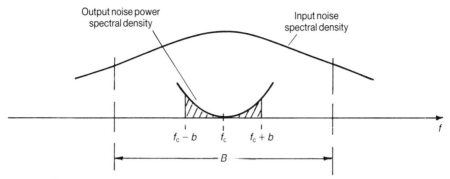

Fig. 4.14 FM input/output noise spectral densities for wideband FM discrimination

The output noise power is proportional to $(f_n - f_c)^2$ which results in an output noise power spectral density as shown in Fig. 4.14.

It can be seen from Fig. 4.14 that at higher frequencies the signal-to-noise ratio is at its worst. This can be improved by the use of emphasis circuits (as discussed later).

The output signal-to-noise ratio is thus:

$$\frac{k_f^2 \overline{m^2(t)}\, 3 V_c^2}{2b^3 N_o}$$

If the modulation is a single sinusoid:

$$k_f^2 \overline{m^2(t)} = k_f^2 \frac{V_c^2}{2} = \frac{f_d^2}{2}$$

Hence:

$$\frac{S_b}{N_b} = \frac{f_d^2 3 V_c^2}{4 b^3 N_o} = 3 \left(\frac{f_d}{b}\right)^2 \frac{B}{2b}\frac{C}{N} \tag{4.5}$$

where $N = N_o B$
B = RF bandwidth
C = input signal power $(V_c)^2/2$

Hence:

$$\frac{S_b}{N_b} = 3 \left(\frac{f_d}{f_m}\right)^2 \frac{B}{2 f_m}\frac{C}{N} \tag{4.6}$$

where $b = f_m$, the maximum modulating frequency.

Equation (4.6) is valid provided demodulation occurs above the threshold. It is usual to have the system working point a fixed value M dBs above the threshold where M is a system margin.

For FDM/FM transmissions the S/N ratio after demodulation is given by:

$$\frac{S_b}{N_b} = 3B \frac{f_d^2}{f_2^3 - f_1^3}\frac{C}{N}$$

where C = carrier power at receiver input (watts)
B = RF bandwidth (Hz)
N = noise power, kTB, in bandwidth B (watts)
f_d = rms frequency deviation/channel corresponding to a test tone at 0 dBmo (Hz)
f_1 = lower baseband channel frequency (Hz)
f_2 = upper baseband channel frequency (Hz)

For a telephony channel:

$(f_2 - f_1) = 3000 - 300$ Hz
$= 2700$ Hz.

This is small compared with the top frequency of the FDM assembly channels so that since:

$$f_2^3 - f_1^3 = (f_2 - f_1)(f_2^2 + f_1 f_2 + f_1^2)$$
$$f_2^3 - f_1^3 = 3bf_m^2$$

where f_m is the top frequency of the FDM assembly and

b is $(f_2 - f_1)$,

then:

$$\frac{S_b}{N_b} = \left(\frac{f_d}{f_m}\right)^2 \frac{B}{b} \frac{C}{N} \qquad (4.7)$$

The value of f_d is the deviation produced by a 1 mW test-tone injected at the zero reference point of the system so that S_b is a 1 mW signal and the quality of the ratio is determined by the noise N_b in pWOP.

For an FDM block of multichannel telephony circuits f_d is the rms deviation corresponding to the complete multiplex signal and the bandwidth B is for the complete FDM/FM transmission. The value for B can be calculated using Carson's Rule, i.e.:

$$B = 2(f_d + f_m)$$

Factors which assist in the calculation of f_d for this arrangement include the following.

Mean power factor (MPF) The mean power of an FDM assembly of n channels at a point of zero relative level has been specified by the CCIR to be:

$-1 + 4\log n$ dBm for $n < 240$
$-15 + 10\log n$ dBm for $n \geq 240$

These power ratios can be converted to a ratio using the expression:

MPF $= a\log(\text{dBm value}/20)$

Multi-channel peak factor (MCPF) The multi-channel peak values have been established to be:

4.7 for $n < 120$
3.16 for $n > 120$

These are ratios and are not in decibels.

The normal routine for this type of system is to determine the performance by means of an equivalent rms test-tone deviation f_d in a single baseband channel rather than the peak deviation in the entire FM assembly ($f_{d.pk}$). The two frequencies are related by the expression:

$$f_{d.pk} = (\text{MPF}) \times (\text{MCPF}) \times f_d$$

Hence, if $f_{d.pk}$ and the number of channels are known, f_d can be calculated for use in equation (4.7).

4.3 Frequency modulation

For SCPC/FDM/FM systems, equation (4.7) is modified to give:

$$\frac{S_b}{N_b} = \frac{3}{2}\left(\frac{f_d}{f_m}\right)^2 \frac{B}{2f_m} \frac{C}{N} \qquad (4.8)$$

where f_d is the rms frequency deviation produced by a single channel.

The difference between equations (4.7) and (4.8) represents the difference between multi-channel and single channel operation. The factor 3 in equation (4.8) is an integration constant for the SCPC condition while the factor 1/2 in that equation is due to the use of a peak deviation value for f_d instead of the rms value used in equation (4.7).

The effects of companding and pre-emphasis could lead to improved values for S_b/N_b. Expressed in dBs, the value for S_b/N_b would become:

$$\frac{S_b}{N_b} = \left(\frac{C}{N}\right) + 20\log\left[\sqrt{\frac{3}{2}}\left(\frac{f_d}{f_m}\right)\right] + 10\log\left[\frac{B}{2f_m}\right] + C + P$$

where C is the companding improvement in dB
P is the pre-emphasis improvement in dB.

Threshold

For an AM system it has been found that at the demodulator output, a required signal is dominant when its peak amplitude exceeds four times the rms noise amplitude, both measurements being taken at the demodulator input. This is the threshold value and has a value of 8 (9 dB) for the signal-to-noise ratio.

In an FM system there are two threshold levels for the demodulation of signals in the presence of noise. The first threshold is the same as for AM, i.e. 9 dBs. The second threshold is defined by the S/N power ratio that gives full FM improvement over AM. This is found to be 2 – 3 dBs greater than the S/N ratio defining the threshold. See Fig. 4.15.

An FM system will only have a better S/N ratio than an AM system, provided:

- the modulation index is greater than unity;
- the amplitude of the carrier is greater than the maximum noise peak amplitudes;
- the receiver is insensitive to amplitude variations.

If the modulation index (m_f) is not greater than unity, the S/N ratio will be less than the reference, but at point A, as the value of m_f is increased, the two are equal.

As m_f is further increased, the S/N ratio improves further until point B (threshold of full FM improvement); when the modulation index equals the deviation ratio (D), the limit for that value of D is reached.

Thus, when the S/N ratio is high there is a considerable advantage in using FM, and the improvement over the AM value, in dBs, increases in direct proportion to the peak frequency deviation.

The difference in value between the working value of C/N and the threshold value is the threshold margin (M).

Pre-emphasis and de-emphasis

It has already been established that the noise effect in an FM system increases with frequency deviation from the carrier. Since FM of constant amplitude results in smaller values of phase deviation at the higher modulating frequencies, while the phase deviation

104 Modulation and demodulation

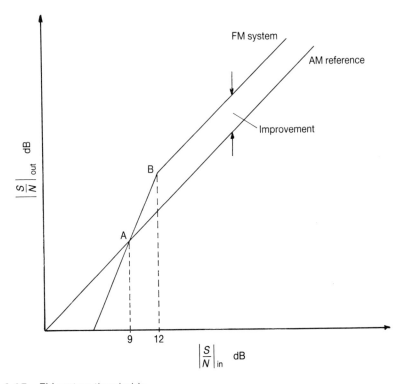

Fig. 4.15 FM system thresholds

of white noise is constant for all frequencies received, the S/N ratio deteriorates at the higher modulating frequencies.

Pre-emphasis is the process whereby at the transmitter, the modulating amplitude is increased proportionally with frequency, above a certain cut-off frequency, by means of a differentiating circuit.

At the receiver, de-emphasis (using an integrator circuit), is used to restore the signal to its original form.

The effect of emphasis on signal amplitude is shown in Fig. 4.16.

From the curve of Fig. 4.16 it can be seen that a 3 dB point occurs at a frequency given by $f_1 = 1/(2\pi T)$ where T is the emphasis time constant. Although pre-emphasis results in greater deviation at higher frequencies, these frequencies have a lower energy content so that, although pre-emphasized, the system is not over-modulated. There is a limit to those frequencies that can be emphasized, hence the need for a cut-off point, given by f_1, below which no emphasis occurs.

The pre-emphasis circuit is placed at the transmitter just before the modulator and, at the receiver, the de-emphasis circuit follows immediately after the discriminator.

Assume the integrator (low-pass filter) de-emphasis circuit in the receiver has an amplitude response:

$$|H_{DE}(f)| = \frac{1}{\sqrt{1 + \left(\frac{f}{f_1}\right)^2}}$$

where f_1 is the 3 dB frequency of the filter.

4.3 Frequency modulation 105

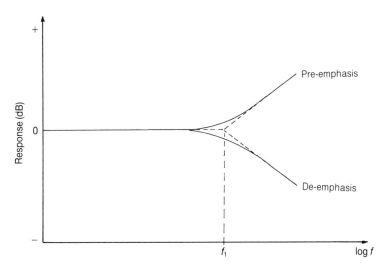

Fig. 4.16 FM pre-emphasis curve

Total noise power output with de-emphasis is:

$$N_b = \int_{-b}^{+b} |H_{DE}(f)|^2 \left(\frac{N_o f^2}{V_c^2}\right) df$$

$$N_b = \frac{N_o}{V_c^2} \int_{-b}^{+b} \frac{f^2}{1 + \left(\frac{f}{f_1}\right)^2} df$$

$$N_b = \frac{N_o f_1^2}{V_c^2} \int_{-b}^{+b} \frac{f^2}{f^2 + f_1^2} df$$

$$N_b = \frac{2N_o f_1^3}{V_c^2}\left[\frac{b}{f_1} - \arctan\frac{b}{f_1}\right]$$

Typically $b >> f_1$ so $\arctan b/f_1 = \pi/2$ which is small compared with b/f_1, hence:

$$N_b = \frac{2N_o f_1^2 b}{V_c^2}$$

and output S/N becomes:

$$\frac{S_b}{N_b} = \left(\frac{f_d}{f_1}\right)^2 \frac{B}{b} \frac{C}{N} \qquad (4.9)$$

Comparing equation (4.9) with equation (4.5) shows that if $b >> f_1$, there can be a dramatic improvement in S/N ratio.

For FDM/FM multi-channel systems, the noise in a telephony channel is proportional to the square of the mid-frequency of the channel. Thus, with a large number of channels the S/N ratio is worse at the top channel compared with all others. In this arrangement emphasis circuits are used with a progressive reduction in the slope of the

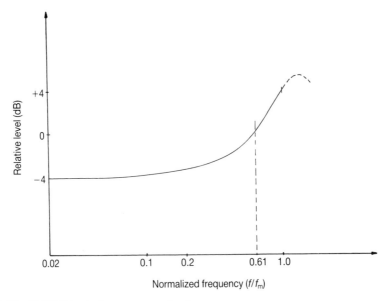

Fig. 4.17 FDM/FM telephony pre-emphasis curve

characteristic over the bottom half of the baseband. The curve of Fig. 4.17 shows a pre-emphasis characteristic for an FDM/FM system as recommended by the CCIR.

Comparison with Fig. 4.16 shows the similarity between the curves. Figure 4.17 has been plotted using relative values with a crossover normalized frequency $f/f_m = 0.61$. This is the frequency where the signal level is unchanged; all signals above that frequency are enhanced by about 6 dB per octave. f_m in this case is the mid-frequency of the top channel in the FDM group. The curve shows that there is an improvement of about 4 dBs in that channel.

Compandor circuit

Companding is a process whereby a voice signal is modified at the sending end in an attempt to improve the signal-to-noise ratio prior to modulation. The inverse of the process must be carried out at the receiver end in order to restore the original speech signal to its correct relative levels. The word *compandor* is a contraction of **com**pressor and ex**pandor**, which refer to the circuits at each end of the link. These circuits perform the task of modifying the speech signal. If the level of gain of the compressor and expandor circuits are controlled by the speech power at a syllabic rate, the compandor is referred to as a syllabic compandor.

In a typical installation the compressor circuit would amplify the high levels of microphone input less than the low levels. This would enable higher modulation levels and improve the signal-to-noise ratio of the transmitted signal. A clipper circuit would limit the amplitude of any transients too fast for the compressor circuit to respond to. The compressed audio signal is then amplified, baseband limited by a filter (to give the required system baseband frequency range and remove out-of-band noise) and then used to frequency-modulate the carrier.

At the receiver, after demodulation and filtering, the received audio signal is processed by the expandor circuit which removes the original compression by amplifying the high levels more than the low levels.

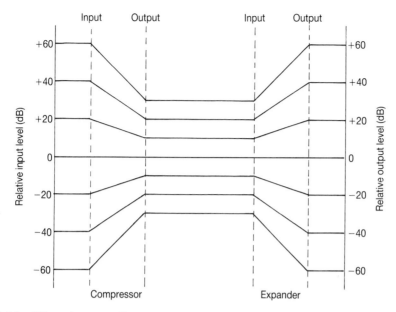

Fig. 4.18 Effect of companding on system power levels

Figure 4.18 shows diagrammatically the effects of companding.

The amount of compression and expansion will depend on the characteristics of the circuit. Figure 4.18 shows, for example, a +20 dB relative input signal level compressed to +10 dB; a −60 dB relative input level is compressed to −30 dB etc. This gives a compression/expansion ratio of 2:1 which is a commonly used value.

The expandor circuit has the added advantage that it will also heavily attenuate input noise making the receive channel sound quieter than would be the case without companding.

The advantage of using a compandor will vary according to its use. Compandors used with an SCPC/FM system give an advantage of 15 – 20 dB.

5
Digital transmission

5.1 Introduction

Messages for transmission are often in digital form, e.g. computer information, telex etc. Voice signals may be represented in digital form by a process of analogue-to-digital conversion. Regardless of whether the message source is digital or analogue, once it is in digital form it may be used to modulate a carrier for transmission. Additionally, if required, an encoder could be used to add redundant digits to the digital signal with the aim of improving the overall quality of the link. The transmission rate of digital signals is known as the bit rate. Demodulation and decoding are required at the receive end, together with digital-to-analogue conversion if required.

The digitization and transmission of video signals would require a very high bit rate. However, video signals exhibit a high degree of redundancy and although a high bit rate is necessary to transmit a video frame, the bit rate needed to transmit the difference between successive frames is much reduced, especially if there is little movement between frames. Standards for digital TV are still not finalized and variations exist between European, US and Japan specifications. However, the Societe Europeanne des Satellites (SES) which operates the Astra satellite system predicts that satellites to be launched in 1994/5 will have a digital TV capacity and that, provided standards are finalized and suitable decoders are ready, digital TV could be realized by 1994. The use of digitized video signals is already a reality for video teleconferencing where the reduced bit rate is possible because of the nature of the video signal, i.e. relatively small movement changes on a frame by frame basis.

The RF bandwidth required by the digital signal, and the power needed to transmit the signal, depend directly on the bit rate. Because of satellite constraints in terms of available power and RF bandwidth it is necessary to limit the bit rate required for the transmission system. A system known as low rate encoding utilizes the known redundancy of speech signals and allows the transmission rate to be reduced.

Advantages of digital transmission include:

- digital systems are less affected by noise, and other interference signals, compared with analogue systems;
- digital systems can transmit both voice and data with equal efficiency;
- digital signals can be encoded for security, and improved quality of the link;
- digital systems can integrate more easily with terrestrial integrated service networks.

A digital SCPC system involves conversion of the analogue voice frequency signal to digital form using one of the many coding techniques. Traditionally for the PSTN, speech is continuously varying waveforms in the audio frequency range 300 – 3400 Hz.

Pulse Code Modulation (PCM) is a technique which samples the speech signal at regular intervals and interprets the sampled analogue level in digital form. The rate of

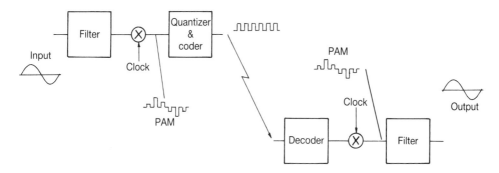

Fig. 5.1 PCM coder/decoder circuit

sampling must be at least twice that of the highest baseband frequency to ensure the speech signal can be reconstructed later with acceptable quality.

PCM systems usually conform to CCITT recommendations with the sampling rate at 8 kHz, with each sample encoded into a 7-bit code giving 128 possible levels of representation. An eighth bit is used to represent the sign of the analogue signal. This results in a bit-rate for the PCM channel of $8 \times 8000 = 64$ kbit/s. Figure 5.1 shows a possible PCM coder/encoder circuit.

Figure 5.1 shows a quantizer encoder circuit, the function of which is to convert the pulse amplitude modulated (PAM) waveform to the uniform amplitude bit stream. The number of steps used (128) to produce the 7-bit code means that there will be a quantized amplitude, represented by the sample binary word, which may not exactly equal the quantity sampled. The difference between the actual sample and its quantized value gives rise to 'quantization noise' since, on reconstruction, the sample will be incorrect by the amount of sampling error. The effect of this could be reduced by increasing the number of sampling levels, and hence the number of bits that represent a sample. Since quantization noise is dependent on the step size between sampling levels, it follows that uniform quantization will give the same level of quantization noise for low amplitude signals as for those with large amplitude. To improve this, non-linear quantization may be employed with smaller steps for low amplitude signals than for large amplitude signals. To achieve non-linear quantization a compandor circuit, similar in principle to that described in Chapter 4, for analogue signals, could be utilized.

Other coding techniques include delta modulation (DM), differential pulse code modulation (DPCM) and adaptive variants such as APCM, ADM and ADPCM. In delta modulation, sampling of the voice signal is at a greater rate than for PCM and use is made of a single bit to track changes in input level from sample to sample. For adaptive DM, the quantizer step size is varied automatically according to the time-varying characteristics of the input signal. For DPCM, the difference between the sample and estimates of it based on earlier samples, is quantized just as for PCM. At the receive end of the link the same predictions must be used as for the transmit end in order to add the same correction.

The coders described so far are all waveform encoders. A vocoder analyses speech in terms of a simplified model of speech production and sends the results of its analysis to the receiver for speech synthesization to reconstruct the original speech

Intelsat use an ADPCM system for their integrated digital communications service (IDR) which produces a 32 kbit/s low-rate encoding (LRE) output from the 64 kbit/s PCM input. Low-rate encoding is desirable since the RF bandwidth occupied by a digital signal and the power required for transmission are directly proportional to the bit-rate. Any reduction in the bit-rate will help increase channel capacity and, since low-rate encoding involves reduction in redundancy in the transmitted signal, its application is well suited to speech or video signals, both of which contain redundant information. The use of Digital Speech Interpolation (DSI) in the IDR system allows the number of channels that can be accommodated to be increased by a factor of up to five compared with the standard PCM 64 kbit/s arrangement. Intelsat refer to the ADPCM/DSI combination as Digital Channel Multiplication Equipment (DCME).

Inmarsat use a system called Adaptive Predictive Coding (APC) for the Inmarsat-B voice channel coding, and Improved Multi-Band Excitation (IMBE) speech coding for the Inmarsat-M voice channel. The IMBE system uses the vocoder principle.

These systems will be described in greater detail in Section 5.3.

5.2 Digital Speech Interpolation (DSI)

This technique utilizes the fact that telephone lines are not occupied 100% of the time. During a conversation there are pauses in the speech and, for a four-wire circuit which provides duplex telephony, one of the transmission paths will be active for only about 40% of the time. In 1959, a system called TASI (Time Assignment Speech Interpolation) was developed by Bell Laboratories. TASI is a high-speed analogue switching system which monitors each user's speech and assigns a channel as necessary. The channel is lost when speech ends and may be used for another speaker. Terminal equipment reassembles the conversation elements into coherent speech without loss of speech quality and without the process being perceptible to the users.

DSI is similar to TASI but operates on the digitized speech signal. The use of DSI increases the channel capacity by a factor of at least two.

Digitally non-interpolated (DNI) channels

A telephony link with DSI may be used for voice signals only since digitized data signals (such as computer data, telex, facsimile etc) may not necessarily be operated satisfactorily with DSI. Any channels which are allocated to digital data signals permanently, without DSI equipment, are known as digitally non-interpolated (DNI) channels.

5.3 Some commercial coding systems

Digital Channel Multiplication Equipment (DCME)

In 1987 Intelsat approved specifications for DCME for continuous and burst mode digital carriers (IDR and TDMA) using a combination of digital speech interpolation (DSI) and adaptive differential pulse code modulation (ADPCM). Some pertinent features of the Intelsat system are as follows:

- a facility exists for 'transparent' 64 kbit/s service, which may be demand assigned or pre-assigned;
- it can accommodate in-band signalling tones, as required for the CCITT signalling system number 5;
- traffic has a multi-destination facility up to a limit of four destinations and can be configured to operate in different modes;

5.3 Some commercial coding systems

- dynamic load control (DLC) prevents excessive DCME load by using a threshold which, when reached, limits the number of calls switched to the system. The threshold level is determined by monitoring the average encoding rate of the voice ADPCM sampler;
- there is adaptive noise insertion, which reproduces at the receive end the background noise sampled at the transmit end, in order to establish subjective voice quality;
- there is an automatic channel check procedure, using test signal transmissions which are checked at the receive end for excessive errors.

Bearer channel The bearer channel format for DCME is shown in Fig. 5.2 for a 2.048 Mbit/s bearer.

Figure 5.2 shows a 125 μs PCM frame with 32 eight-bit time slots. Time slot 0 is used for framing and auxiliary functions while the first 4 bits of time slot 1 are used solely for inter-DCME communications. The last four bits of slot 1 and the remaining slots are used for traffic. The slots are divided into 4-bit nibbles giving 61 nibbles for traffic information.

The satellite channel (SC) is defined in a range of 0 to 127 and is split into the normal range (1 to 61) and the overload range (64 to 83). In the normal range the SC may consist of:

- 3 bits only, with the least significant bit used for the creation of overload channels;
- 5 bits; four from the 'normal' SC nibble and a fifth from a bit-bank (details are shown in Fig. 5.2);
- 8 bits, using two adjacent nibbles.

The SC range 64 to 83 has nibbles formed by the least significant bit (LSB) of some normal range SCs. The overload SC can be 3 or 4 bits wide. The overload range is only used when traffic demands it and hence the overload range SCs can be empty.

Pre-assigned SCs for data must be adjacent to the bit-bank with the spare bits in the

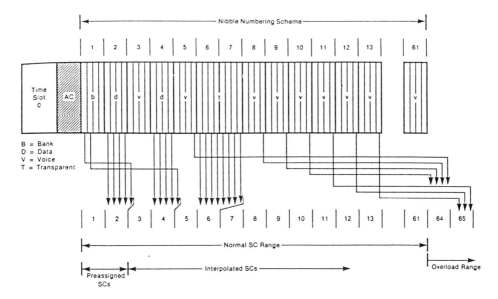

Fig. 5.2 Bearer channel format for DCME (reproduced by permission of *COMSAT Technical Review*, CTR, Vol.22, No.1, Spring 1992, p 29)

112 *Digital transmission*

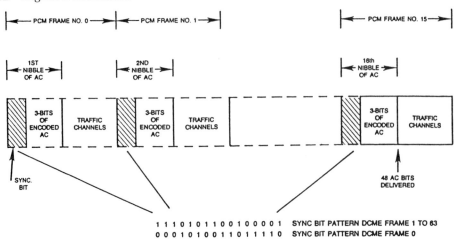

Fig. 5.3 Assignment channel transmission scheme (reproduced by permission of *COMSAT Technical Review*, CTR, Vol.22, No.1, Spring 1992, p 31)

bit bank being available for interpolated (non pre-assigned) data channels. This is shown in Fig. 5.2. Pre-assigned 64 kbit/s channels must occupy adjacent nibbles commencing with an even numbered nibble.

Assignment channel The assignment channel links DCMEs, delivering an assignment message consisting of 24 bits of information every 2 ms. A rate 1/2 Golay code is used and the 48 bits used are sent over 16 PCM frames with 3 bits, plus a synchronization bit, sent every frame. The arrangement is shown in Fig. 5.3.

The 24 information bits of the assignment message is composed of three 8-bit words which form in turn, the SC word, an international channel (IC) word and the data word. The data word in turn is composed of a 4-bit synchronous word and a 4-bit asynchronous word. The information in the asynchronous data word is further multiplexed over a 64 DCME frame multiframe and carries information on alarms, DLC and channel check procedure. The assignment message format is shown in Fig. 5.4.

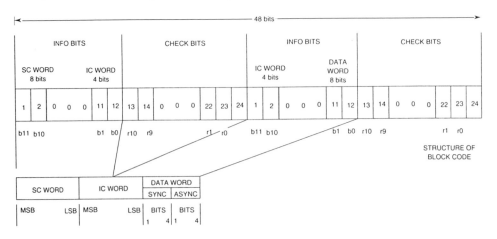

Fig. 5.4 Assignment message format (reproduced by permission of *COMSAT Technical Review*, CTR, Vol.22, No.1, Spring 1992, p 32)

The assignment message is multiplexed over a DCME frame and has the function of notifying the remote DCME of the connections set up in the transmit DCME.

DCME structure On the transmit side, the structure comprises:
- input modules
- transmit channel processing (TCP) function
- coding/mapping unit (CMU)
- assignment channel encoder.

The input modules provide the first level of signal processing on each connected trunk and consist of an activity detector, data/speech discriminator and the signalling detector. The activity detector is an adaptive threshold detector which will be activated typically with a sinusoidal stimulus with a signal-to-noise ratio of 10 dB and duration of 10 ms. The data/speech discriminator determines the type of signal on the trunk using a combination of spectral analysis and 2100 Hz tone detection (this tone is the V.25 answer-back tone used in voice-band data transmission for de-activation of echo control devices). The signalling detector detects the presence of CCITT system number 5 signalling tone (2400 Hz).

The transmit channel processing unit (TCP) contains three main elements; the hangover and signal classification process (HSC), resource allocation and assignment generator (RAG) process and the end processes. The HSC process consists of a state machine in which transitions occur depending on the type of input received. The hangover time varies according to the type of signal. The output from the HSC consists of messages which indicate either a request for capacity or a notification of terminated use. The RAG process receives the message from the HSC processes and generates resource assignment information for both local use and for transmission to the remote DCME. The resource assignment information is logged in the resource map which records the connection and connection type. Messages from the HSC basically constitute a request for assignment action and because the requests are randomly generated a queueing priority is used with the queues being scanned, at the start of a DCME frame, to establish whether there are any stored requests. Prior to servicing a request the bearer bit capacity is checked and, provided capacity is available, an ADPCM encoder is selected from a pool of available encoders in the CMU encoder bank. The end processes take information from the RAG processes and combine it with results of internal routines to provide control information for the CMU. These processes perform the bit association between the banks and the data channels and create the overload channels from the normal SCs. The CMU compresses the N digital input trunks on to the bits of the bearer channel. The output of the encoder consists of 3,4,5 or 8 bit samples and an SC mapping module places the bits of each SC in the appropriate bit locations of the bearer frame in accordance with the map received from the end processes. The assignment channel encoder receives the assignment messages from the RAG processes and encodes the first three information elements in the proper format for transmission.

On the receive side, the structure comprises:
- the assignment channel encoder
- the receive channel processing function (RCP)
- the coder/mapping unit (CMU).

The AC decoder extracts the assignment message information and passes it to the receive channel processing function (RCP). The RCP function takes the assignment information and processes it for the generation of control messages directed to the receive side coder/mapping unit. The RCP also contains receive channel status update and overload decoding process (RUD) and the end processes. The RUD process

114 *Digital transmission*

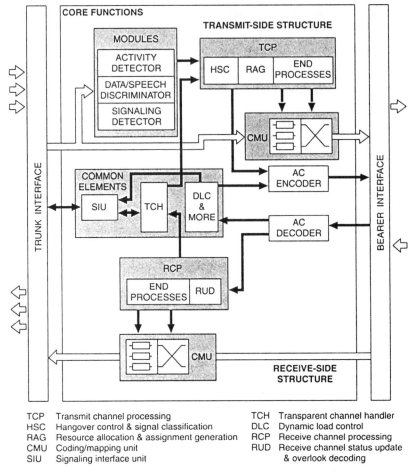

TCP	Transmit channel processing	TCH	Transparent channel handler
HSC	Hangover control & signal classification	DLC	Dynamic load control
RAG	Resource allocation & assignment generation	RCP	Receive channel processing
CMU	Coding/mapping unit	RUD	Receive channel status update
SIU	Signaling interface unit		& overlook decoding

Fig. 5.5 DCME model (reproduced by permission of *COMSAT Technical Review*, CTR, Vol.22, No.1, Spring 1992, p 33)

interprets the received assignment messages and generates information for the end processes. The RUD contains an internal resource map similar to the one described for the RAG process. An important element of the DCME system is that the resource maps in the RUD and RAG processes are identical even while dynamically changing. The RUD has all the information needed for hardware control in the coder/mapping unit (CMU). The end processes drive the CMU hardware and insert a three DCME frame delay. The CMU regenerates the trunk channels from the bits in the bearer channel via ADPCM decoders.

A block diagram of the DCME structure is shown in Fig. 5.5.

Common elements are for functions not specific to DCME and include the transparent channel handler (TCH) and dynamic load control (DLC) facility.

Adaptive Predictive Coding (APC)

The Inmarsat-B system uses 16 kbit/s Adaptive Predictive Coding (APC) as the telephony voice coding technique. There is an optional additional provision for a 9.6 kbit/s APC.

5.3 Some commercial coding systems

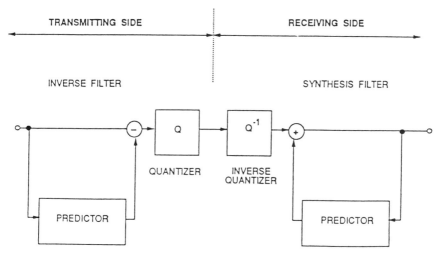

Fig. 5.6 Basic concept of APC coding (courtesy Inmarsat)

The APC technique utilizes an initial estimation of the input speech waveform samples together with a residual (error) signal found by comparison between the estimated and actual speech. The residual signal is quantized and transmitted. At the receiver the quantized residual signal is used to reconstruct the speech signal using a synthesis filter. Figure 5.6 shows the basic concept.

As shown in Fig. 5.6, on the transmitting side the residual signal in the coder is produced using an inverse filter with an all-zero configuration, consisting of a predictor and a subtractor. On the receiving side the decoder uses a synthesis filter with an all-pole configuration, consisting of the predictor and an adder. Quantization of the residual signal is achieved in the coder before transmission and inverse-quantization occurs at the receiving end prior to the synthesis filter. In the coder, the predictor parameters are adaptively controlled on a frame-by-frame basis as the input waveform changes. For speech processing, the predictor function consists of two separate short-term and long-term predictors. Each of the predictors is designed to operate efficiently using known speech characteristics. The main parameters of APC encoding are shown in Table 5.1.

Table 5.1 Main parameters of APC encoding (courtesy Inmarsat)

Coding rate	16 kbit/s	9.6 kbit/s
Input bandwidth	0.3 – 3 kHz	0.3 – 3 kHz
Sampling rate	6.4 kHz	6.4 kHz
Frame length	20 ms	20 ms
Residual signal bits	12.8 kbit/s	6.4 kbit/s
Total side information bits	3.2 kbit/s	3.2 kbit/s

116   *Digital transmission*

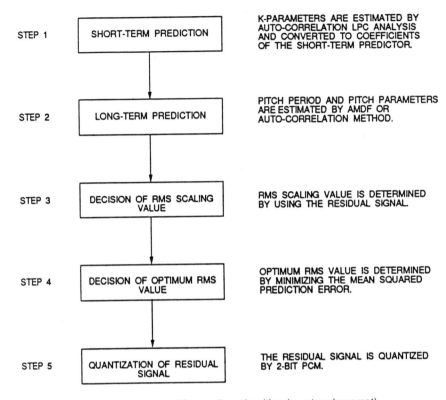

Fig. 5.7 Basic processing flow of APC encoding algorithm (courtesy Inmarsat)

The basic processing flow of the APC encoding algorithm is shown in Fig. 5.7.

The sampled input data is fed to the LPC analyser in which a set of K-parameters (reflection coefficients) is determined by autocorrelation. The resultant K-parameters are quantized, converted to binary code and then converted to coefficients of the short-term predictor.

Next, after the residual signal from the short-term inverse filter has been fed to the long-term inverse filter, the pitch period and pitch coefficients (long-term predictor coefficients) are estimated by an average difference function, or autocorrelation, method.

An rms scaling value is then selected using the residual signal from the long-term predictor.

Next, an optimum rms value is determined by minimizing the mean-square error between the input signal and a locally reconstructed signal.

Lastly, the resulting residual signal from the long-term inverse filter is fed to the quantizer, after subtracting a noise shaping filter output.

APC encoding algorithm.
- *Short-term predictor*   This is defined in terms of a digital filtering system with an output S'_n given by:

$$S'_n = \sum_{i=1}^{N} a_i S_{n-i}$$

where S_n is the input sample at the nth sampling index
a_i are predictor coefficients
N is the order of the predictor (six in this case).
The error power E is given by:

$$E = \sum_{n=1}^{L}\left[S_n - \sum_{i=1}^{N} a_i S_{n-i}\right]^2$$

where L is the frame length and coefficients a_i are chosen to minimize E.

The predictor coefficients are obtained, through the reflection coefficients, as auxiliary parameters. The reflection coefficients (K-parameters) are calculated using autocorrelation and derived by means of Linear Predictive Coding (LPC) analysis. The K-parameters are code/decoded for conversion into the coefficients of the short-term predictor, the all-pole synthesis filter and noise-shaping filter; they are also similarly used in the decoder. The K-parameter magnitude is always less than unity to ensure system stability with the synthesis filters. Leaked predictor coefficients (r) are used to allow for transmission errors and tonal oscillations in the reconstructed speech signal. The transfer function of the short-term predictor is:

$$P_1(Z/r_1) = \sum_{i=1}^{N} a_i r_1^i Z^{-i}$$

where r_1 is a constant and $0 < r_1 < 1$
- **Long-term predictor** This estimates the signal U_n' as:

$$U_n' = P_o U_{n-p}$$

where p is the pitch period
U_{n-p} is previous sample of the long-term predictor input at the $(n-p)$ sampling time.

P_o is a pitch gain parameter indicating a change from frame to frame.

The long-term predictor is inserted in the coder following the short-term predictor. U_n thus corresponds to the residual signal from the short-term predictor.

The coefficient P_o of the long-term predictor having one tap (first order) is defined by:

$$P_o = \frac{\sum_{i=1}^{M} U_i U_{i-p}}{\sum_{i=1}^{M} U_{i-p}^2}$$

where M is the number of samples in a frame.

The pitch period and the coefficient are used for calculating the residual signal from the long-term predictor. The resulting residual signal is fed to a forward adaptive quantizer after subtracting a noise-shaping filter output.

The error power E_T between the input speech signal S_i and the locally reconstructed speech signal R_i is given by:

$$E_T = \sum_{i=1}^{m} (S_i - R_i)^2 \qquad (5.1)$$

where i is the sampling time index in the sub-frame
m is the number of samples in the sub-frame.

An optimum error power can be derived by utilizing several estimated rms powers and selecting one of these to give the minimum value of error power.

A noise shaping filter is used to weight subjectively equation 5.1. So as to shape

118 Digital transmission

the spectrum of the quantization noise, and improve the overall subjective quality, the quantization error is detected and fed to the noise shaping filter. The Z-transform notation for the noise shaping filter is:

$$F(Z) = r_{N2}P_2(Z) + [1 - r_{N2}P_2(Z)]\, P_1(Z/r_1 r_{N1})$$

where r_{N1} and r_{N2} are noise shaping factors for the short-term and long-term predictors respectively, and r_1 is a leakage factor. These factors have values in the range 0 to 1.

APC decoding algorithm At the receiver, the information received is decoded into the K-parameters, pitch period p, the coefficients of the long-term predictor and the optimum rms power etc. The short-term predictor coefficients are regenerated from the decoded K-parameters.

A post noise-shaping filter is included in the decoder to improve speech quality. The post noise-shaping filter $N(Z)$ in Z-transform notation is given by:

$$N(Z) = G_o \left[\frac{1}{1 - P_1(Z/r_1 r_{N3})} \times \frac{1}{1 - r_{N5} P_2(Z)} \right]$$

where G_o is an adaptive level control factor and r_{N3}, r_{N5} are post noise-shaping factors for the short-term and long-term predictors respectively. These factors are in a range from 0 to 1.

G_o is given as the ratio of the input level to the output level across the post noise-shaping filter. G_o is adjusted at every sub-frame. The coefficients of the short-term and long-term predictors are used in the synthesis filters and the post noise shaping filter in the all-pole decoder. Lastly, the post noise shaping filter reconstructs the speech signal from the output sequence of the cascaded synthesis filters.

Frame structure of output data stream The frame format of the output data stream consists of:

- frame alignment and spare bits
- reflection coefficient bits for short-term prediction (K-parameters)
- pitch parameter bits for long-term prediction
- bits of rms value for quantizer step sizes
- scaling bits of rms power
- residual signal bits.

The bit assignments are summarized in Table 5.2 and the frame structure is shown in Fig. 5.8.

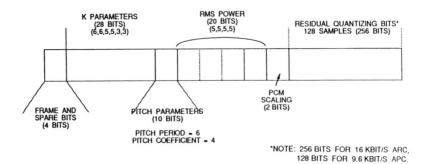

Fig. 5.8 Frame structure of APC output data stream (courtesy Inmarsat)

5.3 Some commercial coding systems

Table 5.2 Parameter bit assignment in frame (courtesy Inmarsat)

Frame alignment and spare bits	4 bits
Reflection coefficients	28 bits
(K-parameters)	1st : 6 bits
	2nd : 6 bits
	3rd : 5 bits
	4th : 5 bits
	5th : 3 bits
	6th : 3 bits
Pitch parameters:	
Pitch coefficient	4 bits
Pitch period	6 bits
RMS power	20 bits
Quantization	5 bit/subframe
PCM scaling	2 bits
Residual quantization bits (16 kbit/s APC)	2 bit/sample, 256 bit/frame
Residual quantization bits (9.6 kbit/s APC)	1 bit/sample, 128 bit/frame

The input baseband signal necessary for the APC algorithm, and the output baseband signal produced by it, are linear PCM signals sampled at 6.4 kHz. If the coder input analogue-to-digital conversion and/or the decoder output digital-to-analogue conversion is performed using standard 64 kbit/s non-linear PCM techniques, with sampling at 8 kHz, it is necessary to convert the 8 kHz samples to 6.4 kHz samples required by the APC algorithm and vice versa. Also, non-linear/linear conversion is required.

Improved Multi-Band Excitation (IMBE) speech coding/decoding

IMBE is used in the Inmarsat-M system operating at 6.4 kbit/s.

The IMBE speech coder, or vocoder, is a model-based system which reconstructs a synthetic speech signal containing the same perceptual information as the original speech signal. The main advantage claimed for this system, compared with other

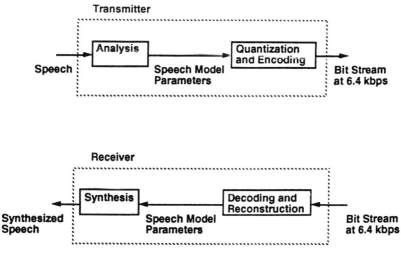

Fig. 5.9 IMBE speech encoder (courtesy Inmarsat)

120 Digital transmission

vocoders, is the use of a sophisticated algorithm for the estimation of speech model parameters and the ability to synthesize the speech signal from these parameters. The basic arrangement is shown in Fig. 5.9.

Multi-band excitation speech model A discrete signal is obtained by sampling an analogue speech signal and passing it through a discrete filter. A window $w(n)$ is applied to the speech signal $s(n)$ to give a windowed speech signal $s_\omega(n)$, where:

$$s_\omega(n) = s(n)\,w(n)$$

The sequence $s_\omega(n)$ is a speech segment or speech frame. The speech signal $s(n)$ is shifted in time to select any required segment. The frame time is 20 ms. A speech segment $s_\omega(n)$ is modelled as the response of a linear filter, $h_\omega(n)$, to an excitation signal $e_\omega(n)$. The main difference between traditional speech models and the MBE speech model is the excitation signal. Rather than using a single voiced/unvoiced (V/UV) decision for each speech segment, as in traditional speech models, the MBE speech model utilizes several non-overlapping frequency bands and makes a V/UV decision for each frequency band. This allows $e_\omega(n)$ for a particular speech signal to be a mixture of periodic (voiced) and noiselike (unvoiced) energy.

For MBE the excitation spectrum comes from the pitch period P_o (fundamental frequency) and the V/UV information by combining a periodic spectrum in the frequency bands declared voiced, with a random noise spectrum in the frequency bands declared unvoiced. The periodic spectrum is generated from a windowed periodic impulse train which is completely determined by $w(n)$ and P_o while random noise is generated from a windowed random noise sequence.

A comparison of traditional and MBE speech models is shown in Fig. 5.10.

The example shown in Fig. 5.10 illustrates the advantage of the MBE speech model in that the V/UV determination is performed in frequency bands and that where the ratio of periodic energy is high the bands are declared voiced; frequency bands where the ratio is low are declared unvoiced.

Speech analysis/synthesis Unlike traditional methods, the IMBE system estimates the excitation and spectral envelope parameters simultaneously, giving a synthesized

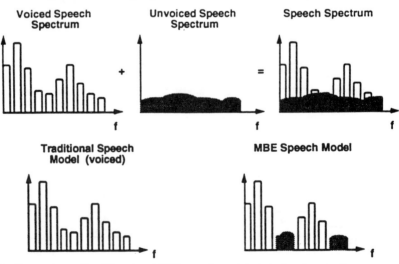

Fig. 5.10 Comparison of traditional and MBE speech models (courtesy Inmarsat)

5.3 Some commercial coding systems

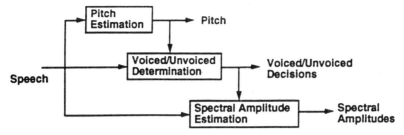

Fig. 5.11 IMBE speech analysis algorithm (courtesy Inmarsat)

spectrum close to the original speech spectrum. Model parameters are estimated by minimizing the error criterion given by:

$$E = \int_{-\pi}^{\pi} |S_\omega(\omega) - \hat{S}_\omega(\omega)|^2 d\omega$$

where $s_\omega(\omega)$ is the Fourier transform of the speech segment $s_\omega(n)$ and $\hat{S}_\omega(\omega)$ is the corresponding synthetic spectrum.

The overall sequential analysis method is shown in Fig. 5.11. The MBE speech model parameters which have to be estimated are the pitch period, the V/UV decisions and the spectral amplitudes which characterize the spectral envelope.

For pitch estimation it is necessary to find P_o corresponding to the current speech frame $s_\omega(n)$. A pitch tracking algorithm considers the pitch from previous and future frames in order to determine the pitch of the current frame. The pitch estimation algorithm involves a two-step procedure whereby an initial estimate is obtained and refined to obtain the final estimate of the fundamental frequency $\hat{\omega}_o$ which has one-quarter sample accuracy. Pitch tracking is used to improve the pitch estimate by trying to limit pitch deviation between consecutive frames. For each speech frame two different pitch estimates are defined, one is a backward estimate which gives pitch continuity with previous pitch frames while the second is a forward estimate to maintain pitch continuity with future frames. Both look-back and look-forward pitch estimates are calculated with suitable algorithms. Pitch refinement is achieved by a pitch refinement algorithm which improves the resolution of the pitch estimate from one half-sample to one quarter-sample.

The voiced/unvoiced (V/UV) decisions $\hat{v}_k$ are found by dividing the spectrum into frequency bands and evaluating the voicing measure, which can be compared with a threshold function. If the voicing measure is less than the threshold function then that frequency band is declared voiced. The width of each frequency band is $3\hat{\omega}_o$ so that each band contains three harmonics of the refined fundamental frequency. This is true of all bands except the highest which can have one, two or three harmonics. If a band is declared voiced, then all harmonics within that band are defined as voice harmonics. Also, if a band is declared unvoiced then all the harmonics in that band are defined to be unvoiced harmonics. Since the number of harmonics in a fixed (3.84 kHz) bandwidth is a function of the fundamental frequency, the number of frequency bands will vary from frame to frame. The IMBE speech coder uses a maximum of 12 V/UV decisions.

The spectral envelope in a frequency band is specified by three spectral amplitudes $\hat{M}_1$ which have values estimated according to whether the band is voiced or unvoiced.

Speech synthesis can be achieved, at the receive end after decoding, by a set of model parameters based on those parameters used in the coder. The reconstructed model

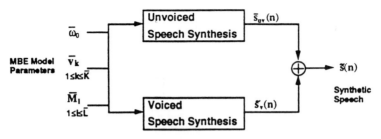

Fig. 5.12 IMBE speech synthesis (courtesy Inmarsat)

parameters are $\tilde{\omega}_o$, $\tilde{v}_k$ and $\tilde{M}_l$ which correspond to the reconstructed fundamental frequency, V/UV decisions and spectral amplitudes respectively. Additionally, a parameter $\tilde{L}$ is generated from $\tilde{\omega}_o$ and defines the number of harmonics in a current frame. Because of quantization noise and channel errors, the reconstructed model parameters are not the same as the estimated values $\hat{\omega}_o$, $\hat{v}_k$ and $\hat{M}$. For each new set of model parameters the synthesis algorithm generates a 20 ms frame of speech $\tilde{s}(n)$, which is interpolated between the previous set of model parameters and the current set. For each new set of model parameters, $\tilde{s}(n)$ is generated in the range $0 \leq n < N$, where N equals 160 samples (20 ms). The synthetic speech signal is divided into a voiced component, $\tilde{s}_v(n)$ and an unvoiced component, $\tilde{s}_{uv}(n)$. These components are synthesized separately and combined to form $\tilde{s}(n)$. This is shown in Fig. 5.12.

6
Digital modulation

6.1 Introduction

A carrier may have the form $V(t) = V_c\cos(\omega_c t + \phi)$, where V_c is the carrier amplitude, ω_c its angular frequency, t is time and ϕ the phase.

For digital modulation the types available are amplitude shift keying (ASK), frequency shift keying (FSK) and phase shift keying (PSK).

Amplitude shift keying can be accomplished simply by the on–off gating of a continuous carrier. The simplest technique is to represent one binary level (binary 1) by a single signal of fixed amplitude and the other level (binary 0) by switching off the signal. See Fig. 6.1.

The absence of a signal for one of the binary levels has the disadvantage that if a fault condition exists it could be misinterpreted as data received. Figure 6.2 shows an alternative method which prevents this disadvantage.

As with speech telephony circuits, the upper sideband and carrier may be suppressed to reduce the bandwidth requirement and concentrate the available power on the signal containing the information.

Frequency shift keying may be used whereby the carrier frequency has one value for a '1' bit and another value for a '0' bit, as shown in Fig. 6.3.

The main difficulty in the use of this FM technique is that the gap between the frequencies used must be increased as the modulation rate increases. Thus, for a restricted channel bandwidth, especially using in-band supervisory signalling, there is a limit to the maximum bit rate that is possible with this technique.

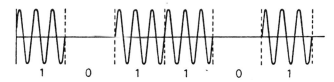

Fig. 6.1 Waveform for ASK using on/off keying

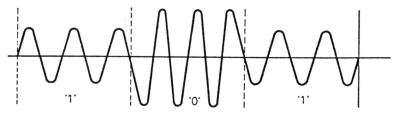

Fig. 6.2 Waveform for ASK using different amplitude signals for the logic levels

124 *Digital modulation*

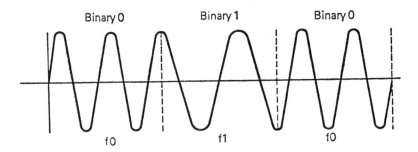

Fig. 6.3 Waveform for FSK

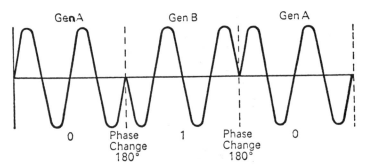

Fig. 6.4 Waveform for PSK

Phase shift keying is a technique in which, using multi-state signalling, the rate of data transmission can be increased without having to increase the bandwidth. Consider Fig. 6.4.

A simple way of appreciating PSK is to imagine the outputs from two generators each operating at the same frequency but out of phase with each other by 180°. The outputs of each generator could be imagined to be switched to represent each binary level.

An alternative modulation technique (not illustrated) is a hybrid combination of ASK and PSK, often referred to as amplitude phase keying (APK). The waveform of Fig. 6.2 illustrates the technique except that at the symbol transitions there is also a phase shift similar to that shown in Fig. 6.4.

If the bit-stream at the modulator input has only two possible values (i.e. 0 and 1) then the system is binary. If the arrangement of bits gives more than two possible values (00,01 etc), then the system is referred to as M-ary where M is a value > 2. Examples of M-ary modulation schemes include quadrature phase shift keying (QPSK), offset QPSK (O-QPSK) and minimum shift keying (MSK).

A digital system is coherent if a local reference is available at the demodulator which is in phase with the transmitted carrier; otherwise the system is non-coherent. Similarly if, at the receiver, a periodic signal is available in synchronism with the transmitted digital sequence, the system is synchronous. If such a periodic signal is not required at the receiver the system is asynchronous.

The choice of a particular system will depend on many factors such as available power, bandwidth requirements, complexity of the equipment required and the effects of the transmission channel on the required signal.

Unlike analogue systems where system performance can be evaluated in terms of the signal-to-noise ratio (S/N), the performance for digital systems is measured by the

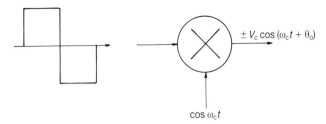

Fig. 6.5 Effect of digital modulation on a carrier using PSK

bit-error rate (BER). The bit-error rate is defined as the number of bits transmitted that are received with errors, compared with the total number of bits transmitted. Typically, the value of BER is specified by:

- average BER over a predetermined period of time;
- proportion of fixed-length time intervals which are either error free or undergo an error rate no worse than a specified value.

One of the most efficient modulation methods is PSK with coherent detection. This technique enables transmission of a constant envelope signal containing the required information as phase transitions—ideal for coherent detection. Only PSK will be considered in this chapter since it is the only system used in the Intelsat, Inmarsat and Eutelsat systems.

6.2 Phase shift keying (PSK)

The simplest form of PSK is binary PSK or BPSK where the digital information modulates a sinusoidal carrier. This is shown in Fig. 6.5.

For phase modulation, the phase of the carrier waveform varies according to changes in modulating amplitude. Generally:

$$\theta(t) = \omega_c t + \phi(t)$$

In BPSK the frequency of the carrier remains constant while the phase shift is one of two fixed values. Consider Fig. 6.6 which shows the effect on carrier phase for a binary 1 and binary 0.

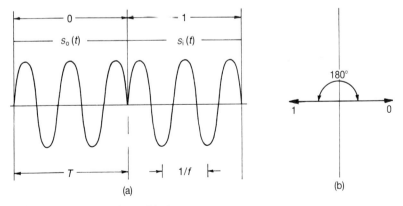

Fig. 6.6 BPSK, (a) carrier waveform; (b) phase states

126 Digital modulation

The two signals used to represent 0 and 1 could be expressed as:

$$s_0(t) = V_c\cos(\omega_c t + \theta_0)$$
$$s_1(t) = V_c\cos(\omega_c t + \theta_1)$$

where $0 < t \leq T$

θ_0 and θ_1 are constant phase shifts.

Using a phase difference of 180°, i.e. $\theta_1 = \theta_0 + 180°$, the equations can be simplified since a phase shift of 180 is simply a reversal of polarity i.e.:

$$s_0(t) = V_c\cos(\omega_c t + \theta_0)$$
$$s_1(t) = -s_0(t) = -V_c\cos(\omega_c t + \omega_0)$$

Figure 6.6 shows the effect on the carrier phase using the time domain spectrum. In the frequency domain the power spectral density of the modulated carrier varies according to the relationship:

$$H(f) = \left|\frac{\sin f}{f}\right|^2$$

This is shown in Fig. 6.7 which illustrates that most of the energy is contained in the major lobe and, for a band-limited system, the bandwidth would be restricted to $2/T$. Since T represents the pulse duration, the bandwidth approximates to the bit-rate.

Restricting the bandwidth will result in the loss of some energy contained in the side lobes. However, the energy in those lobes, removed in frequency from the main lobe, is reduced with increasing frequency.

Incoherent detection for PSK is not feasible since it would be impossible to detect a 1 or 0. If a carrier of the same basic frequency as the transmitted signal is available at the receiver, synchronous demodulation is possible. For the arrangement shown in Fig. 6.8 the output will vary between two levels given by $\pm (V_c T/2)\cos(\theta - \phi)$ according to the

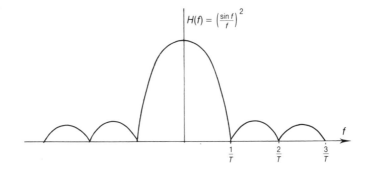

Fig. 6.7 BPSK in the frequency domain

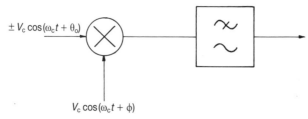

Fig. 6.8 Synchronous demodulation of PSK signals

6.2 Phase shift keying (PSK)

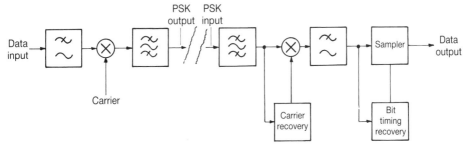

Fig. 6.9 PSK modem block diagram

phase of the input signal. However, with this arrangement it could be difficult to ensure which of the voltage output levels corresponded to the transmitted 1 or 0.

The principles outlined for the modulation/demodulation for a BPSK system can be expanded into the circuit of Fig. 6.9. On the modulator side a filter precedes the modulator in order to shape the modulation waveform and limit the bandwidth of the modulated signal. Post-modulation filtering is used to band-limit the transmitted signal. At the receiver the PSK signal goes through a pre-detector filter which limits the amount of channel noise allowed through to the modulator. A locally regenerated carrier component is extracted from the received carrier via a phase-locked loop (PLL) and assists in the product demodulation process. The demodulated signal passes through a low-pass filter and, via a second PLL, a bit-stream is produced which allows sampling of the signal at the pulse mid-point to reconstruct the original data stream.

As mentioned earlier, there may be a difficulty with this form of demodulation in successfully identifying the correct phase of the regenerated signal for demodulation. The ambiguity could be resolved by the use of a differential PSK system where, instead of the instantaneous phase determining which bit is transmitted, it is the change in phase which carries intelligence. This requires that the original baseband signal (unipolar or bipolar NRZ) waveform be recoded so as to register changes in phase with one logic level and no change in phase with the other logic level, i.e.:

- if a digit changes (from 1 to 0 or vice versa), a 1 is transmitted;
- if no change occurs, a 0 is transmitted.

A basic demodulator arrangement for DPSK is shown in Fig. 6.10. The received waveform is delayed by one sampling period so that the multiplication is the product of a current sample and what is in effect a local oscillator input with the phase of the previous sample input. If both inputs to the multiplier have the same phase the demodulator output is positive; if the phases are different by 180° the demodulator output will be of negative sense.

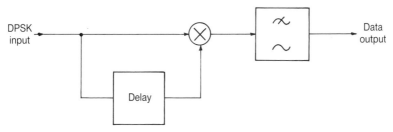

Fig. 6.10 DPSK demodulator

6.3 Quadrature phase shift keying

A QPSK modulated signal can be formed by operating two BPSK modulators in quadrature. If the bit stream is split so that even numbered bits are routed to the in-phase stream and odd numbered bits to the quadrature-phase stream, then:

$d_i(t) = d_0, d_2, d_4$ etc
$d_q(t) = d_1, d_3, d_5$ etc

A possible bit stream input is shown in Fig. 6.11 together with its realization into in-phase and quadrature streams.

It can be seen from Fig. 6.11 that the two derived bit streams have half the bit rate of the input stream.

One method of expressing the QPSK waveform mathematically is to assume the waveform $s(t)$ is realized by amplitude modulating the in-phase and quadrature data streams onto a cosine and sine function of the carrier f_c:

$$s(t) = \frac{1}{\sqrt{2}} d_i(t) \cos(\omega_c t + \frac{\pi}{4}) + \frac{1}{\sqrt{2}} d_q(t) \sin(\omega_c t + \frac{\pi}{4})$$

This could be rewritten as:

$$s(t) = \cos(w_c t + \theta(t)) \tag{6.1}$$

A QPSK modulator is shown in Fig. 6.12.

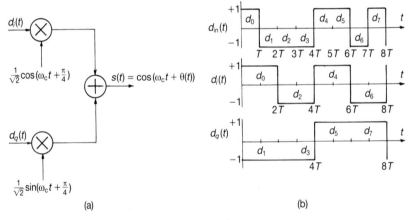

Fig. 6.11 Example of QPSK modulation

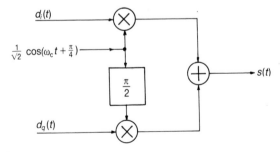

Fig. 6.12 QPSK modulator

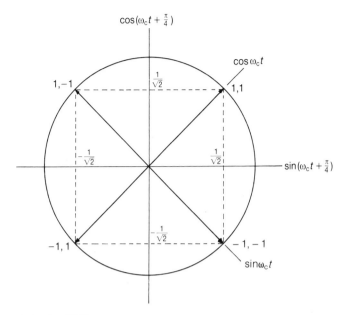

Fig. 6.13 Phase states for QPSK

Pulse stream $d_i(t)$ is shown in Fig. 6.11 as a binary signal with $+1$ representing binary 1, and -1 representing binary 0. This stream will amplitude modulate the cosine function and will have the effect of shifting the phase of the function by 0 or π. This is BPSK. The pulse stream $d_q(t)$ will have the same effect on the sine function producing a BPSK waveform orthogonal to the cosine function. Summing the two orthogonal components results in the QPSK waveform.

Figure 6.13 shows the four phase states assumed for the QPSK signal. The values of $\theta(t)$ from equation (6.1) will be $0°$, $90°$, $180°$ and $270°$. Each phase state depends on a pair of bits. As drawn, the i-channel bits operate on the vertical axis at phase states $90°$ and $270°$, while the q-channel bits operate on the horizontal axis at phase states $0°$ and $180°$. The vector sum of an i-channel and q-channel phase will produce each of the four states shown.

The phase state of the output of the modulator depends on each input channel so that the output state for each signal interval depends on a pair of bits. In this case, the transmission rate depends on a pair of bits, and is measured in terms of symbols per second. For QPSK the power spectrum is the same as for BPSK but since the transmission rate in symbols per second is half the bit rate of BPSK, the bandwidth is halved. Figure 6.7 also represents QPSK except that the time intervals shown are halved.

Because the i and q channels are orthogonal, coherent detection is possible in the receiver with each of the two BPSK signals being detected separately.

Offset QPSK (O-QPSK)

The basic concept of QPSK also applies to offset QPSK, the only difference being the timing of the two baseband signals. In QPSK, as shown in Fig. 6.11, each input pulse has a duration of T seconds and the odd/even streams have a duration of $2T$ seconds. All transitions are aligned in time as Fig. 6.11 shows.

For O-QPSK, the alignment of the odd/even streams is shifted so as to be offset by T

130 *Digital modulation*

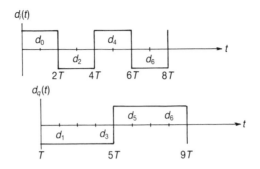

Fig. 6.14 Example of O-QPSK data streams

seconds. The stream for $d_i(t)$ and $d_q(t)$ for Fig. 6.11 has been redrawn in Fig. 6.14 to show the effect.

For QPSK, carrier phase change can occur only once every $2T$ seconds. Depending on the values of $d_i(t)$ and $d_q(t)$ the carrier phase at any instant will be as indicated by Fig. 6.13. At the next interval, if neither stream changes sign the carrier phase is unaltered; if one stream changes sign the carrier phase is shifted by $\pm 90°$; if both streams change sign, the carrier phase shift is 180°. If the QPSK signal is filtered to remove the spectral side lobes, the waveform that results will no longer have a constant envelope. This effect on the QPSK envelope is shown in Fig. 6.15.

If a QPSK signal is passed through a non-linear amplifier, the amplitude variations could cause spreading of the spectrum to restore the unwanted side-lobes, which could cause interference problems. This effect can be resolved either by backing off the power amplifier in the earth station or satellite, or by using O-QPSK.

In O-QPSK, staggering of the odd and even bit streams means that the possibility of a carrier changing state by 180° is eliminated since only one stream can change at any transition instant. With a band-limited O-QPSK, intersymbol interference can cause a small droop in the envelope every $\pm 90°$ transition, but it will not fall to zero as was possible with QPSK. Thus, a band-limited O-QPSK signal passing through a non-linear

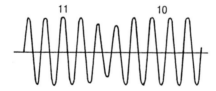

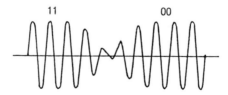

Fig. 6.15 Effect of QPSK phase transitions on QPSK envelope

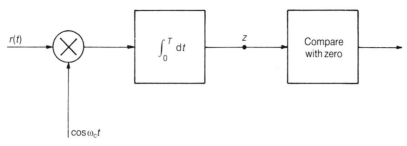

Fig. 6.16 PSK receiver

amplifier will not suffer the out-of-band interference that could be experienced by the QPSK signal.

6.4 Probability of bit error rate

For a PSK transmission in the presence of white Gaussian noise of spectral density $N_o/2$, the input to the demodulator could be represented by $r(t)$ where:

$$r(t) = s(t) + n(t)$$

where $s(t)$ refers to the signal and $n(t)$ refers to the noise.

The receiver circuit used for analysis purposes is shown in Fig. 6.16.
Assume a binary 1 is transmitted corresponding to a value of $s_1(t)$ given by:

$$s_1(t) = V_c \cos(\omega_c t + \pi) = -V_c \cos\omega_c t$$

A comparison is made at the receiver output, at point z, where the output given by:

$$z = \int_0^T r(t) \cos\omega_c t \, dt$$

is compared with zero. If the value at the output is greater than zero then it is decoded as a transmitted 0. If the output value is less than zero it represents a transmitted 1.

For a transmitted 0, the value of $r(t)$ is:

$$r(t) = V_c \cos\omega_c t$$

and the output value of signal is given by:

$$z = \int_0^T r(t) \cos\omega_c t \, dt$$

$$= \int_0^T [V_c \cos\omega_c t + n(t)] \cos\omega_c t \, dt$$

$$= V_c \int_0^T \cos^2\omega_c t \, dt + \int_0^T n(t) \cos\omega_c t \, dt$$

This is a Gaussian variable with mean value $V_c T/2$ and variance given by:

$$\sigma^2 = E\left[\int_0^T \int_0^T n(t) n(\tau) \cos\omega_c t \, \cos\omega_c \tau \, dt \, d\tau\right]$$

132 Digital modulation

$$= \frac{N_o}{2} \int_0^T \cos^2\omega_c t \, dt = \frac{N_o T}{4}$$

The probability density function (pdf) of the Gaussian random noise can be expressed as:

$$p(n) = \frac{1}{\sigma \sqrt{2\pi}} \exp\left[-\frac{1}{2}\left(\frac{n(t)}{\sigma}\right)^2\right] \tag{6.2}$$

where σ is the noise variance.

Since $z(t) = V_c T/2 + n(t)$, then equation (6.2) can be used to determine the probability density functions (pdfs), $p(z/s_1)$ and $p(z/s_0)$:

$$p(z/s_1) = \frac{1}{\sigma\sqrt{2\pi}} \exp\left[-\frac{(z + (V_c T/2))^2}{(N_o T/2)}\right]$$

$$p(z/s_0) = \frac{1}{\sigma\sqrt{2\pi}} \exp\left[-\frac{(z - (V_c T/2))^2}{(N_o T/2)}\right]$$

Figure 6.17 shows the detector probability outputs. The left-hand part of the curve illustrates the probability density of the output for the transmission of $s_0(t)$ and the right-hand part gives the probability density of the output for the transmission of $s_1(t)$.

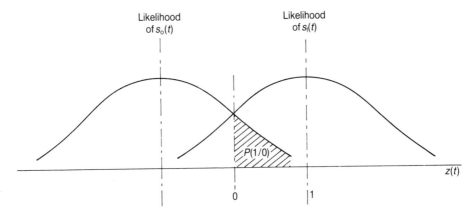

Fig. 6.17 PSK probability density functions

Referring to Fig. 6.17, an error will occur when, say, $s_0(t)$ is sent and channel noise results in the receiver output signal $z(t)$ being greater than the threshold value. This is shown by the shaded area under the curves of Fig. 6.17. The probability of such an error is:

$$P_e = P(e/s_0) = \int_0^\infty p(z/s_0) \, dz$$

$$P_e = \int_0^\infty \frac{1}{\sigma\sqrt{2\pi}} \exp\left[-\frac{(z - (V_c T/2))^2}{(N_o T/2)}\right] dz$$

let $x = (z - (V_c T/2))/\sigma$, or:

$$x = \frac{(z - (V_c T/2)}{\sqrt{N_o T/4}}$$

6.4 Probability of bit error rate

then $dx = dz/\sigma$
and the limit for $z = 0$ is:

$$x = -\frac{V_c T/2}{\sqrt{N_o T/4}} = -\sqrt{\frac{V_c^2 T}{N_o}}$$

This can be written in error function (erfc) form:

$$\text{thus} \quad P_e = \int_{(-\sqrt{(V_c^2 T)/N_o})}^{\infty} \frac{1}{\sqrt{2\pi}} e^{\frac{-x^2}{2}} dx$$

$$P_e = \frac{1}{2} \text{erfc}\left(\sqrt{\frac{V_c^2 T}{N_o}}\right)$$

$$P_e = \frac{1}{2} \text{erfc} \sqrt{\frac{E_b}{N_o}}$$

where the average energy per bit is $E_b = V_c^2 T$.
Figure 6.18 shows the curve for E_b/N_o for bipolar signalling.

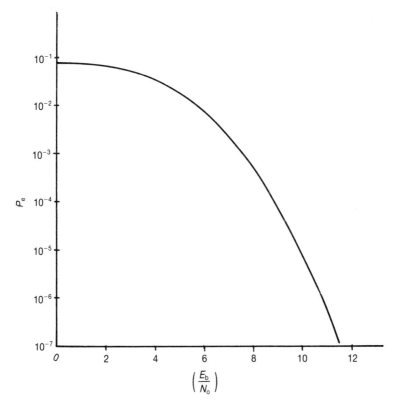

Fig. 6.18 Bit error probability curve for PSK

134   *Digital modulation*

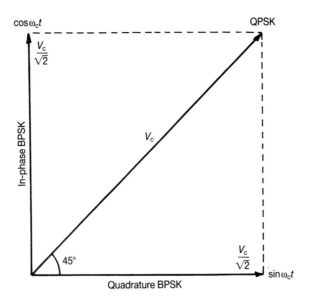

Fig. 6.19   Representation of QPSK using quadrature BPSK components

The parameter E_b/N_o can be expressed in terms of the average signal power (C) to average noise power (N), i.e.:

$$E_b/N_o = (CT)/N_o = C/(RN_o)$$

Using logs,

$$E_b/N_o = (C/N_o) - R$$

where C = average modulating signal power in watts
       T = time duration for one bit, in seconds
       $R = (1/T)$ = bit rate in bit/s.

Calculations have been made in Chapter 3 regarding required values of C/N_o for a particular system. Since $E_b/N_o = (C/N_o) - R$, it follows that the difference between E_b/N_o and C/N_o is the effective bit rate of the system. The truth of this statement can be confirmed by reference to Table 3.1.

Although not described above, for ease of explanation, optimum signal detection requires a matched filter (correlator) demodulator and in this case the required signal bandwidth is equal to the noise bandwidth. In a practical system the value of E_b/N_o required to achieve a specified BER will be higher than the theoretical value because of equipment limitations.

Figure 6.13 illustrated in vector form the production of a QPSK system using quadrature bit streams. Each of these streams modulates an orthogonal component of the carrier at half the bit rate of the incoming stream. The in-phase stream modulates the $\cos \omega_c t$ term while the quadrature stream modulates the $\sin \omega_c t$ term. Figure 6.19 shows that if the QPSK vector has a value V_c, then each of the quadrature BPSK vectors have a value $0.707 V_c$.

Thus, each of the quadrature BPSK signals will contain half the average power of the QPSK signal. It can be deduced from this that if the QPSK signal has a bit rate of R bit/s and average power S watts, then for each of the quadrature BPSK signals the bit rate will be $R/2$ bit/s with average power $S/2$ watts.

Thus, for BPSK and QPSK the ratio of E_b/N_o is the same. It should be noted, however, that there is a difference between bit rate and symbol rate and hence a difference in the probability of symbol-rate error for QPSK and the probability of bit-rate error for PSK.

7
Coding

7.1 Introduction

Assuming that information to be transmitted is in digital form (where the information could be voice, telex or data), the channel through which the signal passes is likely to cause signal degradation. The noise and/or fading etc, experienced during transmission could increase the probability of bit error at the received end. The data signal may be encoded in such a way as to reduce the likelihood of bit error. This process of coding, packages the bits that contain information with other bits. These other bits are known as redundant bits because they contain no information, but they can assist in the detection and correction of errors.

A relationship between a communication channel and the rate at which information can be transmitted over it has been established by C.E.Shannon. Basically Shannon's rule states that if the information rate of a source is less than the channel capacity, there exist coding techniques which allow the transmission of the information with an arbitrarily small probability of error. The capacity (C) of the channel in bit/s with additive white Gaussian noise is given by:

$$C = B\log_2(1 + S/N) \quad (7.1)$$

where B is the channel bandwidth in Hz
and S/N is the signal-to-noise power ratio within the bandwidth.
This relationship is known as the Shannon–Hartley Law.

It would seem from equation (7.1) that by increasing B, the channel capacity can be substantially increased. However, increasing B will increase N since the noise in the channel is proportional to B; thus, if B increases, S/N decreases. Also from equation (7.1) it can be seen that if noise could be eliminated, the channel capacity would be infinite. Channels are noisy however and assuming white noise with two-sided power spectral density $N_o/2$, equation (7.1) can be written as:

$$C = B\log_2(1 + S/N_oB) \quad (7.2)$$

where S is the signal power in watts
N_o is noise power spectral density in watts/Hz.
Equation (7.2) can be modified to give:

$$C = (S/N_o)\log_2[(1 + S/(N_oB))^{N_oB/S}]$$

and, since:

$$\lim_{x \to 0} (1 + x)^{1/x} = e$$

then:

$$\lim_{B \to \infty} C = \frac{S}{N_o} \log_2 e \tag{7.3}$$

The signal power, $S = E/T$, for M-ary signalling could be written in terms of transmission rate R since:

$$R = (\log_2 M)/T$$

and signal power is given by:

$$S = (ER)/\log_2 M$$

so that equation (7.3) becomes

$$\lim_{B \to \infty} C = \frac{E R}{N_o \log_2 M} \log_2 e$$

If the transmission rate (R) equals channel capacity (C), then:

$$E/(N_o \log_2 M) = 1/\log_2 e = 1/1.44 = -1.6 \text{ dB} \tag{7.4}$$

so that as long as $E/(N_o \log_2 M) > -1.6$ dB, the channel can be used with zero error. The value of -1.6 dB is known as the Shannon limit, below which the value of E/N_o should not fall.

Figure 6.18 showed a plot of probability of bit error P_e against E_b/N_o for coherent BPSK and QPSK. To improve P_e performance, the curve should move to the left, i.e. the value of E_b/N_o should be reduced towards the Shannon limit.

Using binary signalling, where $M = 2$, equation (7.4) reduces to $E/N_o = -1.6$ dB.

Providing the signal data is uncoded, $E = E_b$; for coded signals it is usual to quote E_c for the energy content of the signal over the symbol duration T.

For M-ary signalling where processing can be achieved using more than one bit at a time and $M = 2^n$ (where n is the number of bits processed at a time), an improvement can be obtained in bit error probability by increasing the value of n. This improvement assumes orthogonal M-ary signalling in a channel affected by white Gaussian noise. It should be remembered that to achieve this improvement the bandwidth of the channel would also have to increase.

For multiple-phase M-ary signalling, the reverse of the above is true, i.e. as n increases the error performance curve moves to the right, in the direction of a degraded error performance level. However, in this case increasing n results in a larger bit-rate in a fixed bandwidth.

What the conflicting factors show is that a system designer has to strike a compromise between an acceptable E_b/N_o level and channel bandwidth, taking into account other factors imposed by the system such as power levels achievable, system complexity etc.

One method of approaching Shannon's limit is to use a coding stream which allows for the detection, and correction, of any received errors. A possible coding system is shown in Fig. 7.1. The function of the encoder is to produce the required coding stream which is then used to modulate the carrier. The demodulator has the task of retrieving the transmitted coding stream from the received signal. There are two types of demodulation processes that can be used, namely:

- hard decision, where the demodulator produces its best attempt at representing the transmitted information;
- soft decision, where the demodulator additionally provides extra information on the validity of the hard decisions.

138 Coding

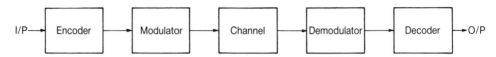

Fig. 7.1 Block diagram of a coding system

The decoder should use the redundancy of the coding stream to establish a valid received sequence.

The types of errors associated with digital transmission are:

- random, where there is no correlation between the symbols in error;
- burst, where a group of consecutive symbols is likely to be in error.

The types of codes chosen may be selected on the basis of how well they deal with each of these errors, or a combination of both of them.

7.2 Block codes

A coding stream could use what are known as block codes. A block code uses input data divided into k information symbols, together with redundant check symbols, to produce a code word of n symbols. An encoder operating in this way would produce a (n, k), or rate k/n, block code.

For a digital system the bit is the basic element and can have the value of binary 0 or binary 1. The bits could be grouped to provide the information that is to be transmitted. For two equal length binary code words there is a distance between them where distance is defined as the number of bit positions in which the two words differ. In general, the distance d_{ij} between two code words s_i and s_j is given by:

$$d_{ij} = (S_i \oplus S_j)$$

where $\oplus$ represents modulo-2 addition, i.e. binary addition without a carry. As an example, if $s_i = 0010011$, $s_j = 1101011$, then the modulo-2 addition of s_i and s_j gives 1111000 which shows there are 4 bits different out of the total of seven bits, i.e. $d_{ij} = 4$.

If the code words used have a minimum distance d_{min} the decoder can detect as many as e errors, where:

$$e = (d_{min} - 1) \text{ bits}$$

The decoder should correct as many as e' errors where:

$e' = (d_{min}/2) - 1$ for d_{min} equal to an even number
$e' = (1/2)(d_{min} - 1)$ for d_{min} equal to an odd number.

This illustrates the difference between detection and correction since if $d_{min} = 4$ for example, three bit errors can be detected but only one bit error could be corrected.

Hamming codes

The Hamming block codes are characterized by:

$$(n, k) = (2^m - 1, 2^m - (1 + m))$$

where $m = 2, 3 \ldots$

These codes have a minimum distance of 3 and thus can detect all combinations of up to two errors per block and correct all single errors per block.

For coherently demodulated BPSK over a channel affected by Gaussian noise, the

7.2 Block codes

symbol error probability in the channel could be expressed as E_c/N_o where E_c/N_o is the code symbol energy per noise spectral density. The relationship between E_c/N_o and the information bit energy per noise spectral density E_b/N_o, could be realized since:

$$\frac{E_c}{N_o} = \frac{2^m - (1+m)}{2^m - 1} \frac{E_b}{N_o} = \frac{k}{n} \frac{E_b}{N_o} \quad (7.5)$$

Hamming codes have values for (n, k) given by $(7, 4)$, $(15, 11)$ and $(31, 26)$. Thus, the value of E_c/N_o for, say, the $(7, 4)$ code is from equation (7.5):

$$E_c/N_o = (4/7)E_b/N_o$$

This shows that the error performance curve for E_c/N_o will move to the left compared with the value of E_b/N_o for an uncoded error performance curve.

Figure 7.2 shows the plot of bit error probability (P_B) against E_b/N_o for coherently demodulated BPSK over a Gaussian channel for different codes. The Hamming codes shown are for (n, k) of $(7, 4)$, $(15, 11)$ and $(31, 26)$.

Extended Golay code

The extended Golay code is produced by adding an overall parity bit to the perfect (23, 12) Golay code. The addition of the parity bit increases the minimum distance d_{min} from 7 to 8 and produces a 1/2 rate code. This code has the advantage of being easier to

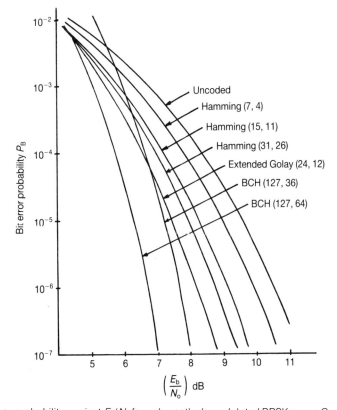

Fig. 7.2 Bit error probability against E_b/N_o for coherently demodulated BPSK over a Gaussian channel for different block codes

140 Coding

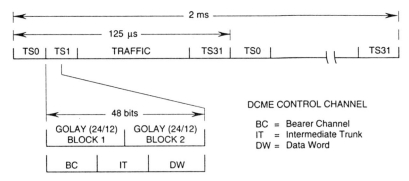

Fig. 7.3 Intelsat IESS-501 frame structure (reproduced by permission of *COMSAT Technical Review*, CTR, Vol.22, No.1, Spring 1992, p 45)

implement than the original 12/23 rate Golay code and is more powerful than the Hamming code described earlier, although at the expense of the need for a greater bandwidth and greater complexity in the decoder. For the extended Golay code with d_{min} of 8, the number of errors that can be corrected is:

$$e' = (d_{min}/2) - 1 = 3$$

The plot of bit error probability (P_B) against E_b/N_o for coherently demodulated BPSK over a Gaussian channel for the rate 1/2 extended Golay code is shown in Fig. 7.2.

From Fig. 7.2 it can be seen that the extended Golay code moves the error performance curve for E_b/N_o still further to the left compared with the value of E_b/N_o for the Hamming codes.

The Intelsat DCME has a 2 ms bearer frame carrying one 24 bit assignment message. A rate 1/2 (24,12) Golay code is applied to each of two 12-bit blocks of information to generate a total of 48 bits. This 48 bit message is sub-divided into 16 groups of 3 bits which are interleaved into the bearer frame at 125 μs intervals. Each 24 bit block is transmitted in 1 ms and 3 bit errors can be corrected within the block. This suggests that the Intelsat DCME bearer frame format can recover from burst errors of less than 256 bits duration on a 2.048 Mbit/s bearer occurring at intervals of 1 ms or greater, and which affect fewer than 4 bits of the 24 bit encoded block. See Fig. 7.3.

Interleaving is dealt with in more detail later in this Chapter.

Coding gain

The difference between the value of E_b/N_o for uncoded and various (n, k) codes over a Gaussian channel can lead to an improvement in the required E_b/N_o to achieve a required error performance for an error-correcting system compared with the uncoded system. This improvement for a given error performance is known as the coding gain and is usually expressed in decibels. The use of a coding system will reduce the energy per bit, as was shown by equation (7.5). It can be shown that there is a threshold below which an improvement does not occur and below the threshold the redundant bits do not provide an improvement in performance. Above the threshold, the reduction in energy per coded bit is overridden by improvement of the coded system to give the coding gain. The coding gain of the Intelsat 120 Mbit/s TDMA system is quoted as 2.5 dB at a value of bit error probability P_e of 10^{-5}.

7.3 Cyclic codes

Hamming and extended Golay codes use parity checks which need to be implemented in the encoder/decoder circuitry. The provision of the required hardware for the parity check could be complex and expensive. A type of parity check code, called cyclic code, is implemented simply by the use of shift registers. Any cyclic codeword can produce another valid codeword by giving an end-to-end shift of one digit, i.e. if a codeword is given by:

$$u_0, u_1, u_2 \ldots u_{n-2}, u_{n-1}$$

then

$$u_{n-1}, u_0, u_1, u_2 \ldots u_{n-2}$$

is another valid codeword.

An (n, k) cyclic code can be generated using a $(n-k)$ shift register and a process of modulo-2 division. A binary number of length k, which consists of n information bits, is followed by $(k-n)$ binary zeros and divided by a binary number of length $(k-n+1)$ bits. Once the process of division takes place, the remainder consists of $(k-n)$ bits which are then added to the n information bits to produce the unique codeword.

Example 7.1 Consider the (7, 4) cyclic code and code word 1000101 which has information bits 1000 and parity bits 101. A shift would give a new codeword 0001011 which corresponds to the information bits 0001. The divisor for this unique set of codewords is 1011, i.e.

```
1 0 1 1 )1 0 0 0 0 0 0
         1 0 1 1
         ─────────
           1 1 0 0
           1 0 1 1
           ─────────
             1 1 1 0
             1 0 1 1
             ─────────
               1 0 1
```

In this example the information bits were 1000 and were added to three zeros to give 1000000, which when divided by 1011 gave a remainder of 101. The remainder replaces the three zeros at the end of the word 1000000 to give 1000101 etc.

The register of Fig. 7.4 is suitable for a (7, 4) cyclic code.

The register initially contains all zeros and the switch is in position 1. The four information digits are shifted into the encoder and as each digit arrives it is routed to the output and added to the value of $S_2 \oplus S_3$ and the sum is placed in the first stage of the

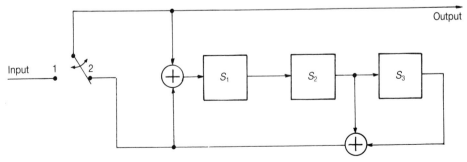

Fig. 7.4 Coder for a (7, 4) cyclic code

register. At the same time, the contents of registers S_1 and S_2 are shifted to the right. Once all the information digits have been entered, the switch moves to position 2 and the register is moved three times to clear it. The sum of S_2 and S_3 will appear at the output during each shift. This is the process described in Example 7.1 for information bits 1000. The sum of S_2 and S_3 added to itself produces a zero which is fed to S_1. After three shifts the codeword is generated consisting of the four information bits and three parity digits. For Example 7.1 the codeword would be 1000101. The register is also cleared ready for the next set of information bits. The process is valid for $2^k = 2^4 = 16$ unique codewords for the (7, 4) encoder.

Addition and subtraction in modulo-2 arithmetic are identical processes and the codeword must be a perfect multiple of the divisor. At the receive end the codeword is divided by the same divisor as used at the transmitting end in order to recover the information. A remainder of zero will indicate that the codeword received is a valid member of the set

Cyclic codes may be represented in polynomial form, i.e. if the elements of the codeword are represented as the coefficients of a polynomial $U(X)$ where:

$$U(X) = u_0 + u_1 X + u_2 X^2 + \ldots + u_{n-1} X^{n-1}$$

and the presence or absence of each term in the polynomial is an indication of the presence of a 1 or a 0 in the codeword.

The degree of the polynomial will be $(n-1)$ if the u_{n-1} component is 1.

If $U(X)$ is a $(n-1)$ degree codeword polynomial, then $U^i(X)$, the remainder resulting from dividing $X^i U(X)$ by $X^n + 1$, is another codeword given by:

$$U^i(X) = X^i U(X) \text{modulo}(X^n + 1)$$

This can be seen by reference to an example.

Example 7.2 Consider the codeword, described earlier, of value 0001011. This can be represented in polynomial form as:

$$U(X) = X^3 + X^5 + X^6$$

To find the value of a codeword corresponding to a single shift, i.e. $i = 1$,

$$XU(X) = X^4 + X^6 + X^7$$

Dividing $X^4 + X^6 + X^7$ by $X^7 + 1$ to obtain the remainder i.e:

$$\begin{array}{r} X^7 + 1 \overline{\smash{)} X^7 + X^6 + X^4} \\ \underline{X^7 + 1} \\ X^6 + X^4 + 1 \end{array}$$

N.B. the remainder is written as $+1$ and not -1 since for modulo-2 arithmetic $+1 = -1$.

Remainder $= 1 + X^4 + X^6$
Hence $U = 1000101$

Compare this result with the one obtained earlier for the same pair of codewords.

A cyclic code can be generated using a generator polynomial. The generator polynomial for a (n, k) cyclic code is:

$$g(X) = g_0 + g_1 X + g_2 X^2 + \ldots + g_r X^r$$

This code is unique and g_0 and g_r must be 1.

7.3 Cyclic codes   143

Every codeword polynomial is of the form $U(X) = m(X)g(X)$ where $U(X)$ is a polynomial of degree $(n-1)$ or less and $m(X)$ is:

$$m(X) = m_0 + m_1 X + m_2 X^2 + \ldots + m_{n-r-1} X^{n-r-1}$$

Since there are 2^{n-r} codeword polynomials and 2^k code vectors then $n - r = k$ or $r = n - k$, and $g(X)$ must be of degree $(n - k)$.

Hence $U(X) = (m_0 + m_1 X + m_2 X^2 + \ldots + m_{k-1} X^{k-1})g(X)$

Utilizing the message digits as part of the codeword can be achieved by shifting the message digits right into the k stages of the register and adding the parity digits in the left-hand $(n - k)$ stages. Multiplying $m(X)$ by X^{n-k} gives:

$$X^{n-k} m(X) = m_0 X^{n-k} + m_1 X^{n-k+1} + \ldots + m_{k-1} X^{n-1}$$

Dividing $X^{n-k} m(X)$ by $g(X)$ gives:

$$X^{n-k} m(X) = q(X)g(X) + r(X) \tag{7.6}$$

where $q(X)$ is the quotient and

$r(X)$ is equal to $X^{n-k} m(X)$ modulo $g(X)$

Using modulo-2 arithmetic to add $r(X)$ to both sides of equation (7.6) gives:

$$r(X) + X^{n-k} m(X) = q(X)g(X) = U(X) \tag{7.7}$$

The left-hand side of equation (7.7) is a valid codeword polynomial and when divided by $g(X)$ there is a zero remainder. The codeword polynomial expands to the form:

$$r(X) + X^{n-k} m(X) = r_0 + r_1 X + \ldots + r_{n-k-1} X^{n-k-1} + m_0 X^{n-k} + m_1 X^{n-k+1} + \ldots + m_{k-1} X^{n-1}.$$

Example 7.3 the generator polynomial $1 + X + X^3$ can be used to generate the parity bits for a message vector 0001. For this case $n = 7, k = 4, n - k = 3$ and $m(X) = X^3$

$$X^{n-k} m(X) = X^3 X^3 = X^6.$$

Dividing $X^{n-k} m(X)$ by $g(X)$ gives:

```
X^3 + X + 1 ) X^6
              X^6 + X^4 + X^3
              ─────────────────
                    X^4 + X^3
                    X^4 + X^2 + X
                    ─────────────
                          X^3 + X^2 + X
                          X^3 + X   + 1
                          ─────────────
                                X^2 + 1
```

Hence $r(X) = 101$ and the codeword becomes 1000101.
Compare this with the previous calculations for the same message word.

Bose–Chadhuri–Hocquenghem (BCH) codes

BCH codes are a class of cyclic codes with a large range of block lengths, code rates, alphabets and error-correcting capability. BCH codes have been found to be superior in performance to all other codes of similar block length and code rate. Most commonly used BCH codes have a codeword block length of $n = 2^m - 1$, where $m = 3, 4 \ldots$

A plot of bit error probability (P_B) against E_c/N_o for coherently demodulated BPSK over a Gaussian channel is shown in Fig. 7.2 for the BCH code (127, 64).

The Inmarsat-A system uses BCH codes for its TDM channels and for the request

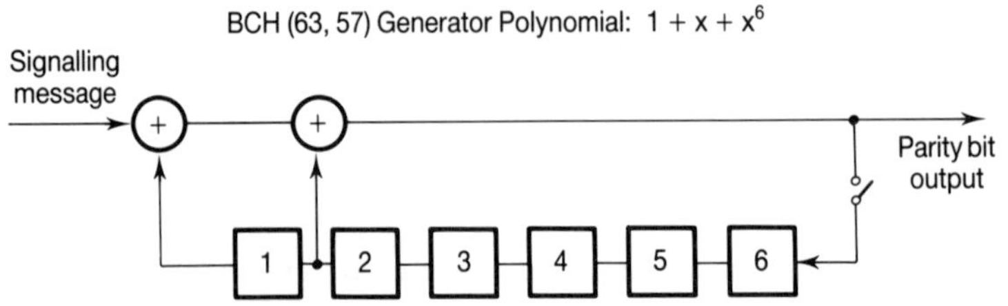

Fig. 7.5 Block diagram of a BCH (63, 57) generator used in Inmarsat-A (courtesy Inmarsat)

Diagram labels: BCH (63, 57) Generator Polynomial: $1 + x + x^6$; Signalling message; Parity bit output; Signalling message parity bit generator implementation.

Notes: 1. Initial state of all registers is 0
2. The feedback switch is closed during the first 57 data shifts and opened for the next six parity shifts during encoding, and no switch is required for decoding. Logic 0s are entered when the switch is open. After shifting 57 data bits plus 6 parity bits during decoding the shift register should contain all 0s if no channel errors have occured.

channel burst. Figure 7.5 shows the encoding arrangement for the BCH (63, 57) code for the TDM channel, with a generator polynomial of $g(X) = 1 + X + X^6$.

The BCH (127,113) code provides two error corrections per block and has been specified for the Intelsat 120 Mbit/s TDMA system.

Reed–Solomon codes

This set of codes has the largest possible code minimum distance of any linear code with the same encoder input and output block lengths. For these codes the distance between two code words is defined in terms of the difference in symbols. The code minimum distance is given by:

$$d_{min} = n - k + 1$$

where n is the total number of code symbols per block
and k is the number of data symbols encoded.

The code is capable of correcting e' errors where, in this case:
$$e' = (1/2)(d_{min} - 1)$$
or
$$e' = (n - k)/2$$

The code therefore requires no more than $2e'$ parity check symbols.

The advantage of Reed–Solomon codes is the reduction in the number of words of n symbols which are codewords, producing a possibly large value of d_{min}.

A Reed–Solomon code may correct e' errors using an alphabet of 2^m symbols

with $n = 2^m - 1$
and $k = 2^m - 1 - 2e'$
where $m = 2, 3 \ldots$

The Reed–Solomon codes may be used for burst-error correction.

Example 7.4 A (31, 23) Reed–Solomon code has $n = 31$ and $m = 5$ (i.e. a codeword of 31 symbols each of which has 5 bits).

Since $k = 23$, $e' = 4$. Thus, the code can correct up to 4 symbols per block. The code could cope with random errors, allowing up to 4 bit errors in the block which would

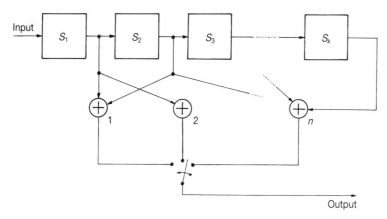

Fig. 7.6 Arrangement of a general convolutional encoder

affect, at worst, four symbols. Burst-errors of length 4 will, at worst, affect only two symbols allowing further burst error correction.

7.4 Convolutional codes

This type of code is not a block code. Instead of parity bits being added to form a block of data, the parity bits are calculated over a longer span of bits to form a continuing message sequence. A convolutional code can be described in terms of integers n and k. Integer n is a measure of the number of coded bits that go into making the sequence while k is a representation of the span of bits forming the message. The factor k is known as the constraint length.

A general convolutional encoder is shown in Fig. 7.6.

A shift register of constraint length k is connected to n modulo-2 adders. The register connections to the adders are random as shown. Whenever a new bit is shifted into the encoder, all bits in the register are shifted right and the output of the n adders is sampled sequentially to produce the code symbols or bits. This code stream can then be used to modulate a carrier for transmission over the link. Multiple input groups of bits are possible but most encoders have input message bits shifted in to the encoder one bit at a time. Under these circumstances there are n coded bits for every message bit and the code rate can be written as $1/n$.

Figure 7.7 shows a convolutional encoder with constraint length $k = 3$. There are 2 modulo-2 adders so that the code rate is 1/2. The input bit m placed into the first of the shift registers causes the bits in the registers to be moved one place to the right. The output switch samples the output of each modulo-2 adder, one after the other, to form a bit pair for the bit just entered. The connections from the register to the adders could be one, two or three connections for either adder. The choice depends on the requirement to produce a code with good distance properties.

The outputs u_1 and u_2 from the adders will have the form:

$$u_1 = S_1 \oplus S_3$$
$$u_2 = S_1 \oplus S_2$$

Suppose a data word 1 1 0 is to be placed into the register, with the leftmost symbol entered first. The output sequence in terms of u_1 and u_2 as the input word is shifted through the register is shown in Table 7.1.

146 Coding

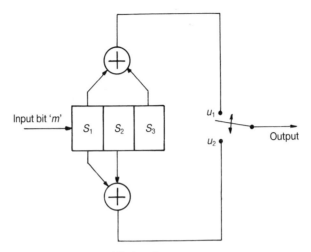

Fig. 7.7 Convolutional encoder, rate 1/2, $k = 3$

Table 7.1 Convolutional encoder output sequence

Register contents			Output word	
			u_1	u_2
1	0	0	1	1
1	1	0	1	0
0	1	1	1	1
0	0	1	1	0
0	0	0	0	0

The output sequence from Table 7.1 is thus:

$$11\ 10\ 11\ 10\ 00$$

The effective code rate for a 3 bit input and a 10 bit output sequence is 3/10, not 1/2 as the 2 bit output for 1 bit input would suggest. The reason is the necessity for pushing the last data bit through the register, which requires four more output bits than the expected six. If the message length is increased for this encoder it is still only four extra output bits so that if the input data is 250 bits, the output is 504, at a rate close to 1/2.

Polynomial representation

A convolutional encoder can be described using a set of n generator polynomials, one for each of the n modulo-2 adders. Each adder has a degree of $(k - 1)$ or less and the coefficient of each term in the polynomial is 0 or 1 according to whether a connection is absent or not between the shift register and each modulo-2 adder. For the circuit of Fig. 7.7, the generator polynomials can be written as:

$$g_1(X) = 1 + X^2$$
$$g_2(X) = 1 + X$$

The output sequence becomes:

$U(X) = m(X)g_1(X)$ alternating with $m(X)g_2(X)$
If the input word is 1 1 0 it can be expressed in polynomial form as $m(X) = 1 + X$.
The output sequence for $m(X)$ is thus:

$$m(X)g_1(X) = (1 + X)(1 + X^2) = 1 + X + X^2 + X^3$$
$$m(X)g_2(X) = (1 + X)(1 + X) = 1 + X^2$$
$$U(X) = (1, 1) + (1, 0) X + (1, 1) X^2 + (1, 0) X^3 + (0, 0) X^4$$
$$U = 1\ 1\quad 1\ 0\quad 1\ 1\quad 1\ 0\quad 0\ 0 \qquad \text{as before}$$

The example shown above has assumed that the contents of the register were initially zero prior to the shifting of the first bit of the required data message. Reference to Table 7.1 will show that if the bits in the register prior to the arrival of the first data message bit were 1s then the output of u_1 and u_2 would both be 0. This would also be the case in the second line of Table 7.1 and the modified output stream would be 0 0 0 0 1 1 1 0 0 0. It can be deduced from this that the output data stream depends not only on the input data but the contents of the register prior to the arrival of that data. Usually, flushing bits are placed at the end of a data stream to clear the registers ready for the next input sequence.

Tree diagram

Convolutional codes can be decoded using a tree diagram. The tree indicates the output sequence for every possible input sequence. Since the input sequence can consist of N information bits, it would seem that the tree would need 2^N branches to cover all possibilities. However, since the constraint length of the encoder is k, the tree is simply repetitive beyond that number. A tree representation for an encoder of 1/2 rate, and constraint length 3 is shown in Fig. 7.8.

The repetitive nature of the tree can be seen from Fig. 7.8 since with $k = 3$ in this arrangement, the branches become repetitive beyond the third level, corresponding to the third input data bit.

The rule for finding a codeword sequence depends on the status of the input data bit and can be explained as follows:
- if the first input data bit is 1, the branch word is found by moving downwards to give an output 1 1; an input data bit equal to 0 requires an upward movement to give an output 0 0;
- if the first input bit is 1 and the second input bit is 0, the second output branch word is 0 1 etc;
- if the first input bit is 1, the second bit is 0 and the third bit is 1, the output branch word is 1 0 etc.

Suppose an input data sequence is 1 1 0 0 1, then assuming the initial contents of the encoder register were zero, the output is given as 1 1 1 0 1 1 1 0 1 1. This path has been shown using a heavy line in Fig. 7.8.

Trellis diagram

The tree diagram of Fig. 7.8 has all nodes labelled and at each node there are two possible paths. The values assigned to each path are complementary, i.e. if an input of 0 at a node yields an output of 1 0 then an input of 1 at that point yields an output of 0 1. This is because each of the two output bits is a sum that includes the most recent bit;

148 Coding

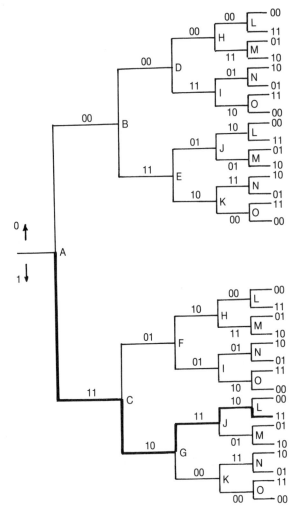

Fig. 7.8 Tree diagram for encoder, rate 1/2, k = 3

hence if all previous bits remain the same while the most recent bit changes from 1 to 0, each output should change. Because of the repetitive nature of the tree beyond level k (3 in this case), it can be seen that there is duplicity in the labelling of the nodes. Once beyond level 3 the output branch words are no longer influenced by the original first input bit which, by now, has been shifted out of the last register. The input sequences will generate the same branch words after level 3 and any two nodes having the same label at an instant in time can be merged since all succeeding paths will be the same. To exploit this repetitive structure of the tree, a trellis diagram can be drawn. The trellis corresponding to the tree of Fig. 7.8 has been drawn in Fig. 7.9.

The labelling on the trellis has been included, to correspond to the labelling on the tree diagram, and the output branch words are indicated on each branch. At each node the trellis splits into two paths, the upper path representing a 0 input and the lower path representing a 1 input. It can be seen from the trellis diagram that all nodes in a horizontal row exhibit equivalent behaviour, giving the same output at each node for a

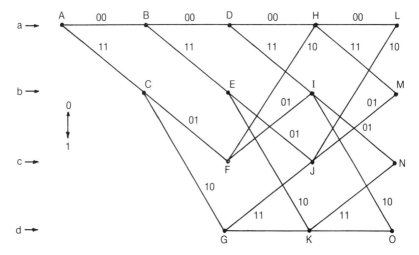

Fig. 7.9 Trellis diagram for encoder, rate 1/2, $k = 3$.

given input sequence. There are four levels of nodes represented and these correspond to the 2^{3-1} possible encoder states. The trellis of Fig. 7.9 assumes a repetitive nature after the depth $k = 3$ is reached and after this point each of the states can be entered from either of two preceding states.

7.5 Decoding

Block codes

The simplest means of decoding block codes is by a method of correlation whereby the decoder makes a comparison between the received codeword and all permissible codewords, selecting that word which gives the nearest match. Decoding of such codes will also depend on whether error detection or error correction is required. Decoders of block codes generally cannot use soft decision outputs from the demodulator unlike the decoders for convolutional codes.

Convolutional codes

The complete transmission loop requires a convolutional encoder followed by modulation, transmission channel, demodulator and decoder. The effect of the transmission channel on the signal, and the probability of detection of a 1 or 0 in the presence of Gaussian noise, has been discussed before in Chapter 6, Section 6.4. The demodulator output can be configured to give a hard decision regarding whether the signal is 1 or 0. The process of decoding then depends on the two state inputs it receives. An alternative demodulator configuration allows quantization of the predicted level, which gives the decoder more information regarding the probable state of the demodulator output. For example, if 3 bit ($2^3 = 8$ levels) quantization occurs then 0 0 0 would suggest a firm valuation of the level received as a 0. On the other hand 0 0 1 suggests the 0 is received close to the threshold and this valuation as a 0 is made with less certainty.

The reason for quantization is to provide the decoder with more information in order to recover correctly the transmitted information with better error performance probability. For a Gaussian channel, 8-level quantization gives an improvement in S/N of about 2 dB compared with 2-level quantization. The use of higher levels results in soft decision decoding and can provide the same bit error probability as hard decision decoding with 2 dB less E_b/N_o for the same overall performance. A disadvantage is the requirement for a more complex decoder.

One of the most common decoder processes uses the Viterbi convolutional decoding algorithm. Basically, this algorithm is equivalent to comparing the received bit sequence with all possible transmitted sequences and choosing that sequence which provides the closest match. The way the algorithm operates can be seen by reference to the trellis diagram. Assuming a 1/2 rate encoder of constraint length 3, the first step is to find the Hamming distances between the initial node and each of the four states, three levels deep into the trellis. Each of the four nodes can be reached from only two preceding nodes, hence eight paths can be identified. The Hamming distance must be found for each path. Referring back to the trellis of Fig. 7.9, this has been redrawn in Fig. 7.10 and labelled with the Hamming distance between the received code symbols and the corresponding branch word from the encoder trellis. The message sequence considered is 1 1 0 0 1 giving an output sequence of 1 1 1 0 1 1 1 0 0 0.

The Hamming distances are estimated according to the difference between the branch code word and the output code word at that instant. For example, from point A to B the branch code word is 0 0 while the required output word is 1 1 giving a Hamming distance of 2. Also, the transition from A to C gives a branch code of 1 1 which corresponds to an output word of 1 1, hence the Hamming distance is 0. This process is repeated for all transitions.

The basis of Viterbi decoding is to establish which of any two paths that converge into a single state can be eliminated in the attempt to determine the optimum path. The total path length can be established by finding the cumulative value of the Hamming path distances. The computation is done for each of the 2^{k-1} nodes at any transition time and repeated at the next transition time. As an example, point D could be reached via ABD

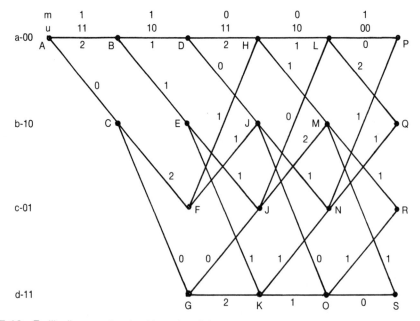

Fig. 7.10 Trellis diagram showing Hamming distances

7.5 Decoding

in a distance of 3, point E via ABE in a distance of 3, point F via ACF in a distance of 2 and point G in a distance of 0. Point H could be reached via ABDH in a distance of 5 or via ACFH in a distance of 3. If there are two input sequences that arrive at the same node, the longer path will be eliminated leaving the surviving path. Point J could be reached via ABEJ in a distance of 4, or via ACGJ in a distance of 0. Hence, the path ABEJ would be eliminated. If two paths have equal distances then one will be chosen arbitrarily. The process of path determination continues, as described, to advance deeper into the trellis and to make decisions on the input data bits by eliminating all but the one selected path.

A disadvantage of the Viterbi algorithm is that although error probability decreases exponentially with the constraint length, the decoder complexity increases exponentially with constraint length. The use of soft decision decoding is easily incorporated, providing enhanced error correction capabilities compared with hard decision decoding.

Sequential decoding

A sequential decoder may be used for convolutional decoding and operates in a similar manner to the Viterbi decoder. A sequential decoder, on receipt of the incoming codeword sequence, will penetrate into the tree according to a decision made regarding the best path to follow. Using a trial and error technique the decoder will progress as long as the chosen path appears correct, otherwise it will backtrack to try a different route. Either soft decision or hard decision decoding is possible with the sequential decoder, although soft decisions would considerably increase the computational time and storage space required.

Suppose the encoder of Fig. 7.7 has been used with the input sequence 1 1 0 0 1 giving an output sequence $U = 1\ 1\ 1\ 0\ 1\ 1\ 1\ 0\ 1\ 1$. The decoder has details of the encoder tree (as shown in Fig. 7.8) and will use the received signal to make decisions regarding the path to take. For any node the decoder will generate both paths and will follow the path that agrees with the received code. If the received sequence is correct the decoder will penetrate the tree in the manner shown by Fig. 7.8. If the received sequence is incorrect, the decoder will start at node A and generate the two paths from that node and will follow the path that corresponds to the received sequence symbols at that time. If there is not an agreement the decoder will follow the most likely path but will keep a count of all the disagreements between the received symbols and branch words on the path followed. Where two paths appear equally likely, the decision to follow one of the paths is an arbitrary one. At each node the decoder generates new branches and makes a comparison with the next received sequence symbols. The process continues deeper into the tree with the decoder keeping a tally of the disagreements. If the number of disagreements exceeds a certain number the decoder decides that it is on an incorrect path and backtracks in order to try another path, at the same time keeping track of the discarded paths in order to avoid re-using them. The sequence is best explained by means of an example.

Example 7.5 Suppose for the sequential decoder the received sequence is incorrect with the third and sixth bits in error compared with the transmitted sequence. The message sequence is thus 1 1 0 0 1 and the (incorrect) received sequence is 1 1 **0** 0 1 **0** 1 0 1 1.

The tree diagram of the encoder of Fig. 7.7 has been drawn in Fig. 7.11 to show the decoding sequence. It is assumed that a disagreement count of 3 is the limit before backtracking occurs and that where alternate paths are equally possible the decoder will follow an input bit one path. The current disagreement count is shown on each path in Fig. 7.11.

152 *Coding*

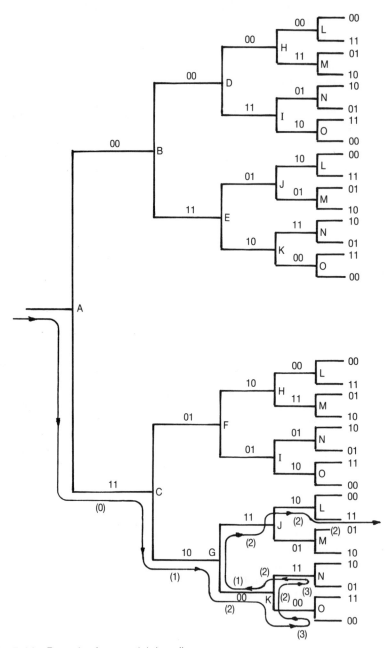

Fig. 7.11 Example of sequential decoding

- At node A, symbols 1 1 are received and compared with the branch words leaving that node. The decoder decides the most likely path is via branch 1 1 and moves via that path to node C.
- The decoder receives symbols 0 0 and compares them with the available branch words leaving that node (0 1 and 1 0). There is no 'best path' and the decoder arbitrarily follows the input bit one (or branch word 1 0) path reaching node G. The disagreement count registers 1.
- The decoder receives symbols 1 0 and compares them with the available branch words leaving that node (1 1 and 0 0). There is no 'best path' and the decoder arbitrarily follows the input bit one (or branch word 0 0) path reaching node K. The disagreement count registers 2.
- The decoder receives symbols 1 0 and compares them with the available branch words leaving that node (0 0 and 1 1). There is no 'best path' and the decoder arbitrarily follows the input bit one (or branch word 0 0) path reaching node O. The disagreement count registers 3.

   However, a disagreement tally of 3 is the set level for backoff so the decoder returns to node K and the disagreement counter is reset to 2. The decoder follows the alternative path from node K via the branch path 1 1 reaching node N. However, this branch word differs from the received word by 1 and the disagreement counter is again set to 3 and the decoder backs off returning to node K, resetting the disagreement counter to 2.

   Since both paths from node K have been tried, the decoder returns to node G, resetting the disagreement counter to 1.
- At node G the decoder follows the untried path (branch word 1 1) to node J. Because the input word at node G (1 0) differs from the path word (1 1), the disagreement counter is set to 2.
- The decoder receives symbols 1 0 and compares this with the branch words leaving that node. The decoder decides the most likely path is via branch 1 0 and moves via that path to node L.
- The decoder receives symbols 1 1 and compares this with the branch words leaving that node. The decoder decides the most likely path is via branch 1 1 and moves via that path to complete the decoding process.

As shown on Fig. 7.11, the decoder will produce a correctly decoded message of 1 1 0 0 1.

A major advantage of sequential decoding is that the number of states examined is independent of constraint length, allowing the use of large constraint lengths and low error probability. A disadvantage is the need to store input sequences while the decoder searches for its preferred route through the tree. If the average decode rate falls below that of the average symbol arrival rate there is a danger that the decoder cannot cope, causing a loss of input information.

Interleaving

Channel impairments such as fading and multipath which exist on a satellite channel cause statistical dependence among successive symbol transmissions, i.e. errors tend to occur in bursts. Fading may be a slow variation compared with the time occupied per symbol. Multipath is caused by signals arriving at the receiver via two or more paths of different lengths; this results in phase differences between received signals causing received signal distortion. Since most block and convolutional codes are designed to resist random independent errors, the effects caused by fading and multipath would render the system inoperative. The process of interleaving causes bursts of errors to be spread out in time and these can thus be decoded as if they were random errors. The interleaver acts as a sort of time-division multiplexer with the symbols of one codeword mixed in time with the symbols of other codewords.

Suppose a codeword is made up of seven symbols and there are seven such codewords in a block (for block codes) or constraint length (for convolutional codes). Each codeword is made up of symbols An_1 to An_7 where A can have the value 1 to 7, i.e.:

$$n_1, n_2, n_3, n_4, n_5, n_6, n_7, 2n_1, 2n_2, 2n_3, 2n_4, 2n_5 \ldots 7n_5, 7n_6, 7n_7$$

If the symbols were interleaved such that the pattern transmitted is of the form:

$$n_1, 2n_1, 3n_1, 4n_1, 5n_1, 6n_1, 7n_1, n_2, 2n_2, 3n_2 \ldots 4n_7, 5n_7, 6n_7, 7n_7$$

then if an error burst occurs that lasts for seven symbol periods, only one symbol from each of the original codes would be affected. Once received and de-interleaved, the original codewords would appear, each with a single bit error. If each codeword possesses a single error-correcting capacity the words should be decoded satisfactorily. Without interleaving the seven-symbol burst could have destroyed one or two of the original codewords.

For block interleaving the coded symbols are received from the encoder in blocks. The interleaver permutes the symbols and sends the rearranged symbols to the modulator. The usual form of permutation is the use of an array (matrix) where the blocks are

n_1	$2n_1$	$3n_1$	$4n_1$	$5n_1$	$6n_1$	$7n_1$
n_2	$2n_2$	$3n_2$	$4n_2$	$5n_2$	$6n_2$	$7n_2$
n_3	$2n_3$	$3n_3$	$4n_3$	$5n_3$	$6n_3$	$7n_3$
n_4	$2n_4$	$3n_4$	$4n_4$	$5n_4$	$6n_4$	$7n_4$
n_5	$2n_5$	$3n_5$	$4n_5$	$5n_5$	$6n_5$	$7n_5$
n_6	$2n_6$	$3n_6$	$4n_6$	$5n_6$	$6n_6$	$7n_6$
n_7	$2n_7$	$3n_7$	$4n_7$	$5n_7$	$6n_7$	$7n_7$

Fig. 7.12 Interleaver array for seven codewords of 7-symbol length

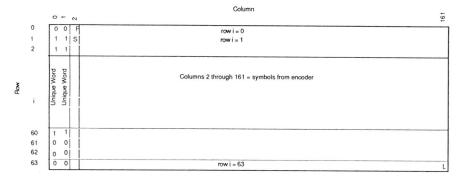

Fig. 7.13 Inmarsat-C interleave matrix (courtesy Inmarsat)

entered by filling the columns of the array and, once the array is full, the symbols are fed to the modulator one row at a time. At the receiver, symbols would be entered into a similar array by rows and taken out to the decoder on a column by column basis. Figure 7.12 illustrates an array of dimensions 7 × 7 filled with seven codewords each of seven symbol length. Codeword 1 (n_1 to n_7) fills column 1, codeword 2 ($2n_1$ to $2n_7$) fills column 2 and so on. The output from the interleaver would be ($n_1, 2n_1, 3n_1, 4n_1, 5n_1, 6n_1, 7n_1$) followed by: ($n_2, 2n_2, 3n_2, 4n_2, 5n_2, 6n_2, 7n_2$) etc.

The memory requirement at the receiver would be for XY symbols (where X = the number of columns, Y = the number of rows). Typically, however, since the array needs to be mostly filled before it can be emptied, a second array of XY symbols is additionally employed so that it can be filling while the first array is being emptied. The use of interleaving with a single error correcting code requires that the number of columns (X) must exceed the expected burst length. The number of rows (Y) depends on the coding scheme used. For block codes the value of Y should be greater than the block length while for convolutional codes Y should be greater than the constraint length. If this requirement is implemented then a burst of length X will cause no more than a single error in the block codeword (for block codes) or a single error in the constraint length (for convolutional codes).

The Inmarsat-C system uses the interleaving process. For this system data is transmitted in blocks using 640 bytes. A 1/2 rate convolutional encoder produces 640 × 8 × 2 symbols (10240 symbols) which are then passed to an interleaver. The interleaver matrix is shown in Fig. 7.13.

For this matrix the first two columns are identical and permanently contain two unique word patterns. The remaining 160 columns contain the 10240 symbols from the encoder starting at column 2 through to column 161 sequentially. In Fig. 7.13 'F', 'S' and 'L' refer to the position of the first, second and last symbols from the encoder. After assembly, the interleave block is transmitted on a row by row basis with the symbols in each row sent in ascending column order, i.e. unique word symbols are transmitted first. However, rows are not transmitted in sequential order; they are sent according to a permuted sequence as follows.

If the rows in the interleave block are numbered from $i = 0$ to $i = 63$ sequentially as shown in Fig. 7.13 and the transmitted order is from $j = 0$ sequentially through to $j = 63$, then i and j are related by either:

- $i = (j \times 39)$ modulo 64 (suitable at the transmitter)
- $j = (i \times 23)$ modulo 64 (suitable at the receiver)

This process is shown in Fig. 7.14.

156 Coding

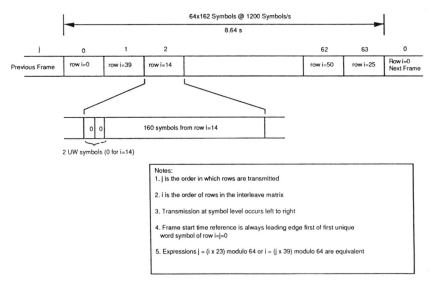

Fig. 7.14 Inmarsat-C forward link serial TDM transmission
(courtesy Inmarsat)

From Fig. 7.14 it can be seen that row $i = 39$ is transmitted as row $j = 2$. This can be deduced from the above since if $i = 39$, $j = (39 \times 23)$ modulo 64.

or $\qquad j = 897$ modulo 64
$= \qquad 1\ 1\ 1\ 0\ 0\ 0\ 0\ 0\ 0\ 1 \qquad (897)$
$\oplus$
$\qquad\qquad 0\ 0\ 0\ 1\ 0\ 0\ 0\ 0\ 0\ 0 \qquad (64)$
$= \qquad 1\ 1\ 1\ 1\ 0\ 0\ 0\ 0\ 0\ 1 \qquad (1)$

Thus, $i = 39$ is transmitted as $j = 1$ (the bits representing values of 64, 128 and above can be ignored in the resultant since both i and j can only have values from 0 to 63, i.e. the final value is given by the right-hand six bits only).

For $i = 50$, $j = (50 \times 23)$ modulo 64

or $\qquad j = 1150$ modulo 64
$= \qquad 1\ 0\ 0\ 0\ 1\ 1\ 1\ 1\ 1\ 1\ 0 \qquad (1150)$
$\oplus$
$\qquad\qquad 0\ 0\ 0\ 0\ 1\ 0\ 0\ 0\ 0\ 0\ 0 \qquad (64)$
$= \qquad 1\ 0\ 0\ 0\ 0\ 1\ 1\ 1\ 1\ 1\ 0 \qquad (62)$
$\qquad\qquad\qquad\qquad\qquad\qquad\qquad\qquad$ etc.

The unique word pattern prior to permuting is given by:

($i = 0$) $\qquad$ 0 1 1 1 $\quad$ 1 0 1 1 $\quad$ 1 0 1 0 $\quad$ 1 0 0 1
$\qquad\qquad$ 0 1 1 0 $\quad$ 1 0 0 1 $\quad$ 0 0 0 1 $\quad$ 0 1 1 1
$\qquad\qquad$ 0 0 1 1 $\quad$ 0 0 1 0 $\quad$ 1 1 1 0 $\quad$ 1 0 0 1
$\qquad\qquad$ 1 0 1 1 $\quad$ 1 0 0 0 $\quad$ 1 1 1 0 $\quad$ 1 0 0 0 $\quad$ ($i = 63$)

and after permuting the unique word sequence becomes:

($j = 0$) $\qquad$ 0 0 0 0 $\quad$ 0 1 1 1 $\quad$ 1 1 1 0 $\quad$ 1 0 1 0
$\qquad\qquad$ 1 1 0 0 $\quad$ 1 1 0 1 $\quad$ 1 1 0 1 $\quad$ 1 0 1 0
$\qquad\qquad$ 0 1 0 0 $\quad$ 1 1 1 0 $\quad$ 0 0 1 0 $\quad$ 1 1 1 1
$\qquad\qquad$ 0 0 1 0 $\quad$ 1 0 0 0 $\quad$ 1 1 0 0 $\quad$ 0 0 1 0 $\quad$ ($j = 63$)

This sequence can be used for receiver synchronization.

7.6 Error correction

Forward Error Correction (FEC)

Forward Error Correction (FEC) coding, which is the result of convolutional coding, is used in Inmarsat systems for some voice and telex channels and signalling channels and in Intelsat IBS and IDR services. Inmarsat-B, for example, uses a convolutional encoder of constraint length 7 and an 8-level soft-decision Viterbi decoder. The coding rate is either 3/4 or 1/2, see Table 3.1. For voice channels the rate 3/4 code is used and is derived by puncturing the rate 1/2, $k = 7$ convolutional code. The generator polynomials for the rate 1/2, $k = 7$ convolutional code are:

$$g_1 = 1 + X^2 + X^3 + X^5 + X^6$$
$$g_2 = 1 + X + X^2 + X^3 + X^6$$

The encoder logic, including the punctured operation is shown in Fig. 7.15.

In the 1/2 rate code, the relationship between three input bits and the six output bits is given by:

Input bit time	1	2	3
Output sequence	$g_1 g_2$	$g_1 g_2$	$g_1 g_2$

The rate 3/4 coded sequence is obtained by eliminating two bits from each block of six output bits from the rate 1/2 convolutional encoder as follows:

Input bit time	1	2	3
Output sequence	$g_1 g_2$	g_1	g_2

The receiving end is informed of the sequence of the four output bits, which make up the punctured code block, by arranging for the first boundary of the punctured code in each 80 ms transmission frame to coincide with the first bit that follows the last bit of the Unique Word or Framing Bit pattern.

At the beginning of each voice activated burst the shift register of the FEC encoder is set to the all-zero state. The Viterbi decoder at the receive end assumes this initial state at the start of each burst.

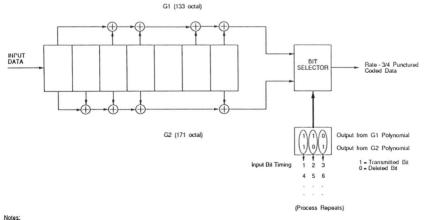

Notes:
- The first bit in each transmission frame is the output from the G1 polynomial
- All bits are transmitted for the rate -1/2 code

Fig. 7.15 Inmarsat-B convolutional encoder and code puncturing logic (courtesy Inmarsat)

Automatic Request Repeat (ARQ)

Forward Error Correction (FEC) requires only a one way transmission link since the message contains parity bits used for detection and correction of errors. Automatic Request Repeat (ARQ) on the other hand requires a two-way link since a receiver, detecting an error, does not attempt to correct it but simply requests the transmitter to retransmit the message. ARQ systems are basically of three types.

- Stop and wait ARQ. Each message block is transmitted and the transmitter waits for an acknowledgement before transmitting a further block. A half-duplex link is required (i.e.,transmission on the link is possible in both directions but not at the same time). If a message block is correctly received the next message block is transmitted; if the block received is in error the transmitter will retransmit that block.
- Continuous ARQ with repeat. The transmitter sends continuously and the receiver acknowledges continuously. Any message block not correctly received causes the transmitter to return to the block in question and re-commence continuous transmission from there. A full-duplex link is necessary (i.e. the ability to transmit in both directions simultaneously).
- Continuous ARQ with selective repeat. In this arrangement only the block received in error is retransmitted and the transmitter continues from where it left off instead of repeating any subsequent correctly received messages. Again a full-duplex link is necessary.

A major advantage of ARQ compared with FEC is that decoding equipment for error correction can be simpler and the redundancy in the total message stream is less. ARQ efficiency is good for low error ratios but for high error ratios requiring retransmission of a large number of message blocks the system becomes inefficient. A disadvantage of ARQ is the variability of the delays experienced from end-to-end of the link and the possible requirement for large data stores for incoming data blocks.

Continuous ARQ may be used in conjunction with FEC to provide a hybrid system. Such an arrangement could be used to provide feedback information to the transmitter regarding slow variations, such as fading. The Inmarsat-C system uses packets of data and every packet transmitted contains a 16-bit checksum field. The receiver completes an expected checksum for each packet and compares this with the actual packet received in order to verify that the packet has been correctly received. ARQ is used if the packet received is in error.

7.7 Pseudo-noise

A pseudo-noise (PN) generator will generate a set of cyclic codes with good distance properties. The name of the sequence is given because the sequence, although deterministic, appears to have the properties of sampled white noise. A PN sequence is easily generated using shift registers and has a correlation function that is highly peaked for zero delay and approximates to zero for other delays. The PN sequence, being deterministic, is useful for synchronization purposes between a transmitter and receiver.

Consider the circuit of Fig. 7.16. The circuit has three shift registers, a modulo-2 adder and a feedback path between the modulo-2 adder and the input to the register. The contents of the register may be shifted to the right and at each instance of time when this occurs the contents of S_2 and S_3 are modulo-2 added with the result sent back to S_1 as the next input.

Suppose the register of Fig. 7.16 is initialized with any 3-bit number other than 0 0 0.

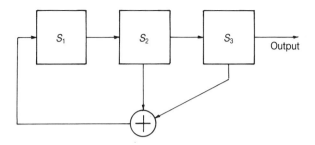

Fig. 7.16 Pseudo-noise generator

For example, let the initial sequence be 1 0 0. It follows then that the circuit of Figure 7.16 will produce a succession of register states given by:

$$1\ 0\ 0\quad 0\ 1\ 0\quad 1\ 0\ 1\quad 1\ 1\ 0\quad 1\ 1\ 1\quad 0\ 1\ 1\quad 0\ 0\ 1\quad 1\ 0\ 0\ldots$$

Since the final state in this sequence is the same as the first state the sequence repeats after $2^3 - 1 = 7$ shifts. An n-stage shift register, where n is the number of stages, would repeat after $2^n - 1$ shifts.

The output is taken from register S_3 and for the above example the output sequence is:

$$0\ 0\ 1\ 0\ 1\ 1\ 1\ldots$$

where the leftmost bit is the first bit of the output sequence. It can be observed from the output sequence that any other output sequence of the same cyclic form could be produced for a different initial register state, i.e. if the initial register state was 1 1 0, the output sequence would be:

$$1\ 0\ 0\ 1\ 0\ 1\ 1\ldots\quad \text{etc.}$$

Comparing this output sequence with the original output sequence, certain uniform properties can be seen. For example there are always four 1s in the sequence and the distance between them is 4. This will be true of all possible output sequences.

The autocorrelation function $R(\tau)$ of a periodic waveform $x(t)$ with period T_0 is given by:

$$R(\tau) = \frac{1}{K}\frac{1}{T_o}\int_{-\frac{T_o}{2}}^{\frac{T_o}{2}} x(t)x(t+\tau)\,dt \quad \text{for } -\infty < \tau < \infty$$

$$\text{where}\quad K = \frac{1}{T_o}\int_{-\frac{T_o}{2}}^{\frac{T_o}{2}} x^2(t)\,dt$$

For $x(t)$ representing a PN code, each fundamental pulse is referred to as a chip. For a PN waveform of unit chip duration and period p chips (where $p = 2^n - 1$), the normalized autocorrelation function is given by:

$$R(\tau) = \frac{N_A - N_D}{p}$$

where N_A = number of like digits in the sequence and a sequence shifted by n pulses; and

160 Coding

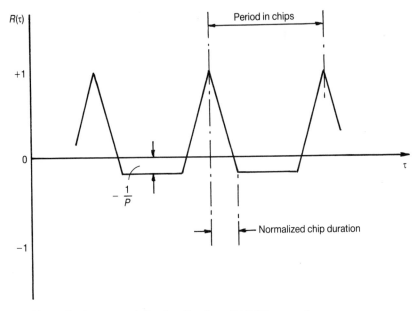

Fig. 7.17 Normalized autocorrelation function for a 7-bit PN generator

N_D = number of unlike digits in the sequence and a sequence shifted by n pulses.

The normalized autocorrelation function for the sequence generated by the shift register of Fig. 7.16 is shown in Fig. 7.17.

As mentioned in Chapter 3, some Inmarsat systems use a scrambler circuit before FEC encoding and a descrambler at the receive end after FEC decoding (see Fig. 3.9). For Inmarsat-B and Inmarsat-M systems for example, the scrambler/descrambler circuits are PN generators using 15 stages. The polynomial for the scrambler/descrambler is $1 + X + X^{15}$ and the configuration is shown in Fig. 7.18.

The scrambler/descrambler are clocked at the rate of one shift per information bit. The first bit into the scrambler at the beginning of a frame is modulo-2 added with the output of the scrambler shift generator corresponding to the initial state scrambling vector. The initial state of the shift register is set at the beginning of a burst and a frame.

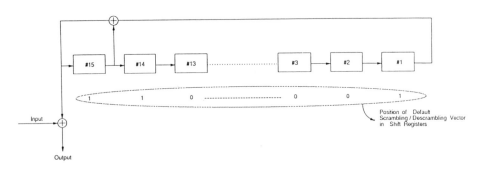

Fig. 7.18 Inmarsat-B scrambler/descrambler configuration (courtesy of Inmarsat)

7.7 Pseudo-noise

Considering the Inmarsat-M system, the initial state of the scrambler shift register at the LES (for the SCPC channel operating in voice mode) is sent to the LES by the MES at the start of a call as part of the call set-up sequence. The MES chooses any initial state (except all zeros) on a random basis for each call and signals this 'scrambling vector' message (1 0 0 0 1 1 0 1 or 8D in hexadecimal form) for implementation at the LES with the least significant bit (LSB) in shift register #1 and the most significant bit (MSB) in shift register #15 of the scrambler. The MES simultaneously sets the descrambler shift registers with the same scrambling vector. For MES to LES channels, a fixed initial state default value of 1 1 0 1 0 0 1 0 1 0 1 1 0 0 1 (6 9 5 9 in hexadecimal form) is used in the MES scrambler and LES descrambler. The position of bits of the fixed default pattern in the shift registers of the scrambler/descrambler is shown in Fig. 7.18.

The Intelsat system uses Digital Channel Multiplication Equipment (DCME) which is applicable to QPSK/TDMA and QPSK/FDMA (IDR) carriers. The intermediate data rate (IDR) system utilizes spectral energy dispersal with data scrambled via a self-synchronizing scrambler conforming to CCITT Rec.V.35. The scrambler consists of a 20 stage feedback shift register with modulo-2 addition to data of the parity of the third and twentieth stages prior to feeding into the first stage of the scrambler register. There is differential encoding for the resolution of phase ambiguity and FEC is employed with a punctured code of rate 3/4, constraint length 7. A bit error in the transmission link will cause a burst of errors in the recovered signal, creating a memory window of 22 bits which has to be flushed before an error event can be regarded as terminated. The output of the scrambler is then passed to the DCME. Details of DCME can be found in Chapter 5.

8
Earth stations

8.1 Introduction

An earth station is that part of the communication link which receives and/or transmits voice, video or data signals via a satellite. The term 'earth station' is used to describe the earth-based terminal although the station may be on a ship or an aircraft. In its most basic form an earth station consists of a transmitter, a receiver, an antenna which is able to transmit and receive signals, channel equipment for signal processing (into whatever form is required by the system used) and an interface to the terrestrial network. A block diagram of the basic system is shown in Fig. 8.1.

In some cases the full basic block diagram will not apply and the station may simply transmit or receive but not both. For example, in a direct broadcast system the earth station will receive only, while for, say, data transmission, the station may transmit only.

Although not shown in Fig. 8.1, an earth station will require sub-systems such as antenna tracking, power distribution and test equipment needed for efficient operation of the system.

The following sections will deal with the earth station equipment in more detail.

8.2 Earth station operating FDM/FM/FDMA

The block diagram of a complete FDM/FM/FDMA earth station operating at 6/4 GHz is shown in Fig. 8.2.

Voice and data signals arriving at an earth station via the public switched telephone network (PSTN) are usually FDM or TDM assemblies. For FDM, not all signals

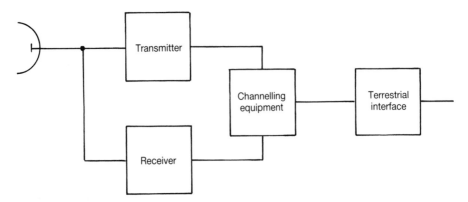

Fig. 8.1 Block diagram of an earth station

8.2 Earth station operating FDM/FM/FDMA   163

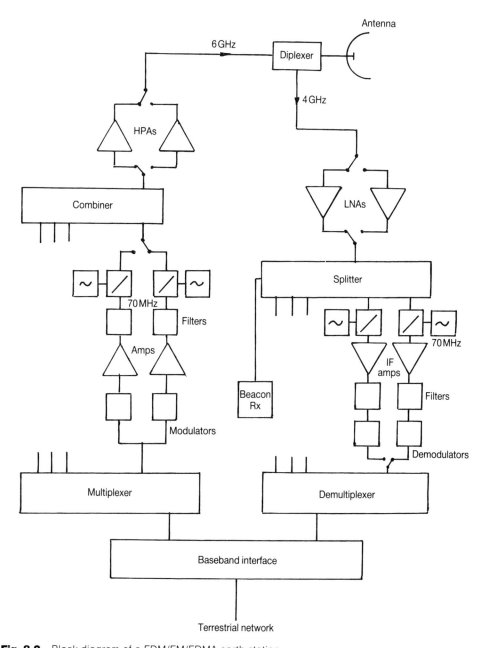

Fig. 8.2 Block diagram of a FDM/FM/FDMA earth station

received are intended for transmission by the earth station and the terrestrial interface may separate the incoming channels back to baseband and reassemble new FDM groups for transmission by satellite. Alternatively it may simply assemble the incoming FDM groups. The station transmits more than one carrier and the incoming groups may be sub-divided in order to be assigned to a particular carrier. Where the incoming traffic is assembled using TDM then the channels must be demultiplexed and reassembled as FDM groups.

If required, the signal may be passed through a pre-emphasis network and an energy dispersal signal added. When FM is used the power transmitted in a given bandwidth varies depending on the modulating signal. With no modulation, the power is radiated at the carrier frequency, whereas with modulation the power is spread over a large bandwidth. As the amplitude of the modulating signal increases the smaller the average transmitted spectral density becomes, i.e. the value in watts/Hz of the bandwidth declines. The spreading of power in order to avoid mutual interference with terrestrial links operating in the same frequency bands is known as energy dispersal. In earth stations the process is achieved by a triangular voltage energy dispersal waveform which produces a uniform spectral density across the RF bandwidth. The energy dispersal signal can be removed at the receive earth station.

The next stage is modulation up to the 70 MHz intermediate frequency (IF). For transmissions using bandwidths up to 36 MHz the IF used is typically 70 MHz with IFs of 140 MHz for bandwidths greater than 36 MHz. Equalizers are included to compensate for variation of the gain and group delay with the frequency. Next, the signal is translated from the IF to the transmission frequency by an up-converter which provides low-level outputs in the 6 GHz frequency band. The 70 MHz channel translates to a separate RF carrier in the transmitted spectrum. In some cases two stages of up-conversion are used, namely 70 MHz and 770 MHz. The use of the 770 MHz second stage IF is to ensure that image frequencies produced by the up-conversion process lie well outside the transmission frequency range and are easily rejected by a transmission bandwidth filter.

The signals are then power amplified up to the required output power levels. The power amplifiers may be narrowband klystrons or wideband travelling wave tube amplifiers (TWTAs). The klystron is normally tuneable over the required band in discrete bandwidth units and, with these devices, each carrier to be transmitted is amplified to the required power level in a separate transmitter and the transmitter outputs are combined prior to the antenna feed connection. Using TWTAs the carriers are combined at low power, as shown in Fig. 8.2, prior to amplification by the wideband power amplifier. Spare high-power amplifiers are normally available as standby units should an operational unit fail.

On the receive side the frequency modulated (FM) carriers received in the 4 GHz range are fed to a low-noise amplifier (LNA). The output from the LNA is split via a branching network into several narrow band carriers, each corresponding to one transponder, and these in turn, are each processed by a down-converter to an IF of 70 MHz. The down converters, while changing the carrier frequency to a lower value that is more suitable for amplification before demodulation, also serve to filter out unwanted carriers from the received signal and allow group delay equalization to be undertaken before demodulation. The output from each down-converter is amplified and then passed to the demodulator which extracts the baseband signal. The baseband is processed for connection to the terrestrial network via the demultiplexing equipment. Included in the baseband processing is the de-emphasis circuit.

The circuit described is for a MCPC/FM/FDMA system. For SCPC/FM/FDMA the elements of the system are broadly similar but SCPC usually has extra functions such as voice-activated carrier switching, companding and automatic frequency control (AFC). The SCPC equipment will have a channel unit which can select one of, say, 1200 carriers spaced at 30 kHz intervals across a 36 MHz IF bandwidth.

On the transmit side the outputs of the channel units are combined prior to IF processing, up-conversion and power amplification. On the receive side the composite signal is passed through the LNA, is down-converted and IF processed before passing to a number of channel units, each of which selects a particular channel for further processing. AFC is required on the receive side where a pilot signal transmitted by a reference station is compared with a stable oscillator frequency. The use of AFC is necessary to compensate for any frequency translation errors that may have occurred in the satellite and the receive side down-conversion.

As well as incorporating companding and voice activation circuits, the channel units could contain echo suppressors and pre/de-emphasis circuits. A basic arrangement has been described in Chapter 3.

For SCPC/PSK the common equipment is similar to that described for SCPC/FM. The voice channel units must include coders/decoders (codecs) with voice switched carriers and synchronizing units that generate the preamble for each speech burst used for carrier recovery and bit timing in the receiver. See Chapter 3.

8.3 Earth station operating TDM/QPSK/TDMA

Using TDMA, the earth station assembles the input signals into a series of bursts which can be interlaced in time with bursts from other earth stations when accessing the satellite. The incoming signal can be organized using FDM or TDM and it is the latter which is assumed here. Figure 8.3 shows a block diagram of a TDM/QPSK/TDMA system.

Voice channels from the International Exchange are passed to the terrestrial interface modules (TIMs) in 30 channel, 2.048 Mbit/s PCM streams. The TIMs convert the input streams from continuous signals to short intermittent sub-bursts at a much higher transmission rate. The efficiency of the digital transmission rate may be improved by the use of digital speech interpolation (DSI) and this can be applied at the TIM (see Chapter 5, Section 5.2). From the TIMs the signals progress to the Common TDMA Terminal Equipment (CTTE) which provides a preamble, providing synchronization and control information, and the sub-bursts into a burst. The CTTE also controls the timing of the burst transmission using a reference burst received from the satellite. The earth station burst modulates a 140 MHz carrier using quadrature phase shift keying (QPSK) with the modulated burst then upconverted to the transmitter frequency (6 GHZ or 14 GHz). The HPAs amplify the burst to a level suitable for transmission.

The receive sub-system passes the low-level modulated RF signal to the low-noise amplifier (LNA) which amplifies it to keep the carrier-to-noise (C/N) ratio at a value which provides an acceptable error rate. The downconverter translates the RF to an IF frequency and passes it to the demodulator for extraction of the wanted information. The demodulator output contains symbols based on the received IF carrier. The probability of symbol error will depend on parameters such as the C/N of the modulated carrier, the satellite channel and the detection scheme used. Because of the probability of errors, the decoder must utilize whatever redundancy was incorporated at the encoding stage to minimize the bit error rate. The probability of bit error rate is discussed in Chapter 6, Section 6.4 while error correction is discussed in Chapter 7, Section 7.6.

8.4 High-power amplifiers (HPAs)

Most of the large earth station HPAs are klystrons or TWTAs. The TWTA is a wide band device (typically 500 MHz at 6 GHz) and can be used to amplify a number of carriers within that bandwidth although if more than one carrier is used the output power of the TWTA must be backed-off (i.e. reduced) in order to prevent intermodulation

166 *Earth stations*

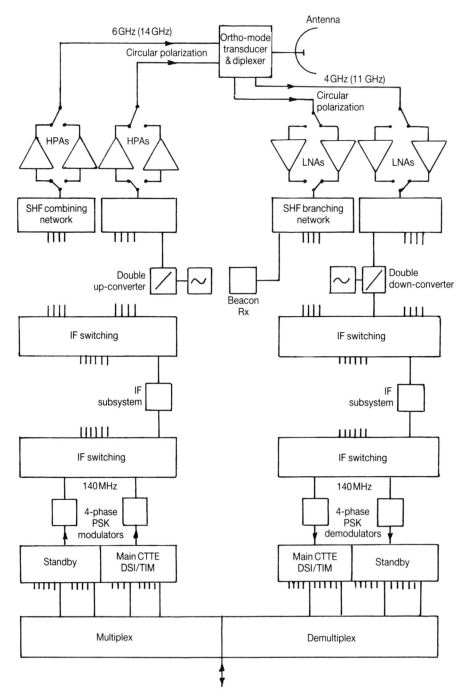

Fig. 8.3 Block diagram of a TDM/QPSK/TDMA earth station

8.4 High-power amplifiers (HPAs)

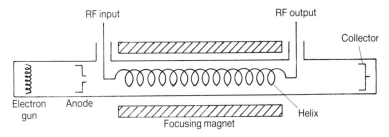

Fig. 8.4 Principle of the TWT

effects. The klystron on the other hand is a narrow-band device (about 40 MHz at 6 GHz) although they may be tuned over a range of 500 MHz. Klystrons are generally cheaper and simpler than TWTAs. The use of TDMA with switching during a TDMA frame between transponders occupying different frequency bands (as in Intelsat and Eutelsat systems) necessitates the use of TWTAs because of the wide bandwidth requirement.

Small earth stations with only a few voice channels could operate with solid-state power amplifiers (SSPAs) for the HPAs utilizing one amplifier for each channel. The advantage of SSPAs is lower operating voltages than TWTAs and klystrons although the output power levels are restricted. The smaller earth station would operate with a smaller antenna so that for a given EIRP the output power would need to be relatively high.

Travelling wave tube amplifier (TWTA)

A basic arrangement of a travelling wave tube (TWT) device is shown in Fig. 8.4.

In the TWT an electron gun fires a tightly focused electron beam along the tube; the focusing on to a collector is achieved by means of concentric magnets. A wire helix, known as the slow-wave structure, has the radio frequency signal entered as an input at one end. A travelling magnetic wave is produced which moves down the tube, interacting with the electron beam. The dc energy of the electron beam is converted into RF energy by velocity modulation of the beam, producing a progressive increase in the strength of the RF signal in the helix. The RF output is taken from the other end of the helix as shown in Fig. 8.4.

Disadvantages of the TWT include its weight, physical size and input power requirement. The major disadvantage, however, is the non linearity of its input/output characteristic. Figure 8.5 gives a typical input/output characteristic for a TWTA showing the saturation effect which occurs at the maximum available power output. The axes are labelled in dB relative to the maximum value at saturation.

When the amplifier has only a single carrier, as in the case of TDMA, it can be handled close to saturation without serious impairment. When the amplifier has to amplify two or more carriers, as for FDMA, SCPC and partial transponder TDMA, the non-linear response near to saturation causes intermodulation distortion. Under these circumstances the amplifier input back-off is increased resulting in an output power back-off from the saturated level to a point on the curve which produces a more acceptable level of non-linearity. In the case of multiple carrier amplification a lower output power is used compared with that shown in Fig. 8.5, due to the power loss caused by intermodulation products falling out of the band and hence not being available as output power.

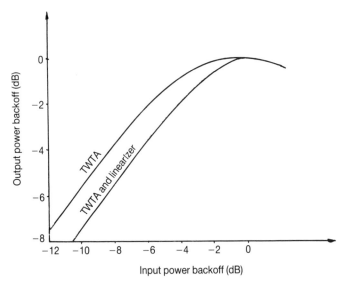

Fig. 8.5 TWT power input/output characteristic

Reduction in input power (input back-off) causes a reduction in output power (output back-off) which, in turn, will affect the overall C/N for a link. Figure 8.6 shows the up-link and down-link thermal carrier-to-noise density ratios plotted against input power back-off. Also included in Fig. 8.6 is the plot of $(C/N)_{IM}$ to show the effects of intermodulation distortion on the overall C/N ratio for a FDM/FM/FDMA link. The curve shows that the effect of $(C/N)_{IM}$ is reduced as back-off is increased. The $(C/N)_u$ value increases linearly while $(C/N)_{dw}$ increases linearly with a slowing down of the rate

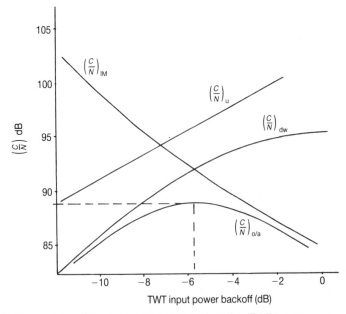

Fig. 8.6 Carrier-to-noise (*C/N*) characteristics plotted against TWT input power

of increase as saturation is approached. The composite effect of $(C/N)_u$ and $(C/N)_{IM}$ is responsible for the well-defined optimum value for the overall carrier-to-noise ratio $(C/N)_{o/a}$.

It is possible to improve the linearity of the TWTA and reduce the amount of output back-off required. The TWTA device exhibits amplitude modulation (AM) non-linearity and this in turn produces frequency and phase modulation (PM) non-linearities, which together produce the intermodulation distortion. Amplitude variations can exist even when a constant amplitude modulation technique is being used. Such variations may be caused by ripples in the passband of filters and, for multiple carrier applications, the effect of many different incoherent carrier frequencies causes variations in the amplitude of the composite envelope. A non-linear network with inverse characteristics to those of the TWTA can result in a more linear TWTA characteristic up to saturation. This curve is also shown in Fig. 8.5. The effect of the linearizer could reduce the amount of back-off needed to restrict intermodulation between multiple carriers by up to 3 dBs.

TWTAs are available with bandwidths of 500 MHz or more and with maximum power outputs of up to 10 kW for earth station use.

8.5 Low-noise amplifiers (LNAs)

The receive side of the earth station is critically affected by the noise temperature of the system. Chapter 2 has shown that the temperature of the receiver stage, antenna and the feeder stage contributes to the G/T factor for the station. The LNA noise temperature is a major contributor to the total system noise and the amplifiers should have low noise temperatures in order to achieve suitable values for G/T.

The two main types of LNA in use are the parametric amplifier (paramp) and gallium arsenide field effect transistor (GaAs FET). Both types may operate at ambient temperature or be thermoelectrically cooled using Peltier-effect diodes; additionally, the paramp may be cryogenically cooled using liquid helium. Typical noise temperatures for LNAs operating at 4 GHz are as follows:

cryogenically cooled paramp	20 K
thermoelectrically cooled paramp	30 – 40 K
uncooled paramp	50 – 60 K
thermoelectrically cooled GaAs FET	50 – 60 K
uncooled GaAs FET	60 – 120 K

The figures for noise temperature at 11 GHz and above are correspondingly higher than those quoted.

The cryogenically cooled paramp has been used since the start of satellite communications but suffers from the need for external equipment and high cost. Latterly, the GaAs FET has found favour because of its relative cheapness, reliability and stability. Later development of the GaAs technology has produced high electron mobility (HEM) devices. These devices, because of the increased electron mobility, have lower thermal generation values and hence lower noise temperature when employed as an LNA stage.

8.6 Antennae

For earth station use, antennae have reflector configurations which produce narrow beams. For transmission, RF energy from a feeder system is fed to the reflector which focuses the radio waves into a narrow beam. For reception, the reflector dish collects RF energy and reflects it as a tightly focused beam on to the feeder which carries the signal to the first amplifying stage of the receiver, which is a low-noise amplifier (LNA).

An important characteristic of the antenna is its effective area, since this will, in part, determine the gain (G) of the antenna. The system noise temperature T_s comprises the noise temperature of the antenna, the LNA and the feeder system (see Chapter 2). The antenna temperature depends on sky noise (which in turn depends on the antenna elevation angle, with noise decreasing with increasing elevation angle) and terrestrial noise. Sky noise is received via the antenna main beam while terrestrial noise is received via the sidelobes. The type of reflector configuration will affect the values of G and T_s achievable and G/T_s is used as a system ratio, which is a critical parameter in the receive mode (see Chapter 2).

The size of an earth station antenna is usually determined by the required value for G/T_s and the noise temperature of the LNA. It may be that the gain of the antenna is specified for the transmit mode, giving rise to a larger antenna than that required to meet the specification for G/T_s. With regard to the Inmarsat system a typical antenna size for a coast earth station (CES) operating at 4/6 GHz would be 13 – 15 m diameter and about 1–2 m for a ship earth station (SES) operating at 1.5/1.6 GHz. Typical antenna diameters for Intelsat services vary according to the Standard Earth Station used. As an example, the Standard A station would use an antenna of 15 – 18 m diameter for international voice, data, TV, IBS and IDR services at 4/6 GHz; the Standard B station operating the same services at the same frequency uses an antenna of 10 – 13 m diameter.

Small earth stations such as those for VSATs and television receive only stations (TVROs) require smaller aperture antennae. The wider beamwidth of such antennae eliminates the need for expensive tracking equipment allowing good reception provided the reduction in G/T_s, caused by the smaller antenna, is either sufficient for the purpose or can be compensated by an increase in satellite transmitter power or a reduction in the RF bandwidth. As a refinement, some TVRO stations can have antennae that track in order to access different satellites. The tracking element is not essential for the satisfactory operation of such stations and the majority operate with only manual adjustment on initial fitting.

For mobile earth stations (MESs), such as those used at sea via the Inmarsat system, some form of tracking is essential. Strict specifications are laid down for the MES transmit EIRP and receive G/T_s which must be met before the station can operate satisfactorily within the system.

Small earth stations can be defined as those where the ratio of antenna diameter (D) to signal wavelength (λ) is less than 60. For a ship earth station operating at 1.5 GHz with a dish of diameter 1 m, the ratio of D/λ is approximately 5. For a TVRO station operating at 11 GHz with a dish diameter of 60 cm, the ratio of D/λ is approximately 55. It follows that where the frequency is increased then, assuming the value of G/T_s is satisfactory, the dish diameter can be reduced.

Some small earth station antennae use symmetrical prime focus fed parabolic reflectors although there is a blockage problem with the position of the feed and generation of side lobes which could cause difficulties where stringent sidelobe envelope specifications must be met. The problem can be overcome by the use of offset fed parabolic reflectors and these are gaining in popularity in modern systems. Cassegrain antennae could be used for systems with $D/\lambda > 50$; the use of the sub-reflector could cause blockage unless the sub-reflector is offset, in which case the blockage is eliminated but at the price of increased complexity of design.

Antenna configurations

Horn antennae   These can be used to provide wide beam coverage directly but more usually they are used as primary radiators, or feeds, in reflector systems to give the high

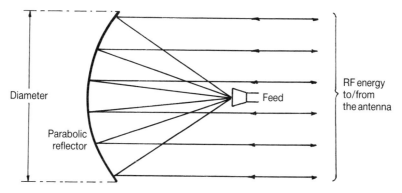

Fig. 8.7 Principle of the reflector antenna

gain/narrow bandwidth required for earth station use. Horns are typically either of rectangular or circular cross-section and may have a flared section which gives a large radiating aperture; this gives improved beamwidth and provides a good impedance match between the waveguide feeder and free space.

Reflector antennae These can be broadly divided into two types; those using a single reflector with horn feed and those using multiple reflectors. In the first category there is the prime focus fed, shown in Fig. 8.7, and the offset-fed parabolic reflectors. The second category contains, among others, the Cassegrain antenna which has a sub-reflector that is hyperbolic in shape and situated within the focus of the main parabolic reflector.

The main advantage of the prime focus fed reflector is its simplicity and low cost. A disadvantage is that, for large antennae, the feeds may be bulky and the distance to the LNA overlong so that the increased attenuation incurred causes a rise in system noise temperature.

The disadvantage of the prime focus feed system restricts its use to antennae of up to about 5 m. diameter. For larger diameter antennae the dual reflector type, such as the Cassegrain, may be used. In this arrangement, the feed is at one focus of a hyperboloid sub-reflector while the other focus of the hyperboloid is coincident with the focus of the main reflector. It can be shown for this arrangement that the use of the secondary reflector causes an increase in the apparent focal length of the antenna compared with

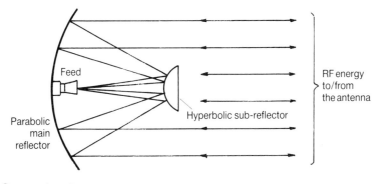

Fig. 8.8 Cassegrain antenna

172 *Earth stations*

Fig. 8.9 Cassegrain Antenna 5 at Goonhilly, UK, used for maritime communications in the Inmarsat system (reproduced courtesy of British Telecommunications plc)

the prime focus feed arrangement. This allows higher aperture efficiencies and lower cross-polarization components. A typical arrangement for a Cassegrain reflector is shown in Fig. 8.8.

For both reflector types described, the antenna efficiency and the magnitude of the sidelobes are affected by aperture blockage due to the primary feed or sub-reflector. There could be an increase in sidelobe levels and a reduction in gain, both causing a degradation in the value of G/T_s.

The Cassegrain antenna with its feed located at the rear of the main reflector and the LNA nearby, requires flexible connections to cross the azimuth and elevation axes. This could be achieved using waveguide rotary joints or flexible waveguide.

An example of an actual installation using a Cassegrain antenna is shown in Fig. 8.9.

Figure 8.9 is the Aerial (Antenna) 5 at Goonhilly UK used for maritime communica-

8.6 *Antennae* 173

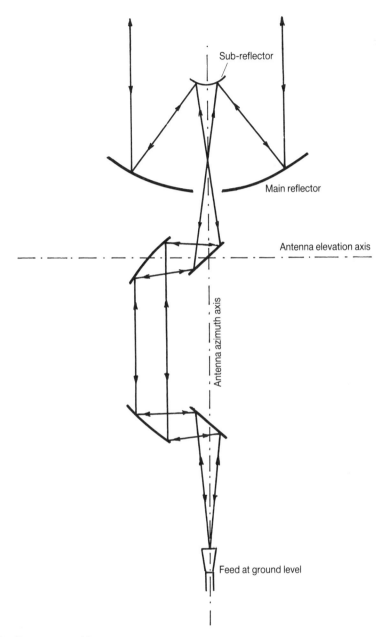

Fig. 8.10 Beam waveguide arrangement

tions in the Inmarsat system, which operates in two frequency bands. Transmission occurs at 6 GHz and 1.6 GHz while reception occurs at 4 GHz and 1.5 GHz. The antenna diameter is 14.2 m and it has a fixed aluminium sub-reflector and a main reflector formed by three rows of aluminium alloy panels with etched surfaces to provide adequate scattering of solar radiation. The reflector backing structure and the

174 Earth stations

support tripod are also constructed from aluminium, which is unpainted so that no maintenance is necessary. The antenna has an elevation over azimuth pedestal with ±135° of azimuth movement, and from 0° to 92° elevation. The drives are from 10 hp SCR (silicon-controlled rectifier) controlled dc motors with two used in each axis in a counter torque configuration in order to prevent backlash. The LNAs (low-noise amplifiers) are accommodated at the rear of the dish in an elevated equipment room which rotates on the azimuth axis. Connections from the LNAs to the moving antenna feed is by flexible waveguide passing through a Teflon weatherproof seal in the wall of the room.

An alternative type of Cassegrain reflector, which has the LNA at ground level and avoids the need for the flexible feed is shown in Fig. 8.10.

This arrangement, known as a beam waveguide, uses a four reflector system with two plane reflectors and two paraboloidal ones. The beam waveguide system has been found to provide a very low-loss path without reducing the antenna gain.

Offset reflector antennae The reflector antennae described in the previous sub-section have the main reflector surface symmetrical about the paraboloid axis; the reflector is thus axisymmetric. The arrangement is simple but suffers from aperture blockage caused by the feed horn or sub-reflector assembly. An antenna with an offset (asymmetric) reflector gives a significant improvement in aperture efficiency and reduces the effect of sidelobes. A disadvantage of the offset reflector is that it is more expensive to produce than the axisymmetric type because of the more complex geometry of the reflector; also, there is a need for a strong mechanical structure to support the offset feed or sub-reflector which could further increase the cost. A basic arrangement of an asymmetric reflector antenna is shown in Fig. 8.11.

The arrangement of Fig. 8.11 shows a asymmetric Cassegrain with a hyperboloid sub-reflector. The axes of the main reflector and the sub-reflector are positioned such that no blockage occurs in the main beam due to the sub-reflector assembly.

Antenna performance

Comparison between types of antenna to assess relative performance can be made in terms of aperture illumination, directivity, sidelobe level, beamwidth and spillover.

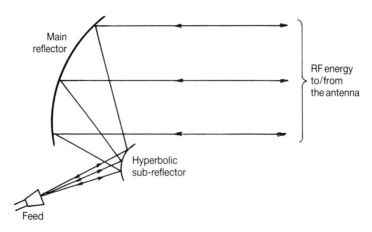

Fig. 8.11 Asymmetric Cassegrain reflector

8.6 Antennae

The gain (G) of an antenna has been shown in Chapter 2, to be:

$$G = \eta 4\pi A/\lambda^2$$

where A = aperture area of the antenna
η = aperture efficiency

The value of η is always less than unity and is often expressed as a percentage. The aperture efficiency is itself composed of many individual efficiencies which together produce the overall value. The efficiency could be expressed as:

$$\eta = \eta_I \eta_B \eta_S \eta_E \eta_O$$

where η_I is the illumination efficiency, which is the result of non-uniform illumination and phase variations in the aperture.

η_B is the blockage efficiency, which allows for the less than complete utilization of the aperture due to blockage by the sub-reflector or feed.

η_S is the spillover efficiency, which allows for a loss of energy due to the sub-reflector and main reflector not collecting all the energy directed towards them.

η_E is surface efficiency and includes variations from uniform phase distribution in the aperture due to reflector profile irregularities.

η_O is ohmic loss due to conducting surfaces in the feed system.

To improve overall efficiency most of the component efficiencies should be increased. However, improvement in one component area may adversely affect another, leading to a compromise. Consider for example the illumination efficiency η_I which could achieve a value of unity if the aperture is illuminated uniformly in both amplitude and phase. This gives good directivity but also a large sidelobe component. The sidelobes can be reduced by tapering the illumination intensity so that it is reduced towards the aperture edge. This has the effect of reducing η_I and widening the width of the main beam. Tapering the illumination does improve spillover efficiency but, since the taper is produced by narrowing the beam of the feed system, which in turn requires a larger feed horn/sub-reflector, the blockage efficiency will decrease. The effects of blockage can be minimized by the use of offset antennae. Most Cassegrain antennae use 'shaped' reflectors which deviate from the true paraboloidal and hyperboloidal shape in order to improve the illumination of the aperture, and hence the efficiency, as well as reducing spillover.

Polarization

A propagated E-M wave is polarized according to the position of the E-field component. The wave may be linearly, circularly or elliptically polarized. Linear polarization is where the E-field is propagated at a constant angle; this angle is usually at 0° or 90° so that the polarization is defined as horizontal or vertical respectively. The H-field is at 90° to the E-field and the plane of the E-M vectors is at right angles to the direction of propagation.

If the E-field vector has a constant amplitude but rotates about the axis of propagation at a constant angular velocity then it will produce a circular locus when projected on a plane normal to the axis of propagation. The wave is then circularly polarized and can be defined as left-hand circularly polarized (LHCP) or right-hand circularly polarized (RHCP) according to the direction of rotation of the E-field vector. A circularly polarized wave is shown in Fig. 8.12.

Cross-polarization defines the use of two polarizations within a system. For a linear polarization system with, say, a horizontal field polarization, the cross-polarization

176 Earth stations

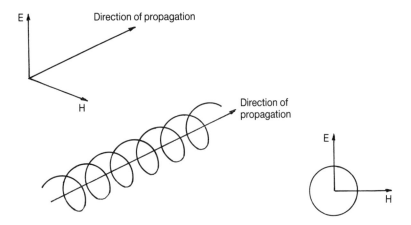

Fig. 8.12 Circularly polarized wave

component would be a vertical field. For circular polarization the LHCP and RHCP are used together. Circular polarization has the advantage that no polarization orientation is required but its performance suffers, due to depolarization, in heavy rain. The antenna must be able to handle both polarizations in isolation. Cassegrain antennae are good in this respect because of their long focal length and axial symmetry. The circularly polarized wave may be considered as the combination of two linearly polarized waves in axis and time quadrature and this phase displacement needs to be maintained over a wide frequency range. The curvature of the antenna could cause variations in the phase relationship causing the polarization to become elliptical. The effect of the ellipticity is to produce a cross-polar component into the co-polar field. Suppose the required polarization is LHCP and suppose the amplitude of the E-field is E_1; a cross-polarization component would be RHCP and could have an E-field amplitude of E_2. An antenna would be affected by the ellipticity and would be able to discriminate against the unwanted polarization according to the ratio:

$$XPD = 20 \log(E_1/E_2)$$

where XPD is cross polarization discrimination.

The ratio between the amplitudes of the E-fields along the major and minor axes of the ellipse is known as the axial ratio and is given by:

$$AR = (E_1 + E_2)/(E_1 - E_2)$$

AR is obviously unity for a circularly-polarized wave.

The effect of a finite value for XPD is to produce a carrier-to-interference ratio that could have an effect on overall performance.

Antenna tracking

With systems operating on a narrow 3 dB beamwidth it is essential that the earth station antenna is pointing at the satellite position within fine limits of accuracy in order to prevent pointing loss. Although placed in a near perfect geostationary orbit, the satellite does move and can drift about its nominal position. This drift is defined (for Intelsat V for example it is within 0.1° N–S and E–W of its nominal position) and can be predicted with accuracy from the earth station. Since the movement of the satellite relative to its nominal position can be comparable with the antenna 3 dB beamwidth, large antennae must be provided with the means of tracking the satellite in order to take account of the satellite drift.

Smaller antennae of just a few metres diameter operate with wider 3 dB beamwidths and the loss of gain caused by satellite drift is not critical. Such antennae can be initially pointed with accuracy and require no further adjustments.

Some of the methods used for large antennae to track the satellite are as follows.

- Monopulse. With this system higher order asymmetric electromagnetic wave modes are generated in the antenna feed when the antenna is slightly off track from the satellite beacon. Extracting these modes at the beacon frequency allows a comparison to be made and error signals to be produced, which are used to activate drive motors which minimize the tracking error. This arrangement uses a beacon frequency and allows automatic tracking without unduly affecting the performance of the communication channel. All satellites have beacons; Intelsat VI for example has three beacons on spot frequencies in the guard band 3.905 – 3.995 GHz and beacons at 11.198 and 11.452 GHz. Monopulse systems are expensive and complicated but are effective for narrow beamwidth large antenna systems.
- Step tracking. With this system the antenna is caused to turn a small pre-determined distance in one direction. If the received signal strength increases the system infers that the movement has been in the correct direction and makes a similar further move. A decrease in signal strength suggests an incorrect move and the antenna is turned in the opposite direction. The system will eventually home in on the correct pointing direction. The use of step tracking may be accomplished in isolation or as a back-up to an automatic system in larger antennae.
- Conical scan. With this system the receive frequency beam is rotated about the axis by a small angle. If the received signal has constant strength, the satellite is on axis whereas if the signal rises and falls within a rotation, there is a pointing error. The antenna can be programmed to track to a new pointing position according to the received signal state.
- Programmed. As mentioned earlier in this section, the satellite drift from its nominal geostationary position is small and since its movement is also slow it can be predicted. The antenna can thus be programmed to track the satellite. Despite high cost, computer control is provided on many large antennae, in addition to manual tracking facilities, to provide the capability should it be needed.

8.7 Monitoring and control

Large earth stations have complex equipment with redundant elements capable of being switched in to service in the event of on-line equipment failure. The process of checking on the essential elements of the system to determine the operating conditions is part of an overall system of monitoring and control which also includes:

- collection of data to provide a record of system status:
- provision of an indication of data, including fault and alarm conditions, to system operators:
- automatic or manual switch over from a failed equipment to standby equipment:
- provision of control data to provide antenna tracking:
- provision of control data to power amplifiers for traffic assignment.

178 Earth stations

The earth station usually has a control centre where the monitoring and control facility is centralized. Facilities also exist for communication to other earth stations within a system network. Monitoring and control systems are complex and expensive but if well designed they can ensure the smooth and efficient operation of an earth station with the minimum likely disruption to the service.

9
Communications satellites

9.1 Introduction

A communications satellite may be broadly sub-divided into two parts:

- the bus, or space platform;
- the communications sub-system, or 'payload'.

The space platform provides the support services, or sub-systems, for the communications payload and must maintain the satellite in its correct orbit. The bus sub-systems in turn can be broadly sub-divided as follows.

- Structure. To support the sub-systems during launch and throughout orbit lifetime.
- Propulsion. For the maintenance of orbital position and initial orbital displacement.
- Attitude control. To keep antennae directed to required locations on earth and solar arrays pointed to the sun.
- Primary power. To provide all satellite power requirements.
- Thermal control. For the maintenance of correct operating temperatures for all sub-systems throughout the lifetime of the satellite.
- Telemetry, Tracking and Command (TT&C). To monitor satellite status, orbital parameters and control satellite operation.

In the hostile environment of space with no apparent gravity, much reduced atmosphere and wide ranging temperature variations, the satellite sub-systems mentioned above must support the communications payload during the launch and programmed life of the satellite. It is the province of the satellite communications systems engineers to design the sub-systems bearing in mind not only the requirements for launch and maintenance for the orbital lifetime, but also the commercial aspects of cost and performance. The design criteria must minimize the mass and power consumption bearing in mind that the mass of the communications payload is only about 15% of the total while the communications sub-system takes about 70% of the total power. Table 9.1 (page 190) indicates a mass and power summary for the Intelsat VI satellite.

These days it is common to base satellite design on a proven bus system and adapt it as necessary for whatever communications sub-system, or systems, it is required to carry. The high cost of development, launch and orbital maintenance must be offset by the commercial gain made by selling the communication capacity to customers either on a lease basis or for simply charging on a time-used basis.

9.2 The support sub-systems

Structure

The structure is designed to give mechanical support for all the sub-systems and also withstand the shocks and stresses of launch and orbital life. Most structures are either

180 *Communications satellites*

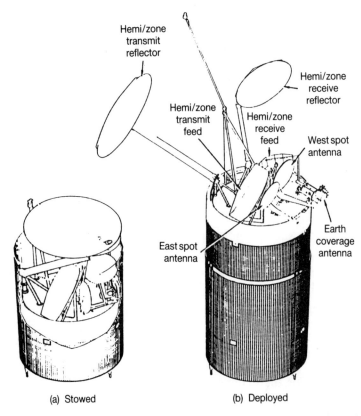

Fig. 9.1 Intelsat VI satellite in its (a) stowed; and (b) deployed condition (reproduced by permission of *COMSAT Technical Review*, CTR, Vol.21, No.1, Spring 1991, p 110)

box-shaped or cylindrical and are designed so as to be able to fit within the launch vehicle. The structures may have elements folded, or stowed, for launch purposes, which may be extended once in orbit. Figure 9.1 shows, as an example, the Intelsat VI satellite in both its stowed and deployed conditions.

Intelsat VI has a cylindrical structure which is an example of a spin-stabilized satellite. Such an arrangement is a dual-spin system where the lower portion of the satellite, which includes the solar arrays, spins while the upper portion, containing the communications payload and antennae, is de-spun by counter-rotation to keep the antennae pointing towards the earth. The box-like structure, as used for example on Inmarsat INM2 and Intelsat V, is a three-axis (or body) stabilized satellite with attitude control about three axes, namely yaw, pitch and roll. Figure 9.2 illustrates the Inmarsat-2 satellite in its deployed mode.

The solar arrays on the 'wings' of the Inmarsat-2 satellite of Fig. 9.2 will have been folded for the launch and opened to its final form for orbital deployment.

Propulsion

The maintenance of correct orbit, attitude control, and possibly the assistance of the satellite into its final orbit, is the function of the propulsion sub-system. The sub-system is often referred to as the reaction control system (RCS).

Initially RCS systems used cold gases such as nitrogen and hydrogen peroxide but these have been replaced by monopropellant or bipropellant systems. The monopropellant hydrazine (N_2H_4), with the assistance of heat or a suitable catalyst, decomposes into nitrogen and ammonia, providing a specific impulse of about 225 s. The specific impulse can be increased by the use of heaters after the decomposition by the catalyst; this increases the exhaust velocity and minimizes the amount of hydrazine needed for a given task. Such an arrangement is known as electrothermal hydrazine thrusters (EHTs). Thrusters are used to position the satellite for attitude control. Intelsat V, for example, has ten working thrusters (and some standby units) with thrust values ranging from 0.3 N to 22 N. There are catalytic and EHT thrusters with the latter used to provide the higher specific impulse needed for N–S stationkeeping.

Latterly, bipropellant systems are increasingly being used, with the most common fuel being monomethyl hydrazine (MMH) with nitrogen tetroxide (N_2O_4) as an oxidizer. The use of these systems allows higher values of specific impulse (typically 290 s) and because of the higher thrusts they can also be used to provide apogee boost. The use of a common system for two purposes can have a significant impact on savings in mass. Intelsat VI uses two 490 N thrusters for apogee boost and six 22 N thrusters for stationkeeping, with enough fuel provided for 10 years of operation.

Attitude control

The main function of attitude control is to keep the communications antenna pointing at the earth and the solar panels towards the sun. A satellite in orbit is affected by disturbances caused by gravitational forces exerted by the sun, moon and planets, solar pressure and changes in the earth's magnetic field. All of these effects can vary as the satellite moves around the earth in its orbit causing nutation (wobbling) of the satellite, which must be mechanically controlled.

The gravitational variations cause orbital changes which alter the inclination of the satellite to the earth's equatorial plane. If unchecked, this inclination would change at a rate of about 0.9° per year and N–S stationkeeping is required to correct for it. Because the earth is not perfectly spherical, the satellite undergoes acceleration towards a stable point of maximum gravitational attraction in the geostationary orbit at longitude 75° E or 105° W. This drift must be corrected at regular intervals in order to keep the satellite on station to within 0.1° of its nominal longitude, as required by the Regulations of the International Telecommunications Union (ITU).

The attitude control system can be seen in block diagram form in Fig. 9.3.

There are two types of attitude control system:

- spin-stabilized in which the body of the satellite containing the solar cells is spun with the spin axis normal to the orbit plane. The communications system and antenna assembly is mounted on the top of the cylinder assembly and is de-spun to keep the antenna pointing earthwards. This arrangement can be seen by reference to Fig. 9.1 which shows the arrangement for the Intelsat VI satellite.
- Three-axis, or body-stabilized in which one or more momentum wheels or three reaction wheels (momentum wheels spin in one direction whereas reaction wheels can spin in both directions) provide the control. A body-stabilized satellite has three orthogonal axes which, in common with the motion of a ship, are labelled roll, yaw and pitch. The axes are shown in Fig. 9.4.

The effect of the momentum wheels on any axis is such that when the wheel is accelerated or decelerated there is a change in angular momentum of the wheel, which is proportional to the applied (motor) torque. This will cause an equal and opposite

182 *Communications satellites*

Inside Inmarsat–2

This cutaway drawing of Inmarsat-2 and its communications payload, by renowned technical artist Tim Hall, provides a unique insight into the workings of a modern spacecraft.

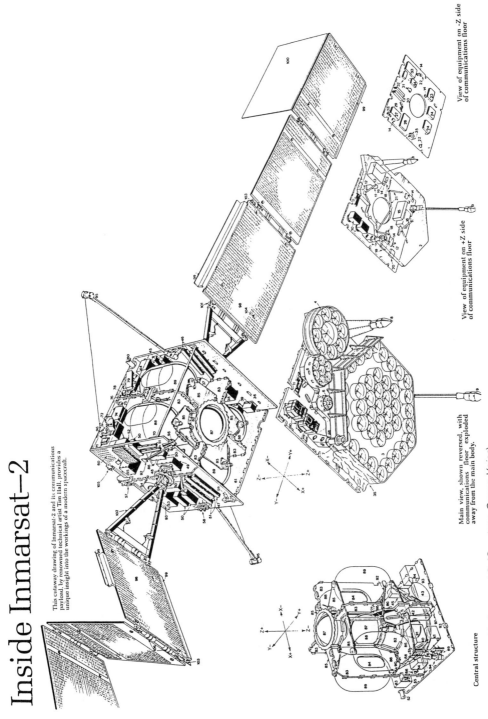

Fig. 9.2 Inside Inmarsat 2. (Courtesy *Ocean Voice*)

9.2 The support sub-systems 183

KEY – INMARSAT-2 SATELLITE

COMMUNICATIONS FLOOR

1. Floor structure of aluminium skins and honeycomb core
2. Antenna farm platform (aluminium skin and honeycomb core)
3. L-band transmit antenna
4. L-band receive antenna
5. C-band transmit antenna
6. C-band receive antenna
7. Tracking, telemetry and command (TTC) fill-in antenna
8. Fixed omni antenna
9. TT & C omni antenna
10. Infra-red 2 axes earth sensors
11. Isolators, LP filter and test coupler
12. Nutation sensor package
13. Payload interface unit
14. Switches
15. Power combiner
16. L-band output band pass filter (BPF)
17. Power monitor
18. Low pass (LP) filters
19. Isolators
20. Sum port monitoring
21. Struts (10) connecting communication floor to antenna farm platform
22. Directional filters
23. L-band receiver
24. Channelization assemblies
25. C-band driver linearizer
26. Upconverter ALC
27. Channel filter
28. C-band output band pass filter (BPF)
29. C-band input filter
30. BPF receivers
31. Signal divider
32. Switch driver
33. Command attenuator (ALC)
34. L-band pass filter (BPF)
35. Multi-layer insulation (MLI)

COMMUNICATIONS PAYLOAD

Y+ WALL
36. Wall structure (aluminium with honeycomb core)
37. Solar array drive mechanism (SADM)
38. Driver amplifier linearizer L-band (FSDAL's)
39. TT & C C-band transponders
40. Battery
41. Central interface unit (CIU)
42. Array switching regulator (ASR)
43. Fixed momentum wheel (FMW)
44. Battery control and interface unit (BCIU)
45. Electronic power conditioners L-band (EPC's)(3)-#4 (1), #5 (2)
46. Travelling wave tubes L-band (TWT's)

Y– WALL
47. Wall structure (aluminium with honeycomb core)
48. SADM
49. PSDAL's
50. EPC's (5)- #1 (2), #2 (2), #3 (1)- (C-band 2 off, L-band 3 off)
51. TWT's (C-band 2 off, L-band 3 off)
52. Battery
53. BCIU
54. FMW
55. Control law electronics (CLE)
56. Thruster module 1A/1B
57. Thruster module 2A/2B
58. Thruster module 3A/3B
59. Thruster module 6A/6B
60. Thermal control mirrors

SPACECRAFT SYSTEMS

X– WALL
61. Wall structure (aluminium with honeycomb core) in sections, upper access panels & lower service module
62. Sun acquisition sensor (SAS)
63. Actuator drive electronics (ADE)
64. Battery discharge regulator (BDR)
65. Pyro safe and arm connectors
66. Battery safety connectors
67. TT & C Connector
68. Power subsystem (PSS) connectors
69. Thruster module 5A/5B

X– WALL
70. Wall structure (aluminium with with honeycomb core) with two sections, upper access panels & lower service module
71. Attitude & orbit control system connector (AOCS)
72. Earth sun sensor (ESS)
73. Gyro
74. BDR's
75. Battery safety connector
76. Clearance hole for relief valve
77. Pyro safe/arm connectors
78. Automatic hold up circuit (AHC)
79. Battery reconditioning unit
80. SAS
81. Thruster module 4A/4B

CENTRAL STRUCTURE AND GENERAL

82. Central structure of carbon fibre reinforced plastic
83. Shearwalls – aluminium with honeycomb core
84. Cutouts for tanks in +X and –X shearwalls
85. Upper tank support panels (aluminium with honeycomb core)
86. Lower tank support panels
87. Pressurant tanks – spherical (2) (Helium)
88. Fuel tanks (2) (Mono methyl hydrazine)
89. Oxidant tanks (2) (Nitrogen tetroxide)
90. Tank support struts (8)
91. Apogee boost motor
92. Struts (6) central structure to +Y and –Y panels
93. Closure panels (2), –Z
94. Z telemetry fill-in horn
95. Stability booms (2)
96. Stability boom hold-down points (2)
97. Electrical umbilical connectors (2)

SOLAR ARRAY

98. Solar panels – single sided panels (array is on a 5 deg eccentric axis to satellite body)
99. Thermal fins
100. Attitude and orbit control subsystem (AOCS) flap
101. Hinges
102. Yoke
103. Solar array sun sensors (SASS)
104. Hold down points on solar panel
105. Hold down points (4) two not shown, on –Y and +Y panels
106. Flexible interpanel harness

184 Communications satellites

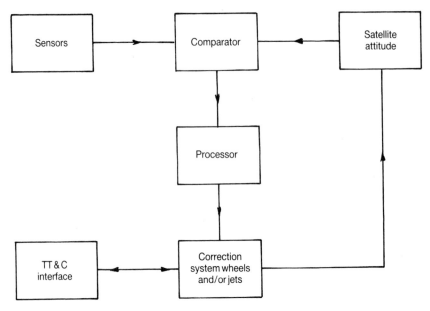

Fig. 9.3 Block diagram of attitude control sub-system

torque on the body of the satellite, causing the satellite to rotate about that particular axis. If the speed of the wheels reaches a limit, thrusters may be used to reduce the wheel speed. Alternatively, reaction wheels may be used instead of thrusters.

The sensors shown in Fig. 9.3 detect the attitude of the satellite and, by comparison with a reference, error signals may be produced which allow corrections to be made. The corrections can be realized by changes in speed of the momentum wheel, by the use of thrusters, or both.

The sensors used have resolution values which will determine the eventual pointing accuracy of the satellite. The sensors are most commonly sun sensors, which have the advantage of requiring no power, or earth sensors which operate in the visible or

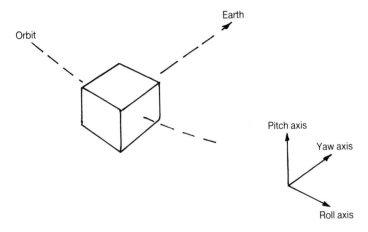

Fig. 9.4 Axes for a body-stabilized satellite

infra-red region of the spectrum. Alternatively, an RF sensor can be utilized which uses the RF beacon frequencies transmitted by an earth station. RF sensors work on the principle that, in a particular plane, two feeds used with a single reflector will produce two different antenna patterns. Only if the beacon signal is received with equal amplitude by both patterns is the boresight axis directed towards the RF signal. Different received signal strengths could be used to produce an error signal to correct the attitude for that plane. For both pitch and roll correction four feeds would be used, two operating in each plane. Inertial sensors (gyroscopes), sun sensors or magnetometers are also possibilities. As an example, Intelsat VI uses a spinning sun sensor and, during an eclipse, one of four spinning earth sensors. A back-up uses the RF beacon system built in to the 6 and 4 GHz antenna systems on the satellite and a beacon signal emanating from one of the ground stations operating with the satellite.

Primary power

For communications satellites, solar power is used to provide primary power. Nuclear power is seen to be advantageous only for interplanetary spacecraft and is not used in commercial satellites because of cost and possible environmental hazards. Solar cells transform solar radiation into electric power but each cell will only provide about 150 mA of current at a voltage of about 500 mV. An array of cells is necessary, connected in series and parallel, in order to achieve the desired operating current and voltage. The average power in the radiation from the sun above the atmosphere of the earth is given by 1.37 kW/m^2. However, the solar cells are relatively inefficient and convert only about 15% of the incident radiation to electric power. This value also decreases with ageing due to bombardment by high-energy particles and damage by micrometeors. Solar cells with coverslides are better able to withstand the high-energy particle bombardment and, at the same time, such coverslides repel infra-red radiation, which does not contribute much to the solar cell output power but which does cause the cell to heat up. The rise in temperature of solar cells also causes a fall in efficiency.

A spin-stabilized satellite such as the one shown in Fig. 9.1 has a solar array made up of two concentric cylindrical panels. The inner, fixed, panel surrounds the heat generating communications equipment and contains a cylindrical band of quartz mirrors, which are the primary means of heat dissipation from the communications equipment. An outer panel is placed over the inner panel during launch and, once in orbit, the outer panel is extended to expose the inner panel to give maximum usage of cells. The arrangement is designed to develop over 2 kW of power for the 10 year planned lifetime of the satellite.

A body-stabilized satellite such as the one shown in Fig. 9.2 has its solar cells on flat panels which can be rotated to intercept maximum solar radiation. Intelsat V satellites have solar 'sails', each of which is 15 feet long, producing greater than 1.2 kW of power. Because the sails have cells which always face the sun, they operate at higher temperature, and hence reduced efficiency, compared with the spin-stabilized satellites where the cells can cool down when in the shadow. The cells also suffer more from high-energy particle bombardment so that thicker coverslides may be necessary for cell protection, with increased mass penalty.

The disadvantage of spin-stabilized satellites is that only about a third of the solar cells are facing the sun at any time and an increase in mass is necessary to obtain the required power level. However, this disadvantage is partly offset, as in Intelsat VI, by a reduction in the mass of the thermal control system and the attitude control system which is generally simpler in such satellites. As already mentioned, Intelsat VI uses concentric cylinders which can increase the surface area after deployment while still

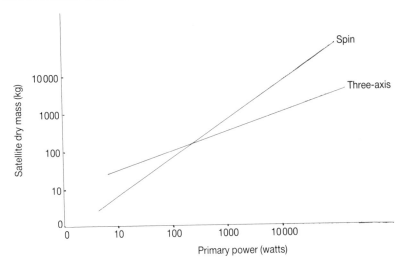

Fig. 9.5 Satellite mass versus primary power curves for spin-stabilized and body-stabilized systems

meeting the stringent launch requirements. Figure 9.5 shows a mass versus power comparison for spin-stabilized and body-stabilized satellites. It can be seen that as newer generations of satellites are built, requiring still more power, the balance may tip towards the use of body-stabilized types. Intelsat VII is of the body-stabilized type structure. The first of the Intelsat VII series (satellite 701) was successfully launched in late 1993.

Geostationary satellites will experience eclipse conditions for a specific number of days per year. An eclipse is caused by the satellite passing through the earth's shadow and, because of the tilt in the earth's axis relative to the geostationary orbital plane, a geostationary satellite will suffer an eclipse for a period around the equinoxes. The geostationary satellite is always in sunlight during the summer and winter solstices but begins to move into shadow 21 days before an equinox, building up to a maximum of 70 minutes of eclipse per day at the equinox, before taking another 21 days to recover to full sunlight. Thus, 84 days per year will see the geostationary satellite suffer an eclipse during which period the solar cells will not supply power and, if necessary, back-up batteries must be used. TV broadcast satellites may be spared the need for battery back-up by locating the satellite to the west of the coverage area where the eclipse will occur after 1 am local time when a shutdown is likely to be acceptable. For voice and data traffic (and in Inmarsat satellites where distress calls may be made) this possibility is not acceptable and batteries must be provided. Early systems used nickel-cadmium (Ni-Cd) cells which have low specific energy (30–40 W h/kg) and needed to be operated with an average depth of discharge of about 40% to maintain life expectancy. Latterly nickel-hydrogen (Ni-H) cells are being used with a higher specific energy (50 W h/kg) operated with a depth of discharge up to 70% without a detrimental effect on life expectancy.

The solar array must be able to supply all normal power requirements and provide a charge for the batteries. A power regulator controls the charging current, and data regarding the solar cells and batteries are supplied to the on-board control system and the control earth station via the telemetry down-link. Typical battery voltages are up to 50 V with capacities in excess of 50 ampere-hours. Intelsat VI for example uses two 44 ampere-hour nickel-hydrogen batteries for eclipse power. Details on Intelsat VI battery charging requirements are shown in Table 9.1 (page 190).

Thermal control

The operating conditions of a satellite in space require careful control of the thermal sub-system in order to maintain thermal equilibrium. Energy is absorbed by the satellite directly from the sun and by re-radiation of the sun's energy from the earth; at the same time heat is generated within the satellite by power dissipation from the electronic equipment, mainly by the power amplifiers. To maintain thermal equilibrium, heat must be removed from the satellite by radiation since in space this is the only mechanism for that purpose. In some cases heat may be generated in an area where radiation of the heat is not practicable and the heat has to be conducted to an area where radiation can occur; this could be achieved by the use of 'heat pipes', where heat generated by a power amplifier could be transferred by fluid convection to a radiator situated at a part of the surface of the satellite that is suitable for radiation of the heat into space. A surface suitable for radiation should have high emissivity ϵ and low absorptivity α since the surface is likely to be in direct sunlight receiving an average value of $1.37\ \mathrm{kW/m^2}$.

During eclipses the thermal balance is altered since the sun's radiation no longer reaches the satellite and the communications capacity could be reduced. To maintain thermal balance, louvres could be used to reduce the radiating surface area and heaters brought into play in order to maintain the correct operating temperature for items like batteries, thrusters and fuel lines.

The two types of satellites used tend to require different solutions for thermal control. A body-stabilized satellite will rotate once daily, relative to the sun, and thus the temperature on the side facing the sun could be much higher than the side facing space. It is these extremes of temperature ($+270\ \mathrm{K}$ to $-270\ \mathrm{K}$) that require thermal control to safeguard items such as batteries, which can only satisfactorily perform over narrow temperature limits, and to keep the fuel from freezing. Because the solar cells of a body-stabilized satellite are carried on the solar sails, the majority of the main body of the satellite can be covered in a thermal blanket which is usually made of synthetic material, in layers, coated with aluminium. Areas which receive little or no solar radiation would not be covered by the thermal blanket and could be used to radiate heat into space. A spin-stabilized satellite will rotate many times per minute and thus the temperature around the periphery of the satellite tends to be evenly spread. On Intelsat VI the dual solar panels are arranged to provide thermal control. Because it houses the communications equipment, the inner panel has solar cells of low absorptivity; additionally, it has a cylindrical band of quartz mirrors used to radiate into space the heat generated by the communications equipment. The outer panel has cells of higher absorptivity allowable because the the panel does not surround heat generating equipment. The lower absorptivity cells also have lower efficiency than the high absorptivity equivalents but this is an acceptable trade-off to achieve the required thermal conditions.

Values of absorptivity to emissivity (α/ϵ) are available for various materials and enable the system designer to choose the most suitable materials for the task of thermal control. Materials with high values of α/ϵ, solar absorbers, include aluminium or aluminium foil (ratio of $2-4$) while a solar reflector may comprise a synthetic material base with a coating of silver or aluminium giving a α/ϵ ratio of the order of 0.05.

Telemetry, Tracking, Command and Ranging (TTC&R)

Once in orbit, the satellite status needs to be monitored and controlled to ensure maximum operational efficiency. The telemetry system is responsible for collecting data on the various satellite sub-systems and returning the data to an earth station where the

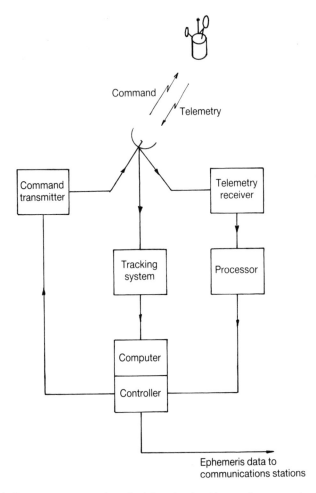

Fig. 9.6 Block diagram representation of a telemetry, tracking and command sub-system

data can be analysed. The command system sends commands to the satellite, possibly as the result of the interpretation of the telemetry data analysis, and for other purposes. Tracking is accomplished by the earth station antenna and ranging is accomplished via a signal transmitted through the command link and returned by the telemetry link. A block diagram representation of a telemetry, tracking and command sub-system is shown in Fig. 9.6.

- Telemetry. This sub-system gathers data from a variety of satellite sensors and sends the information to the ground station by means of a modulated carrier. The type of data collected includes information on the positional state of switches, temperature, pressure, voltages and currents of electronic circuits, outputs of attitude sensors and accelerometers etc. Scientific/experimental data such as radiation levels, magnetic/electric field strengths etc, could also be included. The outputs of the sensors may be analogue or digital but where necessary, analogue data is digitized and the modulation signal is purely digital, with the various signals multiplexed using TDM with PSK as the modulating method. The bit-rates involved in telemetry signals are low,

which allows the use of a small receiver bandwidth at the earth station and thus good C/N ratios, which assist in minimizing errors.
- Tracking. This methodology is used to determine the position of a satellite in space. Angle tracking can be used to determine azimuth and elevation angle from the earth station, using monopulse techniques for example, while ranging uses the time interval taken by a signal to go to the satellite and return. The rate of change of range can be measured either by noting the phase shift of a returning signal compared with the one transmitted or by the use of a pseudorandom code modulation and correlation between transmitted and received signals. Ranging is an important element of a communications system using TDMA but is less important where FDMA is used.
- Command. This sub-system is used during launch to control the firing of the apogee boost motor (ABM), running up momentum wheels, extending solar cells etc. When in orbit the commands are used to control the satellite on-board equipment status such as using thrusters, steering antenna etc. Commands are checked before transmission, to ensure compatibility to known commands, and on reception are stored on the satellite and retransmitted back to the control station via the telemetry link for verification. Once verification is complete an execution signal is sent to the satellite and the command initiated.

The telemetry and command lines are usually operated separately from the communications system although they may use the same frequency band (i.e. 6/4 GHz or 14/11 GHz) utilizing gaps in the communication channel frequencies and using the communications antenna. Alternatively, the TT&C system can use earth-coverage horns operating at specific frequencies in the communications band or at VHF (around 150 MHz) or S-band (around 2 GHz). During launch, however, the antennae are not pointing towards the earth and a back-up system is used via a near omni-directional antenna which allows control of vital satellite functions, such as the ABM, attitude control and orbit control thrusters, until the satellite reaches its operational position. When the required position is reached, the main TT&C system can become operational, although the back-up system can still be used should the main system become defective for any reason.

More information on TT&C can be found in those chapters dealing with operational systems.

9.3 Communications sub-systems

The support sub-system described in Section 9.2 is designed simply to provide a platform in space for the communications sub-system. Table 9.1 shows that the communications sub-system, or payload, represents only a fraction of the total satellite mass although it requires most of the available power. The payload is the *raison d'être* for the satellite and it should produce the revenue for the system operator once the satellite is in operational service.

A communications sub-system is a repeater in space and, in its simplest form, it will perform the same function as a repeater in a terrestrial link, i.e. to amplify a signal before retransmission. A basic arrangement is shown in Fig. 9.7.

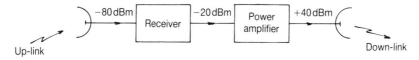

Fig. 9.7 Block diagram of a simplified communications sub-system

Table 9.1 Intelsat VI mass and power summary (reproduced courtesy of British Telecommunications plc)

	kg
Communications antenna	309
Communications repeater	326
Telemetry and command	80
Attitude control	70
Propulsion	120
Power	330
Thermal control	52
Structure	280
Harness	99
Balance	23
Margin	90
Dry spacecraft	1779

	Power (W)		
Subsystem	Solstice	Equinox	Eclipse
Communications	1489	1489	1489
TTC & R	62	62	62
Attitude control	32	32	32
Power electronics	43	43	43
Thermal heaters	31	45	45
Battery charging	34	144	—
Distribution loss	54	58	45
Battery discharge controller	—	—	191
Total	1745	1873	1907
Margin	302	281	—
Panel capability	2047	2154	—
DOD (%)	—	—	64.7

The input signal to the repeater is weak and must be amplified to a suitable level before retransmission without introducing unacceptable levels of noise. Also, the power amplification must be achieved with an acceptable level of distortion introduced by the non-linear nature of the power amplifying device. In the simple circuit of Fig. 9.7 the frequency is assumed to be the same for the receive and transmit sections of the repeater. In practice this would lead to instability and oscillations because of feedback. This can be prevented by translating the frequency of the up-link to a new frequency for the down-link. Frequency bands available for satellite communications have been allocated by international agreement and initially the 6/4 GHz bands were used with 6 GHz for the up-link and 4 GHz for the down-link. The lower frequency was selected for the down-link because it suffers less from attenuation and makes less demands on the satellite power output requirement. Initially, the bands were 500 MHz wide (5.925–6.425 GHz and 3.7–4.2 GHz) but latterly the bandwidth has been increased to 1000 MHz. Full utilization of the increased bandwidth has not easily been accomplished because of the use of part of the band by other services, which could cause interference to the satellite link. Later frequency allocations allowed the use of 14/11 GHz bands and 30/20 GHz. The move to higher frequencies is prompted by congestion in the 6/4 GHz band but is complicated by the fact that higher frequencies suffer more attenuation and the equipment costs are greater. For maritime use the frequency bands 1.6/1.5 GHz (L-band) have been allocated for communication between ship and satellite. Communication between satellite and shore uses the same frequency bands as the fixed satellite service. The use of the L-band frequencies means lower attenuation compared with C-band or Ku-band and allows the ship earth station (SES) to operate with low G/T values comparable with the use of a small antenna. Because of the wider beamwidth associated with L-band antennae, the required tracking of the SES antenna to maintain contact with the satellite is minimized compared with other frequency bands.

Early satellite systems and even some later systems were power limited, i.e. the bandwidth is available for a number of users but, because of the power limitations of the satellite output stage, only a restricted number of users can be accommodated. The other side of the coin is where the output power levels are adequate to support a number of users but the total bandwidth required by those users exceeds the available bandwidth; in this event some users are refused access and the system is bandwidth limited.

Several means are available for increasing bandwidth utilization by bandwidth reuse. This can be achieved by the use of several directional beams at the same frequency (spatial reuse) or orthogonal polarizations at the same frequency (polarization reuse). Several modern systems use these techniques and also use a mix of frequency bands, i.e. 6/4 GHz and 14/11 GHz. Further detail on this aspect can be found in those chapters dealing with particular systems.

The communications repeater may be broadly divided into two categories:

- repeaters which translate the received signal frequency before retransmission without affecting the signal in any other way. Such repeaters are known as transparent;
- repeaters in which on-board processing occurs, in that the received signal is altered in some way before retransmission.

The majority of the contents of this section will refer to transparent repeaters since they form the basis of most existing satellites. Some information will be included, where applicable, on processing repeaters.

Transponders

Transparent repeaters The 500 MHz band available for satellite communication is divided into channels, typically 40 to 80 MHz wide, each of which is handled by a separate repeater known as a transponder (the term originates from aeronautical use where an aircraft, on interrogation by a transmission from the ground, responded with an acknowledgement signal). Each transponder is responsible for a complete signal path through the satellite. A block diagram of a basic system is shown in Fig. 9.8.

The circuit of Fig. 9.8 is referred to as a single conversion stage because it translates from the up-link to the down-link frequency in one step. With this arrangement the

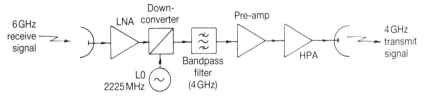

Fig. 9.8 Block diagram of a basic transparent repeater system

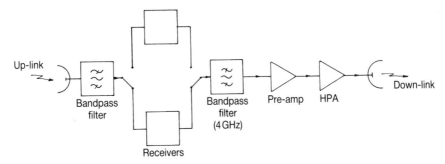

Fig. 9.9 Redundant receiver configuration in a satellite transponder

earlier part of the circuit could be common to all channels within the entire bandwidth; this would comprise a filter (not shown), the low-noise amplifier (LNA) and the down-converter stage. The full bandwidth is then separated into individual transponder channel bands by filters with each channel band amplified by a high power amplifier (HPA). The output of the HPA would then be filtered to eliminate out-of-band signals produced by the non-linearity of that device. The outputs of several channels are then combined in an output multiplexer before feeding to a common transmitting antenna. This type of circuit arrangement is shown in Fig. 9.11. Where a common antenna is used for received and transmitted signals a diplexer is necessary to separate the two signal paths. The use of a diplexer will lead to some leakage of transmitted signal into the receive path but this can be eliminated by the input filter.

Because of the inaccessibility of the satellite from the ground, allowance must be made for equipment failure within the planned lifetime of the satellite. Duplication of active components may be incorporated to allow for system failure. A possible arrangement is shown in Fig. 9.9.

In Fig. 9.9 two-for-one redundancy is shown for the receivers (each receiver consists of the LNA and down-converter) using RF switches, which can be changed by command from the ground. Only the receiver actually routed receives power in order to minimize power requirements and enhance lifetime expectancy. The filters are passive microwave devices with very low failure rates and therefore are not duplicated.

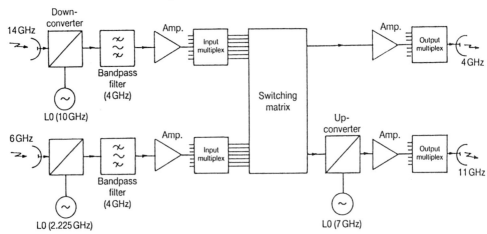

Fig. 9.10 Double conversion technique used in a satellite operating in the 6/4 GHz and the 14/11 GHz bands

9.3 *Communications sub-systems* 193

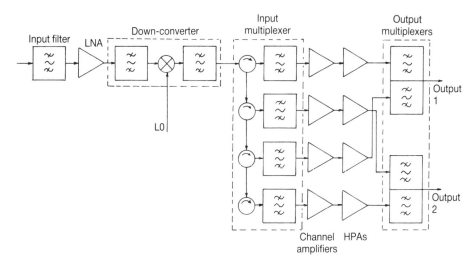

Fig. 9.11 Repeater using input and output multiplexing circuits

Although not shown in Fig. 9.9 redundancy is also likely to be built in for the HPAs using splitter and combiner circuits.

Transponders used in the 14/11 GHz bands are likely to use double conversion as shown in Fig. 9.10.

Figure 9.10 shows a combination of two frequency bands (6/4 GHz and 14/11 GHz) where the frequency conversion of both inputs is to the down-link frequency of 4 GHz. Providing that the channelization and frequency plans of both bands correspond it is possible to use a switching matrix to connect any channel to a down-link frequency of either 4 GHz or 11 GHz. For an output at the latter frequency, a process of up-conversion is necessary following the switching matrix as shown. The advantage of using a common RF of 4 GHz following down-conversion of both input bands is that common circuitry is involved with a simplification of circuit design.

The use of input and output multiplexing circuits is shown in more detail in Fig. 9.11 where a four channel arrangement is considered.

The basic blocks of the repeater of Fig. 9.11 are as follows.

- Input filter. The input filter receives the complete input bandwidth and, where a diplexer is fitted, the filter forms the receive stage of the diplexer. The receive stage is protected from transmit signal leakage. Typically, above 3 GHz a multi-cavity waveguide filter is used where the resonance of the cavities allows the passage of the required band of frequencies through the filter. The characteristics of a filter can vary but the elliptic function filter is popular because of acceptable low passband ripple and good out-of-band rejection. Group delay distortion can be a problem in filters with levels of group delay at the band edges. Group delay depends on the rate of change of phase with frequency and the effect can lead to increased bit error rate with digital signals and phase distortion with an analogue carrier.
- Low-noise amplifier (LNA). The LNA provides amplification, with low noise, of the complete input band and it must possess a very flat passband over the complete frequency range (typically 0.5 dB over 500 MHz) with good linearity to minimize inter-channel modulation effects. Early systems used bipolar transistors and tunnel diodes in the LNA but latterly field-effect transistors (GaAs FETs) have been used.

194 Communications satellites

- Down converter. In this section the incoming signal is translated in frequency to the required output frequency band by mixing with a local oscillator and extracting the difference frequency. The local oscillator frequency may be incremented or decremented in discrete steps to allow matching at various receiver conversion frequencies. After down-conversion the signal is filtered to eliminate oscillator harmonics and higher order mixing products. The down-converter should also possess good linearity to prevent the generation of inter-channel modulation products and its contribution to group delay distortion must be minimized. Typically, at frequencies up to about 20 GHz MIC (microwave integrated circuit) transmission line mixers are used with waveguide devices at frequencies above this value.
- Input multiplexer. This section is basically a collection of filters for specific channels receiving a common input signal and selecting only the channel frequencies required, rejecting all others. The filters require good out-of-band rejection characteristics, and group delay performance, with a very flat in-band passband. The flat group delay response may be realized by extra resonant circuits in the filters or by separate group delay equalizers. The required channel passband may be obtained using quasi-elliptical function filters. Where the design calls for a guard band of one channel bandwidth, filters may use the odd–even method of multiplexing; this calls for two separate paths between the receive antennae and the input multiplexer. Contiguous band multiplexers allow all the channels to be connected to the same antenna. Some systems may use odd–even multiplexers for the input multiplexer and a contiguous type for the output multiplexer. A variety of filter types are available, depending on the frequency used, ranging from surface acoustic wave (SAW) filters, at the lower frequency end, to propagating waveguide filters at higher frequencies.
- Channel amplifiers. These amplifiers are required to provide wideband gain with enough power capability to drive the high power amplifiers that follow. Channel amplifiers above 2 GHz could be a MIC design with input and output isolators to

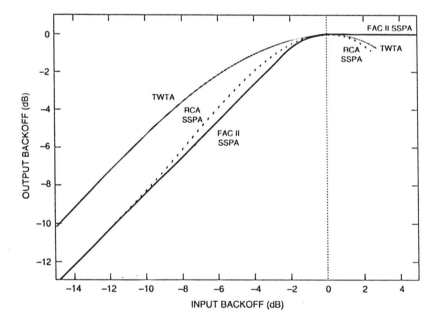

Fig. 9.12 SSPA and TWTA transfer curves (reproduced by permission of *COMSAT Technical Review*, CTR, Vol.20, No.2, Fall 1991, p 302)

9.3 Communications sub-systems

Table 9.2 Maximum phase shift specification with SSPAs (reproduced by permission of *COMSAT Technical Review*, CTR, Vol. 20, No. 2, Fall 1991, p 303)

Relative flux density*	Maximum phase shift		Spec. (deg)
	RCA	FAC II	
0	16.0	13.5	20
3	11.0	6.0	14
6	5.5	3.5	7
9	2.0	2.0	3
12	0.75	0.5	1
14	0.5	0.25	1

* Measured below the flux density that produces single-carrier saturation.

prevent reflective interactions and provide good matching at the input and output ports. Gain and temperature control could be provided.
- High-power amplifiers (HPAs). The majority of applications so far have been with the travelling wave tube amplifier (TWTA), although many communication satellites now use solid-state power amplifiers (SSPAs). Figure 9.12 shows a comparison between the transfer curves of SSPA and TWTA for the Intelsat VI satellites. The SSPA designs were proposed by RCA and Ford Aerospace Corporation (FAC).

Both SSPA curves can be seen to saturate and fall off but the saturation point is not well defined. A reference point was defined by the system engineers for the SSPA which yields an output level nearly equal to the maximum output power without overdriving the amplifier to a point where excessive non-linearity occurs. Table 9.2 shows the SSPA maximum phase shift specification.

Reference to Table 9.2 shows that the two Intelsat VI SSPA designs easily meet the required specification. Comparison with Table 2.1 of Chapter 2 shows that the SSPAs perform significantly better than the TWTA. The results for AM–PM transfer coefficient and C/I specifications are also better using SSPAs than TWTAs. The TWTA is fully described in Chapter 8. The number of TWTAs and/or SSPAs required for a typical communications satellite can be judged by using the Intelsat VI design as an example. At 4 GHz the Intelsat VI satellites incorporate 42 TWTAs with power outputs ranging from 5.5 W to 16 W; the different power levels are required according to which beam coverage the output circuit is connected. At 11 GHz there are 36 devices with power levels from 20 to 40 W. The SSPAs are 1.8 or 3.2 W devices operating at C-band and typically consist of a driver amplifier, a post amplifier, power amplifier stages, a PIN diode attenuator, isolators, a dc/dc converter and bias circuits. Each SSPA uses a pulse width modulation power supply to provide sequential gate and drain bias to the FETs and a gain control voltage to the PIN diode attenuator. Three-for-two redundancy is used for all 4 GHz amplifiers and four-for-two redundancy for the 11 GHz devices.
- Output multiplexer. The output filters are required to reject harmonics, power amplifier generated noise and adjacent channel interference while producing an acceptable pass-band characteristic with low insertion loss. Contiguous band filters allow contiguous channels to be serially multiplexed on a waveguide manifold. Depending on the frequency used, different resonator types are available. Intelsat VI, for example, combines five 72 MHz-wide channels and one 36 MHz-wide channel in the hemi multiplexer with channel filters comprising three physical cylindrical waveguide resonators operating in the dual TE_{111} mode (where each cavity is used twice) achieving an unloaded Q of 10000.

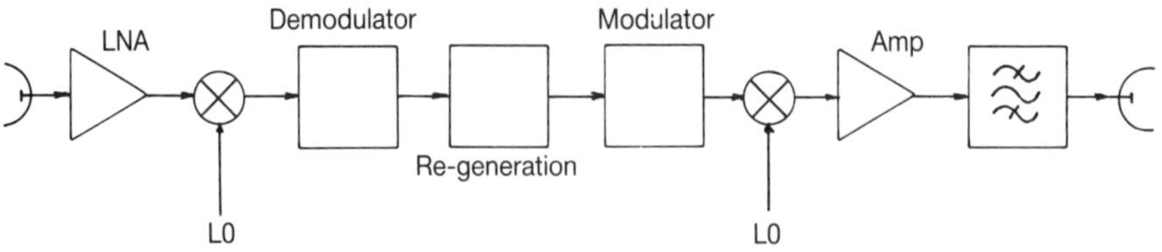

Fig. 9.13 Possible block diagram arrangement of a regenerative repeater

Regenerative repeaters The use of on-board processing can take many forms, including on-board regeneration. The process of regeneration in terrestrial links is used where digital signals are reformed and retimed at regular intervals to maintain the signal characteristics. For a satellite link the regeneration could be carried out in the repeater where the digital signal is recovered by demodulation, restored and retimed before remodulation prior to the down-link. A possible arrangement is shown in Fig. 9.13.

The use of a regenerative repeater allows the total link performance to be measured in terms of the overall bit error rate (BER). Suppose $P_{e(u)}$ is the probability of bit error on the up-link and $P_{e(dw)}$ is the probability of bit error on the down-link. Then the probability of correct reception of the overall link is given by:

$$(1 - P_{e(t)}) = (1 - P_{e(u)})(1 - P_{e(dw)})$$
$$= 1 - (P_{e(u)} + P_{e(dw)}) + P_{e(u)}P_{e(dw)}$$

so that $\quad P_{e(t)} = P_{e(u)} + P_{e(dw)} - P_{e(u)}P_{e(dw)}$

where $P_{e(t)}$ is the error rate of the total link.

If individual error rates are low, the equation reduces to:

$$P_{e(t)} = P_{e(u)} + P_{e(dw)}$$

Thus, if the overall system error rate is to be, say 10^{-5}, then the error rate for each link would be 5×10^{-6} assuming identical rates for each link. In this case, the value of C/N_o required for each link is pretty much the same as the value for the complete link. For a transparent repeater the C/N_o ratio for each link would have to be 3 dB above the total link value.

9.4 Satellite switching

Mention has already been made in this chapter of frequency reuse in order to increase bandwidth availability, and an arrangement using a switch matrix is shown in Fig. 9.10. The concept of switching allows a particular up-beam to be connected to a particular down-beam. Intelsat VI uses static switch matrices (SSMs) and dynamic microwave switch matrices (MSMs). The SSMs allow spot hemi and zone beams to be interconnected in links that are expected to remain in place for a reasonable amount of time. The SSMs are composed of mechanically latching RF switches (C-switches) and are constructed as a $n \times n$ matrix using the C-switches, each of which can be uniquely commanded parallel or crossed to give a very large number of configurations. Redundancy is included to allow the SSM to function correctly despite many switch failures. The dynamic switches are designed with the switched satellite TDMA (SS-TDMA) subsystem. The arrangement provides interconnections among six isolated beams for two independent 72 MHz frequency multiplexed channels at C-band. The dynamic switches allow zone and hemi beams to be interconnected in patterns that can be changed up to 64 times within each 2 ms TDMA frame period. The switching is timed using a time-base generator synchronized with the earth stations. The timing source is contained in a temperature controlled oven and has a frequency correction capability that can be invoked using the telemetry and command system. More information on this type of switching can be found in Chapter 16.

9.5 Satellite antennae

Most satellite antennae are designed to give coverage over a specified restricted area and this causes limitations on antenna gain. Because of the payload restrictions inherent with the launch vehicle, most satellite antennae are small and generally limited to a maximum diameter of about 3 m. For relatively large coverage areas the beam can be produced with a conical horn antenna which, although convenient to use, does exhibit high cross-polar effects which could limit its use in dual-polarization arrangements.

Global beams for Intelsat VI satellites are produced using two simple horns, 0.3 m diameter at 4 GHz and 0.19 m diameter at 6 GHz. The earth edge gain is 17 dBi with a polarization axial ratio of 0.4 dB.

For reduced coverage areas, the smaller beamwidths required preclude the use of horn antennae on their own and reflector antennae are used. The undesirable effects of the use of axisymmetric reflectors, such as gain reduction and aperture blockage, are magnified by the use of small diameter reflectors and offset reflectors are best suited for satellite applications. The requirement to shape the antenna radiation pattern to cover a particular area of interest can be met by:

- shaping the profile of the reflector used with a single feed;
- using a multi-feed array with a conventional offset reflector.

The second of these techniques uses a beam forming network which allows inputs to certain of the feeds to be controlled in order to fix the beam coverage. By varying the amplitude and phase to each feed element the radiation characteristics can be changed to accommodate the required distribution pattern. This type of arrangement is known as a phased array. Such an arrangement will become increasingly important for satellite services to small mobile terminals because of the need for narrow beamwidths and frequency re-use in order to make available the required large number of channels.

On Intelsat VI the antennae for transmit/receive in C-band are parabolic reflectors with offset feed and are designed to provide four co-frequency and co-polarization zone beams with a minimum interbeam spacing of 2.15° with an isolation in excess of 27 dB. The feed consists of an array of 146 dual polarization feed horns with two co-axial connectors per feed. Groups of feeds are excited simultaneously to provide the required beam patterns with sidelobe interference cancelled. The excitation of the feeds for the provision of four zone beams in the Atlantic Ocean Region (AOR) is shown in Fig. 9.14. The actual beam coverage on the surface of the earth is shown in Fig. 9.15.

For the two hemi beams, groups of 79 and 64 feeds are activated.

A similar technique is used on the Eutelsat II satellites.

The use of multiple narrow beamwidth spot beams to cover an area gives rise to increased channel capacity. Provided the multiple beams have minimal overlap, frequency re-use is possible because of spatial and polarization discrimination. For this type of arrangement it is common to use a single reflector with an array of feeds. An alternative to the multi-beam antennae is the use of a single spot beam which can be moved to any part of the total coverage area; interrogation to ascertain user requirements could be by means of a global beam or a second spot beam moving successively across the coverage area.

The system used on the second generation Inmarsat satellites (INM2) is a cup–dipole array. A cup–dipole element comprises a folded dipole, about half a wavelength long,

198 *Communications satellites*

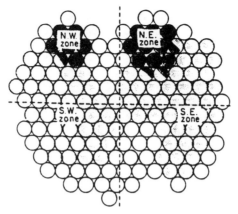

Excitation of 4 zone beams in AOR

Fig. 9.14 Excitation of four zone beams in AOR in Intelsat VI (reproduced courtesy of British Telecommunications plc)

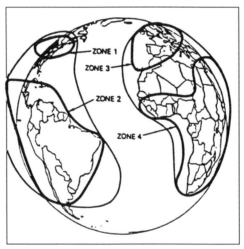

ATLANTIC OCEAN

Fig. 9.15 Intelsat VI Atlantic Ocean coverage (reproduced courtesy of British Telecommunications plc)

enclosed within a reflecting cup structure which has a diameter of several wavelengths. Several cup–dipoles are fixed in a broadside array with a feed through a network of couplers, which function as splitters on the transmit side and combiners on the receive side. The complete antenna farm for an INM2 satellite is shown in Fig. 9.16.

Figure 9.16 shows four sets of antenna arrays, i.e. from shore at 6 GHz, to shore at 4 GHz, to ship at 1.5 GHz and from ship at 1.6 GHz. The transmit antenna to shore and receive antenna from shore are both seven-element broadside arrays while the receive antenna from ship is a nine-element broadside array. The transmit antenna to ship is a 43-element broadside array.

The Inmarsat coverage is global but the use of a broadside array of cup–dipoles allows beam shaping of the coverage to ensure better reception at the edge of coverage.

9.5 Satellite antennae 199

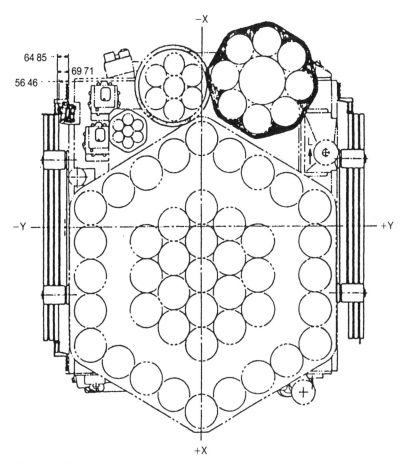

Fig. 9.16 Inmarsat-2 antennae (courtesy Inmarsat)

Satellite G/T

The limit to the gain of satellite antennae combined with the high noise temperatures at their input (caused because satellite antennae 'see' the earth, which has a temperature of about 290 K) may give low values for G/T. A typical value for G/T for ship-to-shore links for Inmarsat first generation satellites is -11.2 dBK, and for second generation satellites it is -12.5 dBK at the edge of coverage. Figures quoted are for the Inmarsat-B voice and data channels.

Typical values for G/T for Intelsat V are $+3$ dB/K for a spot beam transponder operating at 14 GHz and -19 dB/K for a global beam transponder operating at 6 GHz.

Eutelsat II satellites have two dual polarized large reflector multifeed antennae, one used for transmission only (the 'West' antenna) and one for reception and transmission (the 'East' antenna). The 'East' antenna, so called because of its position on the body of the satellite, for reception generates a wide beam giving coverage to all the territories of the Eutelsat member countries. A receive coverage of a Eutelsat satellite in orbital position 10° East is shown in Fig. 9.17.

In Fig. 9.17 the values indicated on the contours are G/T values derived from the antenna iso-gain contours and allowing for a receive system noise temperature of 26.0 dBK (400 K).

200 Communications satellites

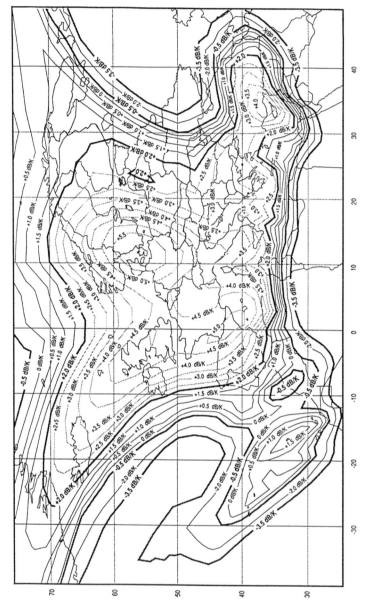

Fig. 9.17 Eutelsat II satellite at 10° East receive coverage (courtesy Eutelsat)

9.5 Satellite antennae

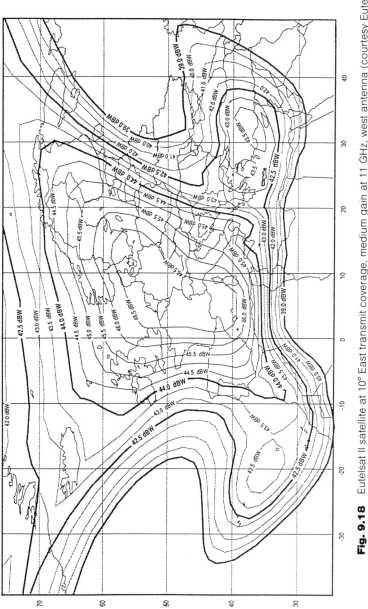

Fig. 9.18 Eutelsat II satellite at 10° East transmit coverage, medium gain at 11 GHz, west antenna (courtesy Eutelsat)

Satellite EIRP

The power output of the satellite transmitter device in saturation could range from about 2 W to 30 W. The figure for EIRP at the edge of the coverage area will depend on the antenna gain. It follows that where necessary the satellite EIRP can be increased by the use of spot beams.

For the Inmarsat system the value of satellite EIRP on the forward link is given as 16.0 dBW while on the return link the value is -8.1 dBW, for first generation satellites, and -3.1 dBW for second generation satellites. Figures quoted are for Inmarsat-B voice and data channels.

For Intelsat VI the minimum value for satellite EIRP at the edge of the coverage area depends on the type of beam used as follows:

hemi and zone, 28 dBW in the 36 MHz channels and 31 dBW in the others;

spot, 44.7 dBW for the 72 and 77 MHz channels and 47.7 dBW for the 150 MHz channels;

global, 23.5 dBW for a specified channel (channel 9) and 26.5 dBW for the remaining channels.

The Eutelsat II 'West' or 'East' antenna can be used for transmission, and a transmit coverage for the 'West' antenna at medium gain at 11 GHz, for the satellite in orbital position 10° East, is shown in Fig. 9.18.

For the transmit coverage the EIRP contours are derived from the antenna iso-gain contours allowing for an input power to the antenna of 15.2 dBW for medium gain coverage (15.4 dBW for high gain coverage). This value of power at the antenna input is a minimum value (end of life, worst path for redundancy).

The transponder output device may be worked at less than saturation in order to limit the effects of device non-linearity. As described earlier, the amount of back-off depends on the type of multiplexing, multiple access method and modulation technique used.

Section Two
COMMUNICATIONS WITH MOBILE EARTH STATIONS

Introduction

Unlike fixed earth station design, the design of mobile earth stations for satellite communications poses unique problems for the project engineer. By definition, a mobile earth station is not stationary and consequently complex satellite aquisition and tracking systems are necessary in order to achieve and maintain the communication link. The physical size of the mobile vehicle must also be considered as this, in turn, will introduce severe constraints on the size of the parabolic antenna and the electronics package. Satellite link budget predictions could well be unattainable if the size of the parabolic antenna is severely restricted. In addition, small mobile vehicles may not be able to provide adequate power to meet transmitter EIRP requirements or even to operate the antenna stabilization and tracking system. This section of the book looks in detail at the International Maritime Satellite Organization (Inmarsat) which provides high quality satellite communications from fixed earth stations via geostationary satellites to ships at sea, passenger aircraft and vehicular mobile earth stations.

10
The Inmarsat Organization

10.1 Introduction

The International Maritime Satellite Organization (Inmarsat) through the establishment and use of several geostationary satellites, covering four ocean regions, provides global communications for mobile users. Whilst Inmarsat was originally established to serve the communication needs of the international maritime industry, the organization's sphere of influence has now been extended to aeronautical and vehicular land based mobile systems. The Inmarsat communications system is the product of extraordinary co-operation between commercial companies, government authorities and telecommunications organizations around the world. These bodies are all part of the Inmarsat system partnership and they include:

- Inmarsat signatories,
- land earth station operators,
- national communications authorities,
- equipment manufacturers, and
- value added service providers.

10.2 The Inmarsat Organization

The Inmarsat organization is a co-operative, commercially orientated enterprise with currently 64 member countries. Inmarsat is headquartered in London, UK. Its prime task is to establish, maintain and operate the satellite system in order to provide global mobile communications.

Inmarsat was created in 1979 in order to provide global communications for the maritime industry. On 1 February 1982 Inmarsat started to provide these services when it took over and expanded the satellite communications system established in 1976 by the US Marisat consortium. The scale of system expansion since 1982 has been breathtaking. Newer and bigger satellites have increased channel capacity by thousands of percentage points. Mobile user terminal fittings of Inmarsat-A have increased from a mere handful, in 1982, to over 16,000 in the period of one decade. A figure which will increase further with the introduction of the new Inmarsat-B system. Inmarsat-C, a system handling data only, has been established, and currently has nearly 3000 mobile terminals accessing the system. It is predicted that there will be some 600,000 Inmarsat-M MESs in use by the year 2005.

In 1992 there were 22 land earth stations (LESs) providing global services on the Inmarsat-A system, and a further 11 are planned for the near future. Eight LESs are providing services in the Inmarsat-C system, with a further 13 planned for the near future. There are ten ground earth stations (GES) providing aeronautical services, with numerous others being planned for the near future. Fig. 10.1 provides information on the status of CES fittings internationally.

COAST EARTH STATIONS

Inmarsat-A

Country	Location	Operator	Coverage Region	Access Code (Octal)	(Decimal)
Australia	Gnangara	IDB Comms Group Inc	IOR	13	11
Australia	Perth	Telstra	IOR/POR	02	02
Brazil	Tangua	EMBRATEL	AORE	14	12
China	Beijing	Beijing Marine Coms & Nav Co	POR/IOR	11	09
Denmark, Finland, Iceland, Norway, Sweden	Eik	Norwegian Telecom	IOR/AORE /AORW	04	04
Egypt	Maadi	National Telecoms Organisation	AORE	03	03
France	Pleumeur Bodou	France Telecom	AORW/AORE	11	09
Germany	Raisting	Fernmeldetechnisches Zentralamt	AORE	15	13
Greece	Thermopylae	OTE SA	IOR	05	05
India	Aarvi	Videsh Sanchar Nigam Ltd	IOR	06	06
Iran	Boumehen	Telecomm Co of Iran	IOR	14	12
Italy	Fucino	Telespazio	AORE	05	05
Japan	Yamaguchi < + >	Kokusai Denshin Denwa	POR/IOR	03	03
Korea, Republic	Kumsan	Korea Telecom Authority	POR	04	04
Netherlands	Burum	PTT Nederland NV	AORE/IOR	12	10
Poland	Psary	Ministry Transp/Marit Economy	AORE/IOR	16	14
Russia	Nakhodka	Morsviazsputnik	POR	12	10
Saudi Arabia	Jeddah	Ministry of PTT	IOR	15	13
Singapore	Sentosa	Singapore Telecom	POR	10	08
Turkey	Anatolia	Comsat Corporation	IOR	01	01
Turkey	Ata	General Directorate of PTT	IOR/AORE	10	08
UK	Goonhilly	BT International	AORW/AORE	02	02
Ukraine	Odessa	Morsviazsputnik	AORE/IOR	07	07
USA	Niles Canyon	IDB Comms Group Inc	POR/AORW	13–1	11–1
USA	Santa Paula	Comsat Corporation	POR	01	01
USA	Southbury	Comsat Corporation	AORW/AORE	01	01
USA	Staten Island	IDB Comms Group Inc	AORE	13–1	11–1

Inmarsat-C

Country	Location	Operator	Coverage	Code
Australia	Perth	Telstra	IOR/POR	302/202
Brazil	Tangua	EMBRATEL	AORE	114
Denmark, Finland, Iceland, Norway, Sweden	Blaavand	Telecom Denmark	AORE	131
Denmark, Finland, Iceland, Norway, Sweden	Eik	Norwegian Telecom	IOR	304
France	Pleumeur Bodou	France Telecom	AORW/AORE	011/111
Germany	Raisting	Fernmeldetechnisches. Zentralamt	AORE	115
Greece	Thermopylae	OTE SA	IOR	305
Netherlands	Burum	PTT Nederland NV	AORE/IOR	112/312
Poland	Psary	Ministry Transp/Marit Economy	AORE	116
Singapore	Sentosa	Singapore Telecom	POR	210
UK	Goonhilly	BT International	AORW/AORE	002/102
USA	Santa Paula	Comsat Corporation	POR	201
USA	Southbury	Comsat Corporation	AORW/AORE	001/101

Fig. 10.1 Inmarsat coast earth stations. (Courtesy *Ocean Voice*)

Inmarsat-M/B

Country	Location	Operator	Coverage	System
Australia	Perth	Telstra	POR/IOR	M/M
Japan	Yamaguchi	Kokusai Denshin Denwa	POR/IOR	M+B/M+B
UK	Goonhilly	British Telecom International	AORW/AORE	M/M
USA	Santa Paula	Comsat Mobile Communications	POR	M+B
USA	Southbury	Comsat Mobile Communications	AORW/AORE	M+B/M+B

Planned

Country	Location	Operator	Coverage	System	Status
Argentina	Balcarce	Comision Nac de Telecom	AORE	A	1993/95
China	Beijing	Beijing Marine Coms & Nav Co	POR/IOR	C	1993
Cuba	N/A	Ministry of Communications	AOR	A	1993/95
Denmark, Finland, Iceland, Norway, Sweden	Eik	Norwegian Telecom	AORW	A	1993
France	Aussaguel	France Telecom	AORW/AORE/IOR	M/B	1993
Germany	Raisting	Fernmeldetechnisches Zentralamt	IOR	A/C/M	1994
Greece	Thermopylae	OTE SA	AORE/IOR	M/B	1995
India	Aarvi	Videsh Sanchar Nigam Ltd	IOR	C	1993
Iran	Boumehen	Telecomm Co of Iran	IOR	C	1993
Italy	Fucino	Telespazio	AORE	C/M	1993/95
Korea, Republic	Kumsan	Korea Telecom Authority	IOR/POR	A/C	1993
Kuwait	Umm-al-Aish	Ministry of Communications	AORE	A	1994
Netherlands	Burum	PTT Nederland NC	AORE/IOR	M/B	1994
Poland	Psary	Ministry Transp/Marit Economy	IOR	C	Planned
Portugal	Lisbon	Comp. Portuguesa Radio Marconi	AORE	C	1993
Russia	Nakhodka	Morsviazsputnik	POR	C	1994
Saudi Arabia	Jeddah	Ministry of PTT	AORE/IOR	A/C/M/B	Planned
Singapore	Sentosa	Singapore Telecom	IOR/POR	C/M/B	1993/94
Spain	Buitrago	Telefonica de Espana SA	AORE	A/C	1993/95
Turkey	Ata	General Directorate of PTT	IOR	C	1993
UK	Goonhilly	BT International	AORW/AORE	B	1993
Ukraine	Odessa	Morsviazsputnik	AORE/IOR	C	1994
USA	Niles Canyon	IDB Comms Group Inc	AORW	M/B	1995
USA	Staten Island	IDB Comms Group Inc	AORE	M/B	1995

Key
AORE Atlantic Ocean Region East
AORW Atlantic Ocean Region West
IOR   Indian Ocean Region
POR   Pacific Ocean Region

Fig. 10.1 (cont.)

Inmarsat is the only provider of mainstream global satellite communication services for mobile users. The telecommunications industry never remains static and Inmarsat is no exception. Additions to the system over the next decade into the new millennium will include some of the following:

- increased channel capacity by the use of Inmarsat-3 (INM3) satellites;
- the expansion of the new systems Inmarsat-B and Inmarsat-M;
- the availablity of position determination and navigation applications using Inmarsat satellites;
- the introduction of the world's first global paging service;
- the introduction of a world-wide pocket-sized telephone service.

The space segment

Whilst it is unlikely that all of the above innovations will be fully available by the year 2000, it is likely that most of them will be in the final stages of development.

Satellites must inevitably form the nucleus of any truly global communications system. As the reader will appreciate, a satellite is an extremely expensive and complex piece of hardware. It is violently hurtled into orbit, and expected to survive in a hostile environment subject to extreme temperature changes and high levels of radiation. However, space technology has progressed rapidly over the past three decades to the point where satellites can be relied upon to operate virtually faultlessly and provide high quality communications throughout their operational lives. The Inmarsat organization bases its earth coverage on a constellation of four prime geostationary satellites covering four ocean regions. Two in geosynchronous orbit over the Atlantic Ocean regions West and East (AORW and AORE respectively), one positioned over the Indian

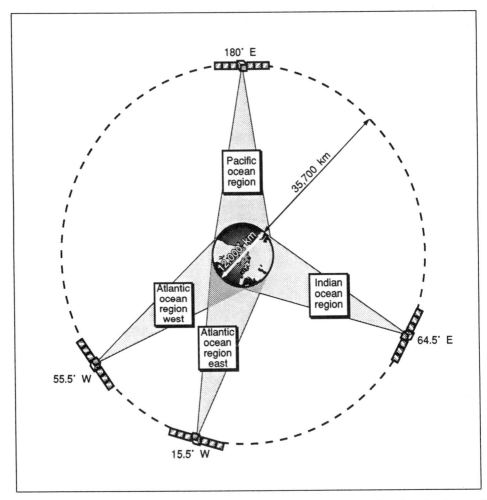

Fig. 10.2 View of the satellites in geostationary orbit above the four ocean regions (courtesy Inmarsat)

10.2 The Inmarsat Organization 209

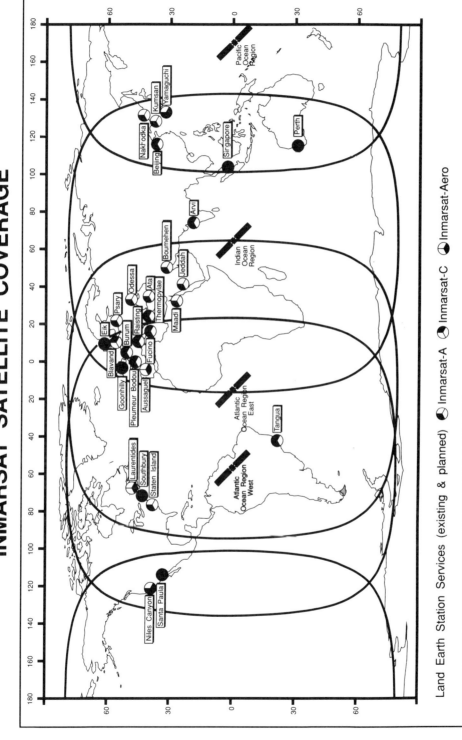

Fig. 10.3 Inmarsat satellite coverage (courtesy Inmarsat)

SATELLITE STATUS				
Ocean Region	Spacecraft	Location	Launch Date	Status
AORW	INM2–F4	55.5W	1992	Operational
	Marecs–B2	58.0W	1984	Spare
AORE	INM2–F2	15.5W	1991	Operational
	Intelsat V–MCS B	18.5W	1983	Spare
	Marisat–F1	106.0W	1976	Spare
IOR	INM2–F1	64.5E	1990	Operational
	Intelsat V–MCS A	66.0E	1982	Spare
	Marisat–F2	72.6E	1976	Spare
POR	INM2–F3	179.0E	1992	Operational
	Intelsat V–MCS D	180.1E	1984	Spare
	Marisat–F3	182.5E	1976	Spare

Fig. 10.4 Satellite status (courtesy *Ocean Voice*)

Ocean region (IOR) and one over the Pacific Ocean region (POR). Several other satellites are maintained in orbit as back-up spares. See Fig. 10.2.

Figure 10.3 shows the footprints projected onto the surface of the earth from the four prime Inmarsat geostationary satellites currently in use. It should be noted that the recommended limit of latitudinal coverage is within the area between 75° north and south. However, as a large percentage of the earth's mobile communication requirements lie within this area, the system is considered to possess a global coverage pattern. The original Inmarsat system was based on a three satellite constellation of AOR, IOR and POR. With the rapid growth of maritime satellite communications traffic in the 1980s and in order to close a small gap in coverage which existed, it was decided that a second satellite would be needed in the Atlantic Ocean region. Consequently, the space segment was expanded to form four ocean regions each with prime satellite coverage. It should be noted that there are considerably more than four Inmarsat operated satellites in orbit. It would be folly indeed to rely upon one operational satellite in each ocean region. In the Summer of 1993 the Inmarsat satellite status was as shown in Fig. 10.4.

In February 1982 Inmarsat took over and expanded the communications system started in 1976 by the US based COMSAT group, Inmarsat's US signatory. The Marisat system was formed by three satellites interlocking three ocean regions. Each satellite was able to provide a mere ten communication channels, a fact which, because of the rapid growth of ship earth station installations, caused queueing to occur, particularly in the AOR. Three Marisat satellites are in fact still in orbit providing back-up facilities as required.

The constant battle to keep the number of available channels ahead of demand led to Inmarsat's decision to lease a 30 channel capacity package, known as the Maritime Communications Subsystem (MCS), on several Intelsat V satellites. These satellites, made by Ford Aerospace, are large capacity vehicles which handle a high volume of the earth's fixed international communications links. In addition, each satellite carries a MCS which is dedicated to mobile communications use and leased from Intelsat by Inmarsat. The Intelsat V-MCS-A satellite was launched in September 1982 with the MCS-B and MCS-C and MCS-D satellites following in 1983 and 1984 respectively.

Capacity was further increased when the 40 channel Marecs satellites were leased in their entirety from the European Space Agency (ESA). These satellites have provided service since the first Marecs A was launched in 1982. Marecs B was lost when the Ariane launch vehicle failed shortly after lift off. The replacement satellite Marecs B2 was successfully launched in 1984.

Because of the unprecedented demand on the system, it soon became evident that, by the end of the 1980s, the demand for channels would outstrip supply. Inmarsat commissioned a number of manufacturers to produce what effectively may be called the second generation satellite system, four of which now form the nucleus of the global system. The second generation of Inmarsat developed satellites, called Inmarsat-2 (INM2) have been designed and built by an international consortium of six companies, headed by British Aerospace, to provide a theoretical 250 channel capacity. The first two, INM2-F1 and INM2-F2 were launched from Cape Canaveral using McDonnell-Douglas Delta-II rockets in October 1990 and March 1991 respectively. INM2-F3 and INM2-F4 were carried into space from the Kourou Space Centre in French Guiana using ESA Ariane rockets, in December 1991 and April 1992.

INM2 satellites incorporate a number of innovations in both vehicle construction and the communications package.

One new idea concerns satellite attitude stability and the way in which it 'sails' on what is euphemistically called the 'solar wind' in order to assist control. The two large solar arrays, which primarily provide all of the power to the electronic systems and subsystems on board the spacecraft, are fitted with extra 'flaps' to provide an increased surface area.

The on-board Attitude and Orbit Control System (AOCS), which is responsible for keeping the spacecraft stable and earth-pointing, gains information from various sensors placed on the body of the spacecraft. This AOCS subsystem then performs calculations for various parameters, resulting in rotational torque control of the solar arrays with respect to each other. The corresponding pressure of the solar wind on these surfaces controls the attitude stability of the spacecraft and maintains the antennae pointing towards the earth. Such control would normally be achieved through the use of several on-board thrusters and the consequent saving of fuel normally needed to perform this function translates into achieving a longer operational life of the spacecraft.

212 *The Inmarsat Organization*

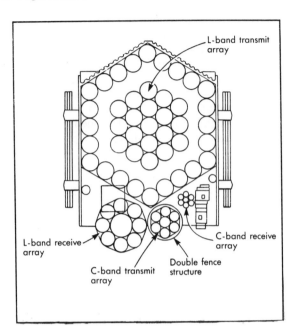

Fig. 10.5 Inmarsat-2 antennae (courtesy *Ocean Voice*)

The most obvious innovation in the communications system is that of the antenna farm. The curious cluster of cups forming each antenna, resembling a fly's eye and known as a cup–dipole array, is a method of reducing the size of the antennae. As an example of this, the large hexagonal cluster of 43 elements comprising the L-band transmit antenna is only 1.7 m in diameter. The antenna, for the Inmarsat-2 satellites are shown in Fig. 10.5.

A conventional parabolic antenna of the same specifications would need to be 30% larger. The antenna is the first direct radiating array on a satellite to use a shaped global beam. The effect is a subtle way to smooth out the power differentials which exist in a conventional beam as the effective received power reduces from the beam centre to its outer edges.

Further innovations in the communications electronics enable the satellite to handle a wide variety of standards of mobile earth stations. Equally radical is the 180 W, L-band high power amplifier, with its four travelling wave tube amplifiers (TWTA), which have been linearized to avoid interference from intermodulation noise and thus improve bandwidth usage. Another first is the use of surface acoustic wave (SAW) filters for channelization, permitting the transponder gain to be set independently for each standard of mobile earth station. The shore-to-ship transponder is divided equally into four sectors. One is devoted to the Inmarsat-A system, while another is used for the new Inmarsat-B service. Unidirectional high speed data occupies a third, while the fourth is dedicated to the expanding Inmarsat-C service plus SAR and aeronautical services. INM2 satellites do, in fact, rotate on their solar panel axis once per day in order that the solar cells continue to face the sun whilst the antenna faces the earth. Power is stored in two nickel cadmium batteries to enable continuous operation whenever the satellite is eclipsed by the earth.

In common with all communications satellites INM2 is basically a signal repeater which is supported by control and environmental 'life support' systems. The satellite

10.2 *The Inmarsat Organization* 213

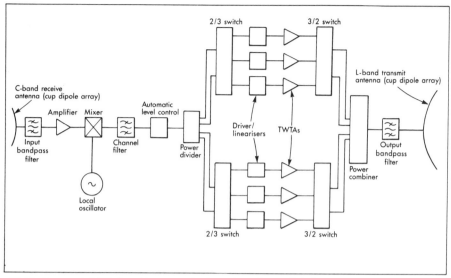

Fig. 10.6 Inmarsat-2 C/L-band transponder (courtesy *Ocean Voice*)

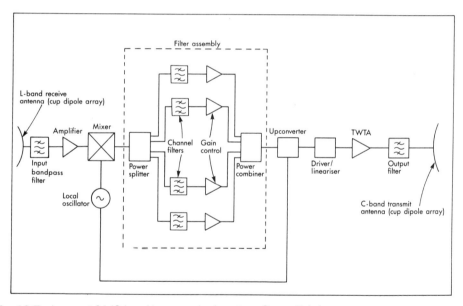

Fig. 10.7 Inmarsat 2 L/C-band transponder (courtesy *Ocean Voice*)

receives information from the land earth station in a directed beam of energy, at approximately 6 GHz in the C-band, converts the information and broadcasts it, at approximately 1.5 GHz in the L-band, in a wide beam over one third of the earth's surface. The link from a mobile earth station (MES) follows the reciprocal path. Figures 10.6 and 10.7 illustrate the C/L and the L/C band transponders which form the nucleus of the system.

The up-link transponder receives information from a MES via nine cup–dipole elements. The minute incoming signal at approximately 1.6 GHz, is carefully bandwidth

214 The Inmarsat Organization

filtered to ensure adequate interference rejection is present before being downconverted to 60 Mhz via a local oscillator and mixer assembly. The intermediate frequency (IF) signal is split four ways using a network of hybrid couplers and the four channels are filtered by surface acoustic wave (SAW) filters. Each channel has adjustable attenuators to allow each channel gain to be varied independently. A further hybrid network recombines the channels. The local oscillator and upconverter now raises the frequency to within the C-band before being applied to the high power amplifiers (HPAs) via an automatic level control circuit. The HPA is a travelling wave tube amplifier (TWTA) which is a non-linear device and, because the composite signal is composed of many FDMA/SCPC carriers, the non-linearity will produce undesirable intermodulation products. These intermodulation products will interfere with the wanted signals and their presence must be suppressed by the use of a linearizer placed before the TWTA.

The TWTA requires low and high voltage power supplies which are provided by an electronic power conditioner (EPC) unit that includes safety devices and regulators in addition to the control interface. Finally, bandpass and harmonic filtering takes place before the C-band signal is transmitted to a LES at approximately 4 GHz. A two-for-one redundancy scheme is used to ensure operational capability in the event of active component failure. Extra switching is provided, in an arrangement known as cross strapping. In this arrangement, cross-connection occurs between the automatic level circuit (ALC) in each of two transponders and the driver stage prior to the HPAs of each of the transponders. Thus, if an ALC and driver both fail, operation can continue regardless of whether the failures are in the same or different transponders. The C/L band transponder operates in a reciprocal way.

The ground segment

Preceding chapters in this book will have made the reader aware that all forms of telecommunications, including communications via satellite, must be rigidly controlled if total chaos is to be avoided. The same applies to the control of satellites in orbit. All satellites, irrespective of their function or nationality, must be rigidly maintained in their

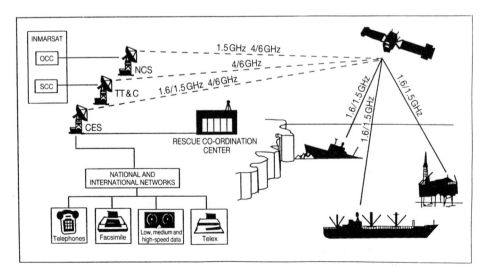

Fig. 10.8 Inmarsat-A ground operations

pre-arranged orbit. To satisfy both of these requirements an extensive ground control network has been established under the ultimate jurisdiction of Inmarsat.

The Inmarsat ground control network is shown in Fig. 10.8. In overall control of the whole network of fixed stations, mobile stations and satellites is the Inmarsat Network Control Centre (NCC), situated in London, UK.

Three Satellite Control Centres (SCC) are responsible for the physical management of the four different types of satellites in use. Marisat and Intelsat V are controlled from the Intelsat and Comsat General SCC in Washington DC. Marec satellites are controlled from the ESA centre in Darmstadt, Germany, whilst the new generation INM2 satellites are controlled from the Inmarsat SCC in London.

Satellite Tracking, Telemetry and Command (TT&C) is obviously of vital importance to ensure efficient satellite management. TT&C for Intelsat V, Marisat and Marecs space vehicles is initiated from their respective SCCs. However, TT&C responsibilities for the new INM2 vehicles has been contracted out by Inmarsat to Telespazio at Fucino, Italy for the AORE and IOR satellites, to the China Satellite Launch & Tracking Control General company in Beijing for the POR craft and to COMSAT in Southbury and Santa Paula, in the USA, for the AORW and back-up services.

In this respect, Inmarsat was again at the forefront of new developments. Although these four TT&C stations are manned by site personnel during normal working hours, the equipment located at the stations is under the direct control of the SCC personnel at the London SCC. All reconfigurations of equipment on site at these stations, on a 24 hour basis, is nominally performed from the London SCC, which also has the ability to repoint remotely the large TT&C antennae.

Each spacecraft is interrogated by an SCC to obtain the following data relating to its physical fitness:

- the operational status of vehicle subsystems;
- satellite orientation in space relative to the sun and the earth;
- a full diagnostic check on all electrical functions;
- the temperature of equipment and surfaces;
- quantity and availability of attitude control fuel.

At first sight this last item may seem to be superfluous. However, the useful life of the satellite is dependent upon its ability to maintain its attitude relative to the earth. The useful life can therefore be equated to the amount of fuel remaining for corrective manoeuvres to be carried out as required. The SCC is able, by the use of telemetry, to maintain a satellite attitude within $\pm 0.1°$.

Whereas the SCC is crucial to space vehicle management, the Network Co-ordination Station (NCS) is crucial to telecommunications services management. A NCS network exists for each Inmarsat service and each ocean region. For the Inmarsat-A service, telecommunications control for the AORE&W is controlled by the Southbury NCS in the USA, and the POR and the IOR by the Yamaguchi NCS in Japan. For the Inmarsat-C service, telecommunications control for the AORE&W is from the Goonhilly NCS in the UK, the POR by the Singapore NCS and for the IOR by the Thermopylae NCS in Greece.

A NCS continuously monitors traffic requests and the flow of telephone and telex traffic through the ocean region satellite for which it is responsible. This service is essential to maintain the correct operation between the mobile and the ground station.

The terrestrial end of the satellite communications link is made via a land earth station (LES) which forms the fixed end of the link. Earth stations currently commissioned by Inmarsat are listed in Fig. 10.1.

A LES is a highly complex and very expensive telecommunications station to construct

216 The Inmarsat Organization

and operate. However, telecommunication services are able to generate large incomes and, as a consequence, new stations are becoming commissioned each year. There is no doubt that the list of LESs, serving the newer Inmarsat systems, will grow rapidly. The functions of a LES are to establish communication channels in response to requests from terrestrial subscribers or mobile stations; to verify and file mobile station identities; to record traffic and process corresponding data and to identify distress priority calls from ships. LES services offered include: automatic calls for telephony and telex, manual operator services, directory enquiries, technical assistance, establish collect and credit-card calls, telegram services, store and forward services, group calls, data communications. A LES is also able to interface to value added services relating to health, navigation and other data.

A full detailed understanding of system operation and the interaction of each component can be gained in Chapter 11, where descriptions of call establishment and procedure are detailed.

An Inmarsat-A land earth station

Land earth stations vary in size and complexity. The Pleumeur Bodou Inmarsat-A/C earth station is shown in Fig. 10.9.

The LES situated at Goonhilly Down in Cornwall, UK, is owned and operated by British Telecommunications (BT), the premier telecommunications provider for the United Kingdom. BT in the UK is an Inmarsat founder member and signatory with an investment share of currently 12.55480%. In addition to offering full Inmarsat-A and C LES facilities, Goonhilly is also the NCS for Inmarsat-C operations for the AORE&W ocean regions and offers the new Inmarsat-B and Inmarsat-M services.

As is the case with many LESs, the satellite earth station at Goonhilly Down provides a huge range of fixed link satellite services and, consequently, the Inmarsat mobile

Fig. 10.9 Pleumeur Bodou Inmarsat-A/C CES (courtesy *Ocean Voice*)

services are but a small part of the overall station. Inmarsat-A (AORW) services are currently based on antenna number 5 (GHY5), a 14 metre Cassegrain dish. A second antenna provides the down-link for the AORE and a spare is provided in order to continue communications during maintenance. A photograph of antenna number 5 is shown in Chapter 8, Fig. 8.8.

The technical side of a typical maritime CES consists of three main features: the antenna assembly, the radiocommunications electronics and the baseband signal processing system.

The parabolic antenna operates in both the L and C frequency bands to and from the satellite. It has previously been stated that the satellite up and down links with a CES occur at approximately 6 GHz in the C-band. However, a CES must also be able to operate in the L-band to enable it to:

- monitor the L-band Common Signalling channel and respond to requests for frequency allocations by the NCS;
- verify signal performance by loop testing between satellite and CES;
- receive the C-to-L band automatic frequency compensation (AFC) carrier.

In order to reduce the frequency errors seen by the SES, AFC is performed by a CES on the shore-to-ship (C-to-L) communications signals. The primary reason for this action is to compensate for the frequency drift which occurs at the satellite and maintain the accuracy of the received frequency at the SES.

The multiple effect of frequency error caused in the satellite transponder and the error caused by the Doppler shift can result in frequency errors at the input to a SES approaching 55 kHz. The CES must include an AFC system whereby the error can be detected and corrected in order that the SES sees the correct L-band frequency from the satellite. To enable corrections to be made, common AFCs are transmitted by selected CESs for each of three sections of each ocean region—Northern, Equatorial and Southern. As an example: for the Northern AOR region the CES at Southbury in the USA transmits the AFC pilot carrier. After monitoring the carrier, each CES is able to correct the C-to-L band frequencies to within +230 Hz relative to nominal and in the L-to-C direction to within +600 Hz.

A typical Inmarsat-A CES antenna would be a Cassegrain dish of approxiamately 14 m diameter. Such an antenna would be able to meet the Inmarsat gain requirements, which are 50.5 dBi and 29.5 dBi respectively. The antenna is designed to withstand high wind speed, typically up to 60 mph in its operational attitude and 120 mph when stowed at 90°. The dish would be steerable ±135° in azimuth and 0° to 90° in elevation. Tracking is either by automatic programme control or operator initiated. A tracking accuracy of 0.01° RMS and a repositioning velocity of 1° per second would be typical parameters for such a parabolic dish. Radiofrequency and baseband processing hardware design varies greatly.

Inmarsat-A RF equipment comprises duplicate C and L-band receive and transmit assemblies. At the level of complexity shown in Fig. 10.10 the RF electronics are fairly straightforward. Intersite links between the RF unit and the baseband units are made at 70 MHz with signals being up or down converted in the RF unit via standard mixing and local oscillator circuits. The 70 MHz to C-band feed for transmission is applied to a 3 kW air-cooled klystron amplifier to drive the antenna. The C-band satellite down-link signal is applied to a low-noise amplifier (LNA) with a 1 dB bandwidth of 500 MHz at 57 K, before being applied to the downconverter.

The L-band transmitter uses duplicated 200 W travelling wave tube amplifiers (TWTA) to provide the low L-band power required for AFC working.

Control and supervisory facilities maintain the equipment in operation. Indications

218   *The Inmarsat Organization*

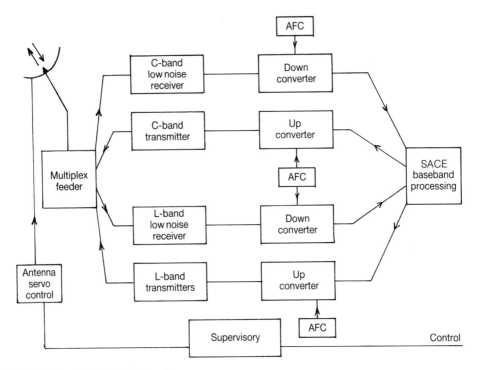

Fig. 10.10 A typical CES RF system

of a path failure of a traffic path and/or equipment are indicated and dealt with automatically. Duplication of equipment is provided on each of the communication chains with automatic path switching. Identified faulty equipment is barred from accepting further traffic and fault reports are generated.

The baseband system shown in Fig. 10.11 is primarily the interface between landlines and the radio equipment. The unit known as the Signalling and Access Control Equipment (SACE) handles all signalling between the CES and mobiles and between the CES and the inland network. Other circuits in the SACE verify the status of mobile users and continuously monitor and route calls. Additionally, the SACE maintains a complete record of each call in order to generate charging information. The SACE consists of:

- a systems control processor based on a mini computer;
- a line control sub-system communications processor;
- a telex switch consisting of a number of micro processors; and
- an electronic four-wire switch which converts all analogue speech signals to digital format before switching and reconverts to analogue after switching.

The systems control processor accepts all incoming call instructions, automatically determines the action required and commands the necessary sub-systems.

The highly complex operation of the CES in response to a request serves to illustrate how the system works. When a ship-to-shore voice call is requested, a message is transmitted by the ship to a specified CES. This would normally be the CES closest to the vessel. The initial message also announces the priority of the call and the type of channel required. The addressed CES receives the call on the Common Signalling

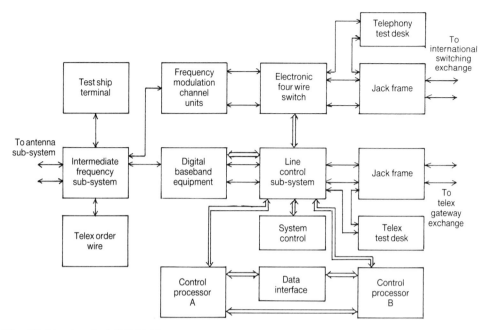

Fig. 10.11 A typical CES SACE system

channel and responds by transmitting a 'request for assignment' message to the NCS. The NCS responds by allocating a satellite voice channel and transmits this information to the ship and the CES. This last transmission occurs in data on the common TDM channel. Both the ship and the CES switch to the allocated duplex voice channel. Once the channel is initialized in this way all further signalling occurs on this channel. The CES SACE now switches the channel through to the national telephone network, of the country of destination, via an international switching centre. The CES continuously monitors the call. At its termination the CES transmits a 'notification of ship clearing' signal to the NCS, which then clears the channel making it available for further use.

LES services

The telecommunication services offered by a LES vary depending upon the complexity of the station selected. As an example of this, a typical LES along with the National Telecommunications Authority could offer a wide range of services from/to the mobile, located in their own ocean region, to any international location. The Inmarsat-A services may include some of the following:

(a) Telephone Distress, urgency, safety and medical assistance calls

- Duplex telephone calls, either automatically or manually connected. Normal telephone calls to business or private subscribers
- Voice messaging. With this service a MES can send, retrieve, reply, redirect and broadcast spoken messages, even out of hours
- National radio paging. The caller is able to page the called party using the National Radiopaging Service of that country.
- Cashless calling. Collect calls or charge cards may be used to pay for telephone calls.

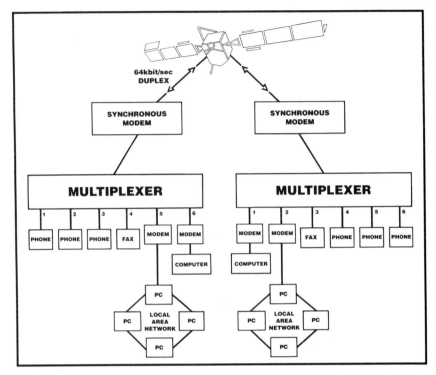

Fig. 10.12 Multiplexing six communications channels onto a single satellite connection using Inmarsat's Duplex HSD service (courtesy *Ocean Voice*)

(b) Facsimile Suitably equipped MESs are able to send fax messages to a business fax number. Additionally, personal or private fax messages may be sent to a LES for forwarding by post.

(c) Telex
- Standard telex point-to-point business connections, where the required telex receive number is not busy
- Store and forward service. The message may be sent to the LES where the message will be stored, processed and forwarded over the telex network. Attempts will be made over the next 24 hour period to deliver the telex message
- Multiple address telex messages. By using the store and forward service it is possible to send the same telex message to a hundred different destinations
- Telex letter and greetings card. Greetings cards designed for special occasions.

(d) Data and information
- Databank. Supplies all the latest weather, financial and sports reports. Also essential maritime information such as navigation warnings and notices to mariners
- Global network services (GNS). GNS offers access to international databases in over 80 countries. GNS gives access to a wide range of subjects including meteorological, geoastrophysical and oceanography. It is also possible to access electronic mail facilities and computing facilities
- Electronic mail. An electronic mail and information management service
- Videotel. This is a Videotext service which is able to provide a vast amount of

information including meteorological, flight and ferry schedules, financial and business information.

Readers should check the documentation provided by LES operators and telecommunication authorities for details of these value added services. It should be noted that enhanced services like those above are not provided by Inmarsat but by an LES operator.

(e) High-speed data services Inmarsat, along with a growing number of LESs, is now able to offer data communications at rates of 56 and 64 kbit/s compared with the much slower standard data rates of 2.4 kbit/s to 9.6 kbit/s. Two services are available, a ship-to-shore simplex demand assigned service and a full duplex 64 kbit/s service. The number of LESs able to offer these services is increasing, although it should be noted that the service is dependent upon an adequate terrestrial link being available from the LES to the destination.

Not all MESs can support the system but some modern equipment may be modified to do so. Multiplexing a number of communications channels onto a single satellite link, and thus greatly improving operational efficiency, becomes a real possibility by using duplex HSD. Fig. 10.12 illustrates the principle.

11
The Inmarsat-A system

11.1 Introduction

The Inmarsat-A system has been the workhorse of marine mobile communications since February 1982 when Inmarsat started to provide satellite communications for the maritime user. Prior to that date a similar service was provided by the US COMSAT organization and was known as Marisat. The number of ship earth station fittings has massively increased since 1976. Demand for satellite channels has increased leading to the launch of newer, larger and more powerful satellites with increased channel capacity. At the same time, electronic engineering has moved ahead to meet consumer demands for better, cheaper and more compact equipment. The net result is that an Inmarsat-A SES of 1994 is but a fraction of the size, weight and cost of that of 1984. The above decks equipment (ADE), including the radome of an SES, has shrunk in size and weight over the last decade to about a third of its original bulk. Owing to high volume sales and mass production techniques, the cost of a 1994 SES has also reduced to a fraction of what it was ten years ago, whilst the facilities it is able to offer have been greatly improved. A decade of development in the field of electronic design is equivalent to several decades in some other disciplines.

Inmarsat specifications provide for three classes of Inmarsat-A SESs as follows.

- **Class 1:** Duplex telegraphy. (Telex)
  Shore-to-ship one-way telegraphy.
  Duplex telephony with and without compandors.
  Shore-to-ship one-way telephony with or without compandors.
- **Class 2:** Duplex telephony with or without compandors.
  Shore-to-ship one-way telephony with or without compandors.
  Shore-to-ship one-way telegraphy.
- **Class 3.** Duplex telegraphy.
  Shore-to-ship one-way telegraphy.

Technical specifications for equipment to be used via Inmarsat satellites are produced by Inmarsat for the guidance of equipment manufacturers. These specifications are extremely complex and are published in the document 'Technical Requirements for Inmarsat-A Standard Ship Earth Stations' (or Inmarsat-B, C or M as appropriate) obtainable from Inmarsat headquarters in London.

SESs must operate, using assigned common frequency channel pairs anywhere between the limits of the two bands: 1535.025 – 1543.475 MHz (down-link, satellite-to-SES) and 1636.525 – 1644.975 MHz (up-link, SES-to-satellite). In order to reduce the possibility of interference between the SES transmitter and receiver stages, duplex communication uses two channels—one in each of the bands—providing a 101.5 MHz spacing between up-link and down-link frequencies. See Fig. 11.1.

There are 339 allocated channels in each band with 25 kHz spacing. Inmarsat channels

11.1 Introduction 223

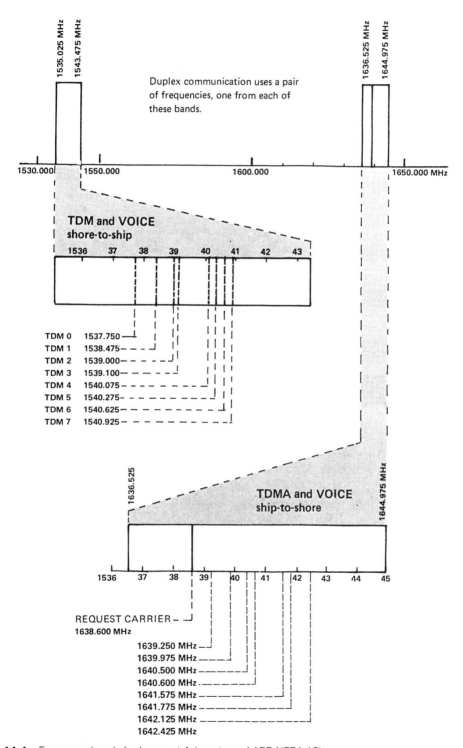

Fig. 11.1 Frequency bands for Inmarsat-A (courtesy of ABB NERA AS)

224 *The Inmarsat-A system*

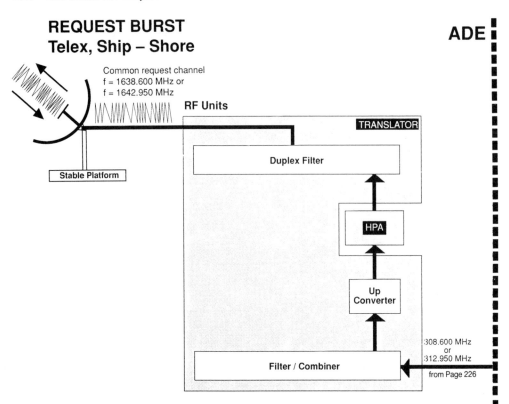

Fig. 11.2 Request Burst Format, ship-to-shore (courtesy ABB NERA AS)

11.1 Introduction 225

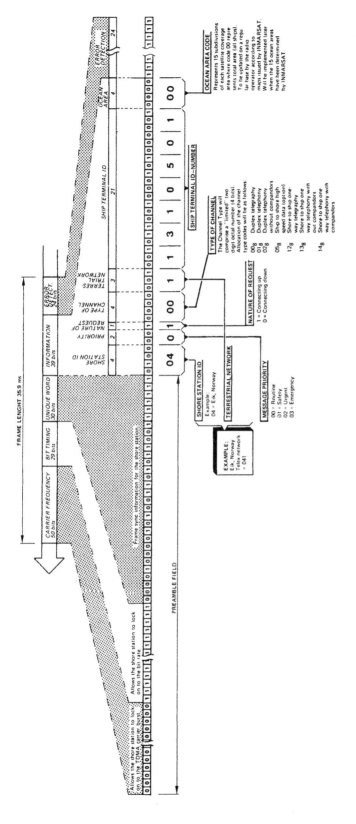

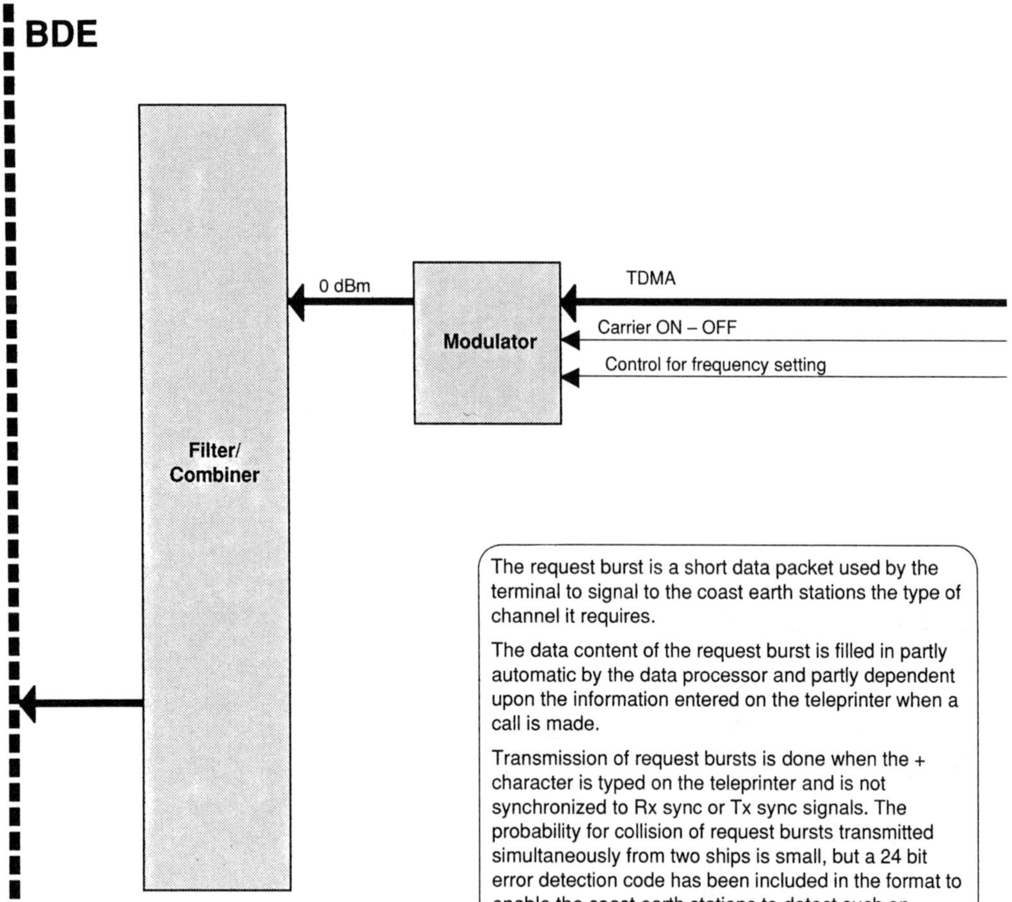

226 *The Inmarsat-A system*

BDE

Filter/Combiner ← 0 dBm ← **Modulator** ← TDMA
Carrier ON – OFF
Control for frequency setting

> The request burst is a short data packet used by the terminal to signal to the coast earth stations the type of channel it requires.
>
> The data content of the request burst is filled in partly automatic by the data processor and partly dependent upon the information entered on the teleprinter when a call is made.
>
> Transmission of request bursts is done when the + character is typed on the teleprinter and is not synchronized to Rx sync or Tx sync signals. The probability for collision of request bursts transmitted simultaneously from two ships is small, but a 24 bit error detection code has been included in the format to enable the coast earth stations to detect such an occurrence.
>
> To protect the request channel from being blocked from interference two frequencies are used and every second request is transmitted on each frequency automatically. Also, to limit the number of request bursts transmitted the terminal will not allow repeated requests more often than one every 10 seconds. Request transmissions is also blocked if the terminal is not receiving the common TDM carries properly.

Fig. 11.2 (cont.)

11.1 Introduction

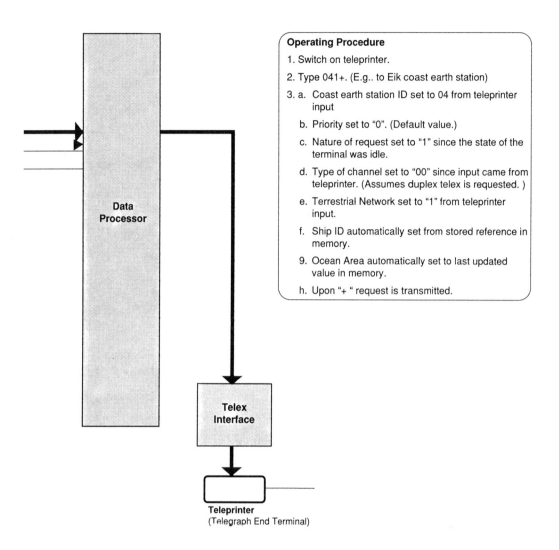

Operating Procedure
1. Switch on teleprinter.
2. Type 041+. (E.g.. to Eik coast earth station)
3. a. Coast earth station ID set to 04 from teleprinter input
   b. Priority set to "0". (Default value.)
   c. Nature of request set to "1" since the state of the terminal was idle.
   d. Type of channel set to "00" since input came from teleprinter. (Assumes duplex telex is requested.)
   e. Terrestrial Network set to "1" from teleprinter input.
   f. Ship ID automatically set from stored reference in memory.
9. Ocean Area automatically set to last updated value in memory.
   h. Upon "+ " request is transmitted.

REQUEST BURST, SHIP – SHORE TELEX
Saturn 3S.90
System Description

228   The Inmarsat-A system

are numbered decimally from 001 to 339 or octally from 001 to 523. Octal channels may carry the notation 'Q'. Two channels are reserved for ship-to-shore requests—ch.124Q and ch.402Q—both of which are monitored by both the CES and the NCS. All assignment messages are transmitted on channels 156Q or 213Q, one of which an SES monitors when in the idle state.

11.2   Details of the Inmarsat-A system

Access control and signalling system

There are two basic types of signalling in the Inmarsat-A system. Out-of-band signalling which is used to set up and control the various types of communication channels via the satellite, and in-band signalling is used for all supervisory and selection signalling once the satellite channel has been assigned.

The out-of-band signalling to the SES is contained in the 'signalling channels' of TDM carriers transmitted by the NCS and the CESs. There is at least one TDM carrier continuously transmitted by each CES. Additionally, there are two CSC carriers continuously transmitted on the selected idle monitoring frequency by the NCS. An SES remains tuned to the appropriate CSC carrier (TDM0 or TDM1) at all times even when the SES is in the idle state. The CSCs are designated TDM0 channel 110 (156Q) and TDM1 channel 139 (213Q) and are modulated with a rectangular BPSK data waveform at 1.2 kbit/s.

A SES transmits a request carrier burst in order to establish a communications link, see Fig. 11.2. This carrier burst is used to obtain the allocation of a communications channel. A SES when initiating calls must use one of the two common request channels alternately for each call, the frequencies of which are; 1638.600 MHz (124Q) and 1642.950 MHz (402Q). Modulation is Coherent BPSK with a data rate of 4.8 kbit/s. NRZ coding format is used to reduce bandwidth.

The EIRP of a single carrier from ship to satellite is 36 dBW and remains constant.

The signalling formats used in the Inmarsat-A system for distress, urgency and safety alerts and traffic are detailed in *Understanding GMDSS*. The following section deals with system signalling for channel requests, duplex telephone and telex (telegraphy) operations.

Communications channels

When a SES is communicating using Telex, it will be tuned to a frequency pair assigned by the correspondent CES. When making a telephone link call the SES will be tuned to a telephone frequency pair assigned by the NCS. Telephone paired carrier frequencies are not associated with a particular CES but are shared as part of a pool and allocated on a demand assignment basis.

Telegraph (Telex) traffic is transmitted in TDMA format. Each TDMA carrier frequency provides up to 22 individual telex channels. There is one TDMA channel frequency associated with each TDM channel generated by a CES, at the TDM frequency plus 101.5 MHz.

- Modulation and data rates are the same as those for the request carrier.
- A frame length is 1.74 seconds. The information frame is 72 bits, 12 characters of 6 bits each.
- The character code used is the ITA2 International Telegraph Alphabet No.2.

Figure 11.3 clearly shows the TDMA structure.

11.2 Details of the Inmarsat-A system

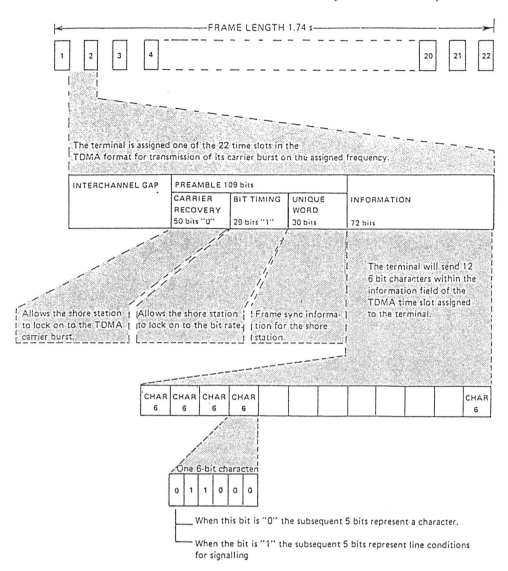

The least significant bit which is here
shown to the left is transmitted first.

Fig. 11.3 TDMA Format, ship-to-shore (courtesy ABB NERA AS)

Telephone channel characteristics:

- Modulation: FM.
- Peak frequency deviation: a channel designated 01, with compandors, is used for speech. The channel with a nominal baseband level of 0 dBm0 produces a peak deviation of 12 kHz in the absence of peak clipping.

A channel designated 02, the customized uncompanded channel, is used for data and facsimile. The compressor is replaced by an amplifier with a nominal gain of 6.5 dB. A nominal baseband level of -6.5 dBm0 will produce a peak deviation of 12 kHz without peak clipping.

- Baseband signal: limited to 300 – 3000 Hz.
- Companding: 2:1 syllabic.
- Emphasis: none.

TDM channel characteristics:

- Data rate: 1.2 kbit/s, ± 1 part in 10^4.
- Frame length: 348 bits (0.29 s)
- Synchronization: 20-bit unique word with a complementary unique word transmitted every sixth frame.
- Modulation: unfiltered constant-envelope rectangular BPSK.

Channel type

Channel numbers are not alterable by the operator and are for information only.

The channel field comprises a two-digit octal number encoded as 4 bits of data. The code identifies the channel type to be used for the communications method requested.

Octal N_8	Channel type
00	Duplex telegraphy. (Telex)
01	Duplex telephony
02	Customized uncompanded channel for data and fax.
03 & 04	Spare
05	Ship-to-shore high-speed data
06 through 11	Spare
12	Shore-to-ship one-way telegraphy
13	Shore-to-ship one-way telephony without compandors. (Slow data or fax)
14	Shore-to-ship one way telephony with compandors. (Voice broadcast)
15 through 17	Spare

RF channels

The SES must be capable of tuning automatically to any one of the 339 (523 octal, suffix Q) paired transmit/receive frequency channels in 25 kHz incremental steps anywhere between the limits of the two bands: 1535.025 – 1543.475 MHz (down-link – satellite to SES) and 1636.525 – 1644.975 MHz (up-link). In the idle state an SES monitors either of the common signalling channels TDM0, channel 110 (156Q), or TDM1, channel 139 (213Q), depending upon how the equipment has been configured, to listen for requests from the NCS. A sample of channels used on the down-link is shown below:

11.2 Details of the Inmarsat-A system

Radio frequency (receive)	N_{10} Decimal	N_8 Octal
1535.025 MHz	001	001
1535.050 MHz	002	002
..	..	..
1537.750 MHz (CSC TDM0)	110	156
..	..	..
1538.475 Mhz (CSC TDM1)	139	213
..	..	..
1543.475 MHz	339	523

To request a communications channel the SES must use alternately one of the two up-link CSC frequencies 1638.600 MHz or 1642.950 MHz, allocated within the transmit band. The SES transmit RF channel frequency is always 101.5 MHz above the receive channel allocated, thus channel pairing is strictly controlled.

Radio frequency (transmit)	N_{10} Decimal	N_8 Octal
1636.525 MHz	001	001
1636.550 MHz	002	002
..	..	..
1638.600 MHz (Common RQ ch.)	084	124
..	..	..
1642.950 MHz (Common RQ ch.)	258	402
1644.975 MHz	339	523

When not actually being used for communication purposes the equipment is left in the idle state, whereby data is being received from a satellite in order to monitor one of the common signalling channels (TDM0 or TDM1) for assignments and to maintain satellite lock.

SES identification

After successful completion of commissioning tests each SES is assigned, by Inmarsat, a unique 21-bit encoded Inmarsat identification number (IMN). The 21-bit code represents a seven digit octal notation number.

$$\text{SES ID} = N_1 N_2 N_3 N_4 N_5 N_6 N_7$$

The first digit of the IMN number identifies the Inmarsat system. A number 1 identifies an Inmarsat-A approved terminal, a 3 denotes an Inmarsat-B terminal, a 4 identifies an Inmarsat-C terminal and a 6 an Inmarsat-M terminal.

Message identification

These numbers are not alterable by the operator and are for information only.
The 6-bit message field comprises two-digit octal numbers which inform the SES of the state of system accessability.

232   The Inmarsat-A system

IDLE STATE

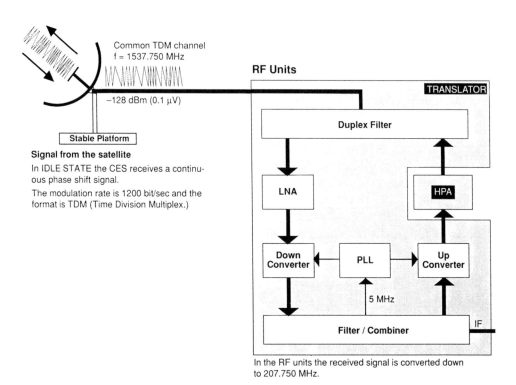

Signal from the satellite

In IDLE STATE the CES receives a continuous phase shift signal.

The modulation rate is 1200 bit/sec and the format is TDM (Time Division Multiplex.)

In the RF units the received signal is converted down to 207.750 MHz.

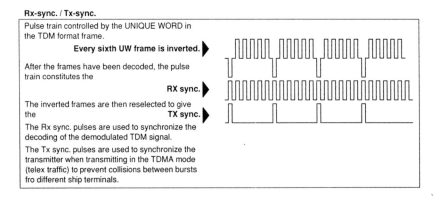

Rx-sync. / Tx-sync.

Pulse train controlled by the UNIQUE WORD in the TDM format frame.

Every sixth UW frame is inverted.

After the frames have been decoded, the pulse train constitutes the

RX sync.

The inverted frames are then reselected to give the **TX sync.**

The Rx sync. pulses are used to synchronize the decoding of the demodulated TDM signal.

The Tx sync. pulses are used to synchronize the transmitter when transmitting in the TDMA mode (telex traffic) to prevent collisions between bursts fro different ship terminals.

Fig. 11.4   Saturn 3S.90 Inmarsat-A SES. Idle State Signalling operation (courtesy ABB NERA AS)

11.2 Details of the Inmarsat-A system

> When the terminal is in idle state the SATURN 3S.90 will continuously track the satellite while monitoring the assignment field of the common TDM carrier for possible assignments addressed to it.
>
> If a match is detected between the terminal's stored address and the address of a particular assignment message the terminal will respond according to the message. It will retune telegraphy or telephony receivers and start the appropriate signalling sequences. After a particular call has been cleared the terminal reverts to the idle mode and continue monitoring the common TDM carrier for new assignments.
>
> The format of the TDM carrier is shown below.

ADE

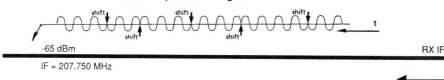

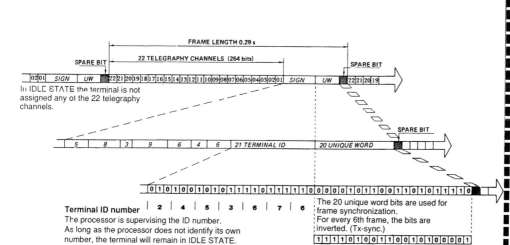

In IDLE STATE the terminal is not assigned any of the 22 telegraphy channels.

Terminal ID number
The processor is supervising the ID number. As long as the processor does not identify its own number, the terminal will remain in IDLE STATE.

The 20 unique word bits are used for frame synchronization. For every 6th frame, the bits are inverted. (Tx-sync.)

234 *The Inmarsat-A system*

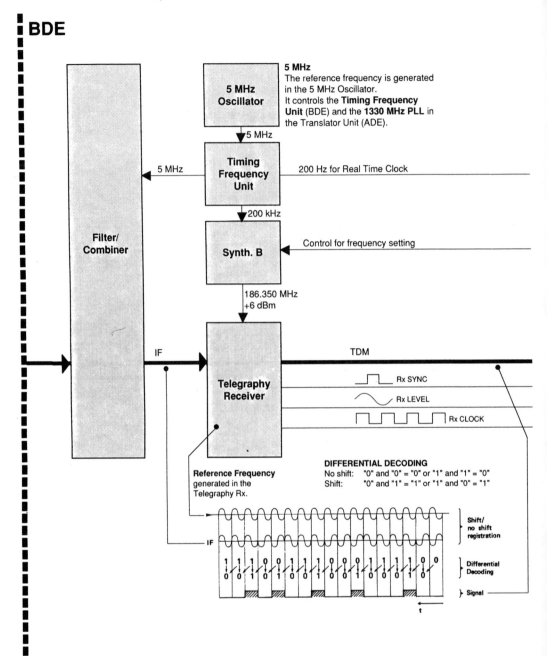

Fig. 11.4 (cont.)

11.2 Details of the Inmarsat-A system

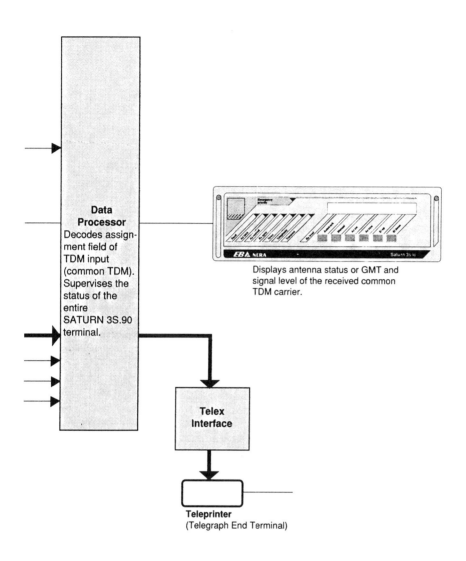

IDLE STATE
Saturn 3S.90
System Description

Octal N$_8$	Message
00	Spare
01	Spare
02	Request not acceptable
03 through 07	Spare
10	Assignment command
11	Acknowledged message. (Request in queue)
12	Channel release command
13	Congestion (full queue or network congestion)
14 through 65	Not relevant to SES
66	Unconditional channel release
67 through 76	Not relevant to SES
77	Idle message

The SES idle state

Figure 11.4 shows how Inmarsat-A equipment operates in the idle state. The unit continuously tracks the selected satellite while monitoring the assignment field of the assigned common signalling channel (previously common TDM channel) for possible assignments addressed to it. If a match is detected between the terminal's stored identification and the address identification of an assignment message, the unit will respond according to the message. It will retune telegraphy or telephony receivers and start the appropriate signalling procedures. Data are modulated, by the CES, onto the CSC carrier using coherent BPSK. The differentially encoded signal shows a phase transition of 180° on a binary one. The data rate is 1.2 kbit/s.

The unique 21 bit station identification IMN number, is the ship's selective calling number which identifies the vessel. This seven digit decimal number is programmed into the computer software of the SES when the installation is commissioned by Inmarsat and cannot be changed by the operator.

The SES continues to monitor the CSC until it identifies its own unique number. Upon receipt of this information, the central processor in the SES BDE commands the SES to enter the call set-up sequence.

The BDE part of the Saturn 3 series in the idle state basically shows the method of decoding differentially encoded data. The intermediate frequency (IF) signal containing information in the form of phase shifts is compared with a locally generated reference frequency in order to reconstitute the information signal, which is then passed to the main processor. Decisions are made by the processor depending upon information received.

Broadcast simplex telex

This signalling is done on the CSC in the direction shore-to-ship only. The format is shown in Fig. 11.5.

The system is used only by Inmarsat for an all-ships call, in order to broadcast service information. Service messages are automatically printed on the SES teleprinter without any action from the operator. Using this channel, Inmarsat can transmit the following calls:

11.2 Details of the Inmarsat-A system 237

- All-Ships Call,
- Ocean Area Group Call,
- National Group Call, or
- Fleet Group Call.

The format of the CSC (previously known as the common TDM carrier) is graphically shown, 22 data slots, signal parameters and the unique word plus spare bit form the 0.29 s frame. Signal parameters are: error detection, 6 bits; shore station information between CES and NCS, 8 bits; channel number, 15 bits; of which nine indicate frequency and six the appropriate time slot; channel type, 4 bits; message type, 6 bits; and the terminal identification IMN number, 21 bits.

The terminal continuously monitors the CSC and decodes all ID numbers. Upon receipt of a valid call the processor decodes the message type, channel type and reprograms timers to assign the correct time slot. It then reads the data in that timeslot. The teleprinter prints the decoded characters received in the time slot at the end of which the mark-to-space transition causes the printer to switch off and the terminal returns to the CSC and continues monitoring.

Duplex telex TDMA format. Ship-to-shore

Figure 11.6 clearly shows how the TDMA format for duplex telex communications is organized. Each of the 22 multiplexed slots within the 1.74 second frame is composed of a 41.4 ms interchannel gap and a 37.7 ms information slot. This slot is constructed as shown with the carrier frequency comprising 50 bits to permit the shore station to lock on to the TDMA carrier burst; the 29 bit timing section permits the shore station to lock on to the bit rate; the 30 bit unique word which follows is used for timing purposes. Finally, the information block of 72 bits contains 12 telex characters of 6 bits each. In this case spelling the ship's name. 1.74 ms later data slot 20 will contain the next 12 characters and so on.

Ship-to-shore telex channels are multiplexed in TDMA format enabling 22 50-baud channels to be transmitted simultaneously on the same frequency. Telex characters are sent in 37.7 ms bursts and 22 bursts are transmitted in every 1.74 s frame. Each burst carries synchronizing bits and unique word bits plus 72 bits (12 telex characters) which equates to a baud rate of 50. The modulation is by 2 phase BPSK.

It may appear that the 'Interchannel gap' of 41.4 ms in the TDMA format between telex channels is excessively large when compared with the 37.7 ms information frame. Obviously, if the gap could be reduced, more channels could be made available within the data frame. The interchannel gap is 41.4 ms which has been calculated as a result of differences in signalling distances between ships, the satellite and a CES. The difference in path distances from the best case, where the SES and the CES have a combined path length of 72,000 km to the worst case where the SES, in high latitudes, and CES have a path length of 78,000 km produces a difference in signal delay of approximately 50 ms. The figure quoted (50 ms) is a very broad indication of the signal delay over 6000 km for a radio wave travelling at 300×10^6 m/s. In practice, the maximum delay likely to occur has been accurately calculated to be 41.4 ms. If the interchannel gap was reduced, ships communicating in this worst case scenario may suffer synchronization problems causing channel time slots in the TDMA frame to overlap with neighbouring time slots. Clearly, this would lead to cross-channel interference. Figure 11.7 illustrates the worst case where two ships are communicating with the CES over very different distances.

Inmarsat-A system access can readily be understood by considering the process of call initialization and clearing and the actions of the SES, CES and the NCS.

238 The Inmarsat-A system

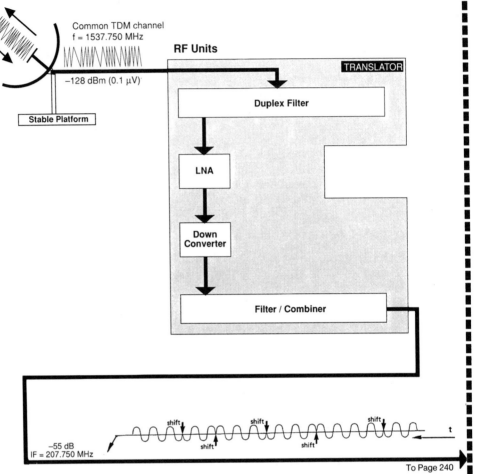

Fig. 11.5 Saturn 3S.90 Inmarsat-A SES. Broadcast, Simplex Telex. TDM Format, shore-to-ship (courtesy ABB NERA AS)

11.2 *Details of the Inmarsat-A system* 239

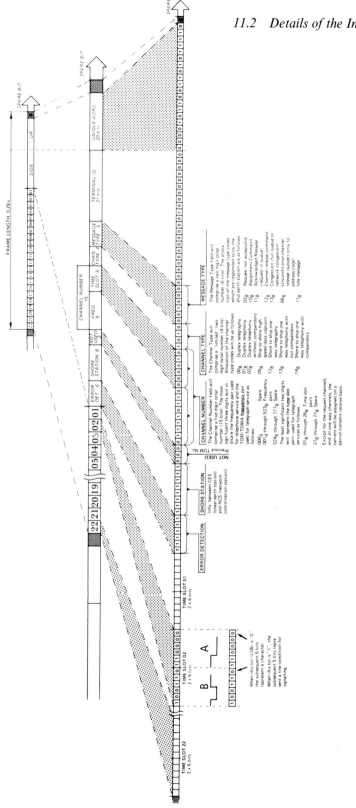

240 *The Inmarsat-A system*

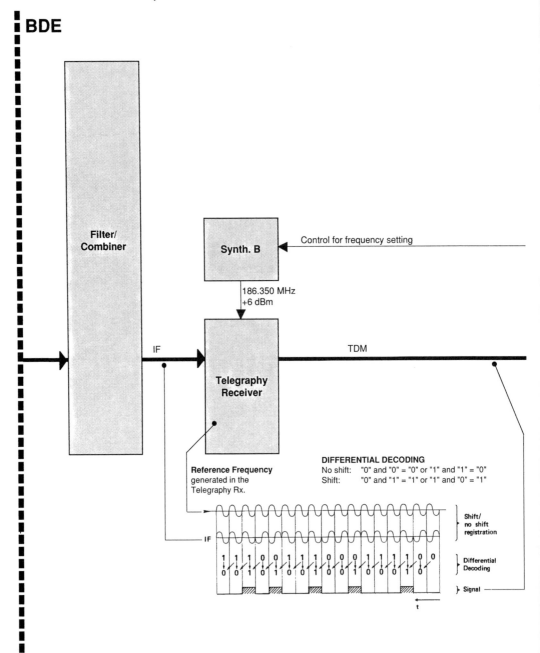

Fig. 11.5 (cont.)

11.2 Details of the Inmarsat-A system

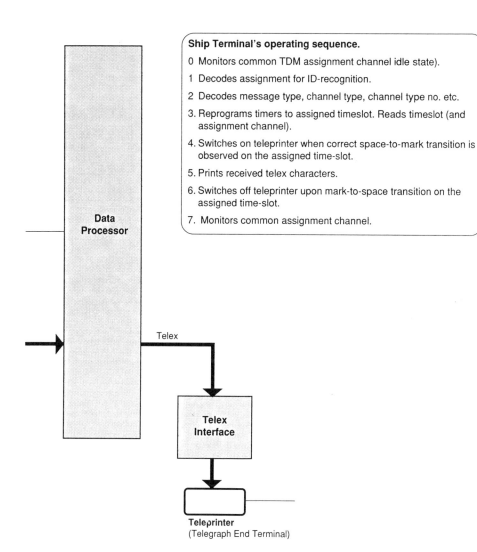

Ship Terminal's operating sequence.

0. Monitors common TDM assignment channel idle state).
1. Decodes assignment for ID-recognition.
2. Decodes message type, channel type, channel type no. etc.
3. Reprograms timers to assigned timeslot. Reads timeslot (and assignment channel).
4. Switches on teleprinter when correct space-to-mark transition is observed on the assigned time-slot.
5. Prints received telex characters.
6. Switches off teleprinter upon mark-to-space transition on the assigned time-slot.
7. Monitors common assignment channel.

BROADCAST, SIMPLEX TELEX
Saturn 3S.90
System Description

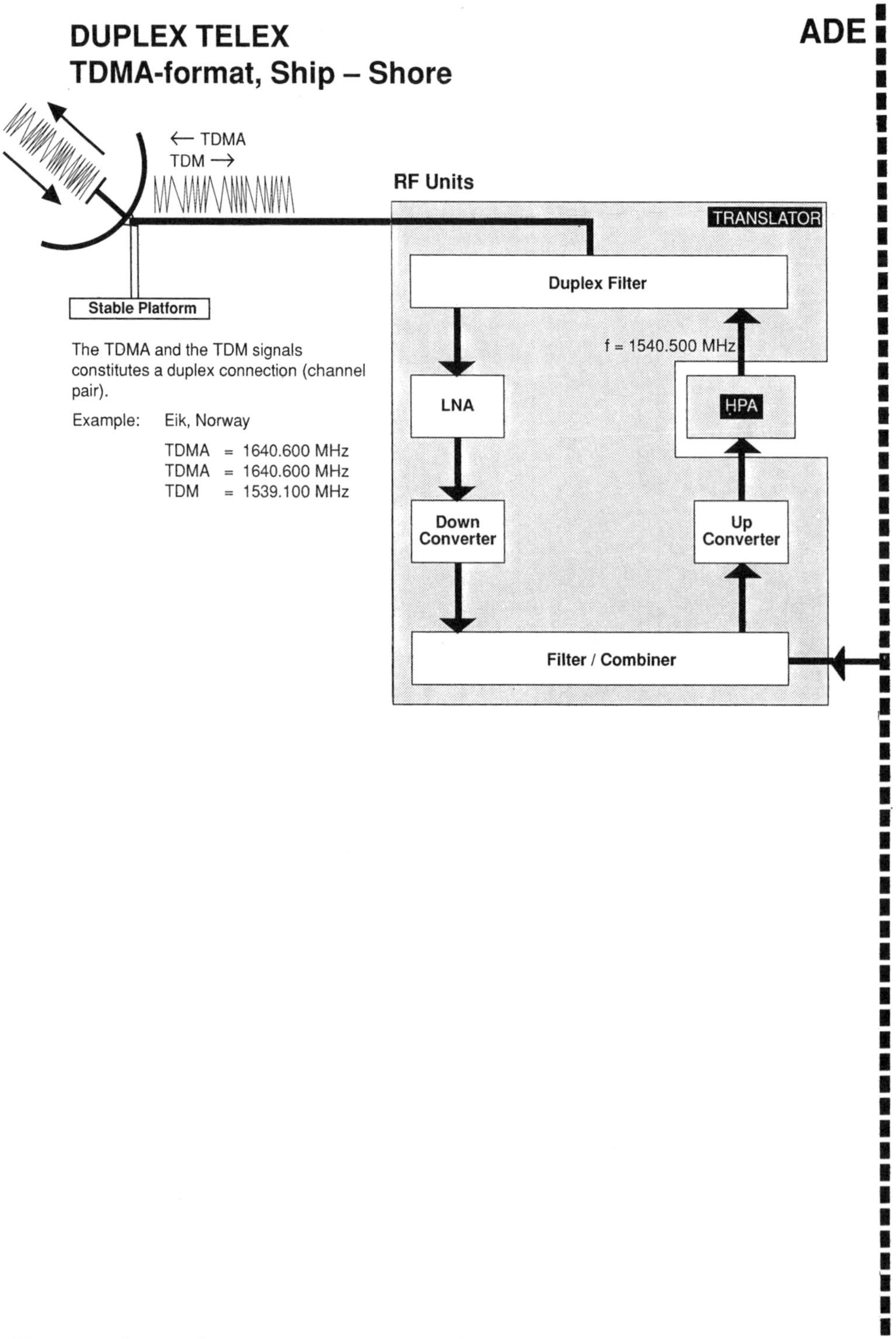

242 *The Inmarsat-A system*

DUPLEX TELEX
TDMA-format, Ship – Shore

ADE

← TDMA
TDM →

RF Units

TRANSLATOR

Duplex Filter

$f = 1540.500$ MHz

Stable Platform

The TDMA and the TDM signals constitutes a duplex connection (channel pair).

Example: Eik, Norway

TDMA = 1640.600 MHz
TDMA = 1640.600 MHz
TDM = 1539.100 MHz

LNA

HPA

Down Converter

Up Converter

Filter / Combiner

Fig. 11.6   Saturn 3S.90 Inmarsat-A SES. Duplex Telex TDMA Format, ship – shore (courtesy ABB NERA AS)

11.2 Details of the Inmarsat-A system

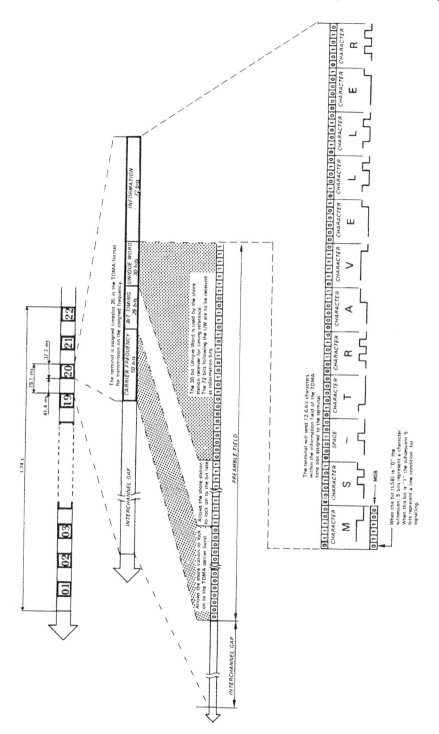

244 The Inmarsat-A system

BDE

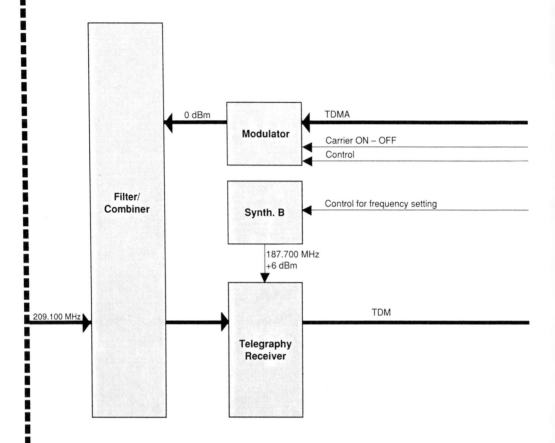

The telex channels in ship—shore direction are multiplexed in a TDMA format, allowing 22 50-baud telex channels to be transmitted virtually simultaneously on the same frequency.

The telex characters are sent in short bursts, each lasting 37 milliseconds, and one burst is transmitted every 1.74 seconds. Each burst carry 12 telex characters which gives a mean baud rate of 50 baud for each user.

The modulation is phase shift keying as for the request burst modulation.

The telex characters in the shore—ship direction is modulated and demodulated in the same way as for telex broadcast, except that the telegraphy receiver is retuned from the common TDM frequency to the TDM carrier frequency of the coast earth station which handles the telex message.

Fig. 11.6 (cont.)

11.2 Details of the Inmarsat-A system

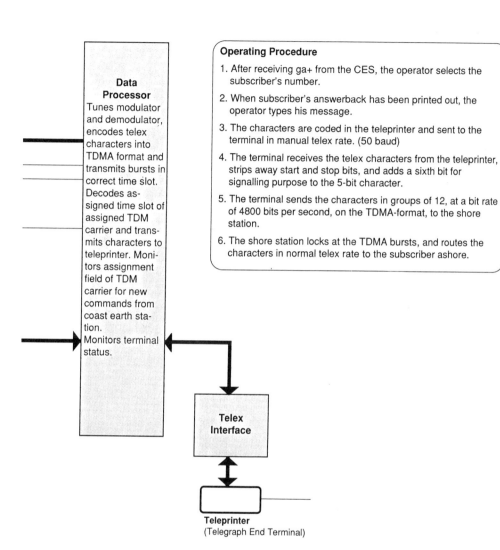

Data Processor
Tunes modulator and demodulator, encodes telex characters into TDMA format and transmits bursts in correct time slot. Decodes assigned time slot of assigned TDM carrier and transmits characters to teleprinter. Monitors assignment field of TDM carrier for new commands from coast earth station. Monitors terminal status.

Operating Procedure
1. After receiving ga+ from the CES, the operator selects the subscriber's number.
2. When subscriber's answerback has been printed out, the operator types his message.
3. The characters are coded in the teleprinter and sent to the terminal in manual telex rate. (50 baud)
4. The terminal receives the telex characters from the teleprinter, strips away start and stop bits, and adds a sixth bit for signalling purpose to the 5-bit character.
5. The terminal sends the characters in groups of 12, at a bit rate of 4800 bits per second, on the TDMA-format, to the shore station.
6. The shore station locks at the TDMA bursts, and routes the characters in normal telex rate to the subscriber ashore.

Telex Interface

Teleprinter
(Telegraph End Terminal)

DUPLEX TELEX, SHIP – SHORE
Saturn 3S.90
System Description

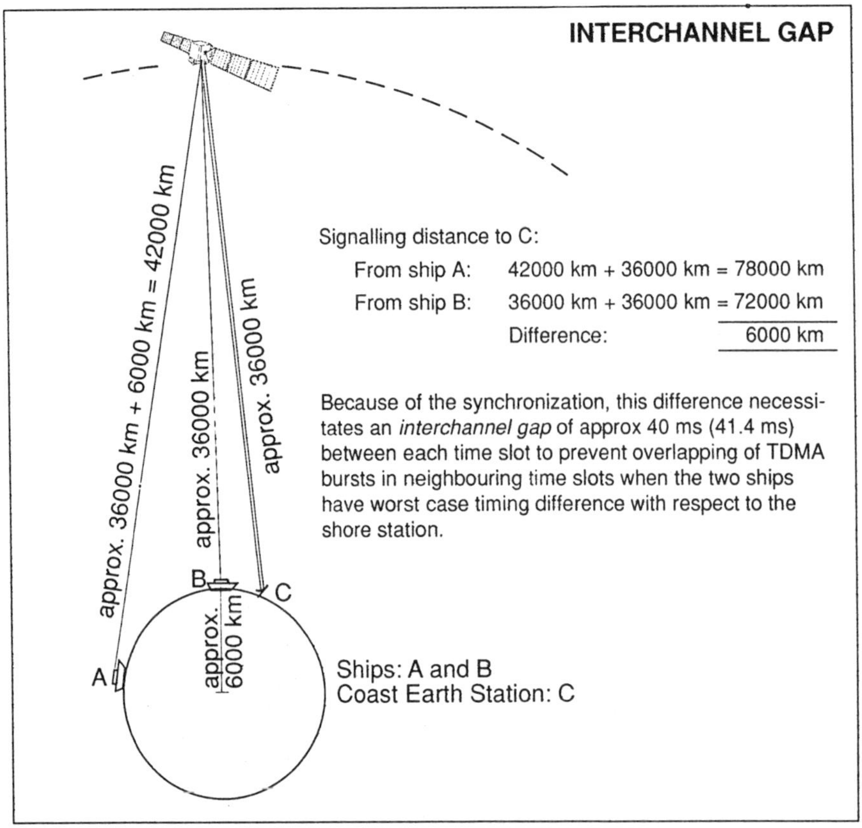

246   *The Inmarsat-A system*

```
                        INTERCHANNEL GAP
```

Signalling distance to C:
 From ship A: 42000 km + 36000 km = 78000 km
 From ship B: 36000 km + 36000 km = 72000 km
 Difference: 6000 km

Because of the synchronization, this difference necessitates an *interchannel gap* of approx 40 ms (41.4 ms) between each time slot to prevent overlapping of TDMA bursts in neighbouring time slots when the two ships have worst case timing difference with respect to the shore station.

Ships: A and B
Coast Earth Station: C

Fig. 11.7   Explanation of the Interchannel Gap (courtesy ABB NERA AS)

Call initializing and clearing procedures

Call initialization and clearing for telephone working   Telephone calls and voice-band data calls (medium and low speed data and fax) are transmitted over the network using FM channels.

It should be noted that the procedure detailed below is fully automatic and only requires the SES operator to input the dialling sequence of numbers to make a call. The description which follows assumes that the SES has been tuned to monitor the CSC TDMO channel 156Q.

The process is started by the operator who selects a CES and initiates a request data burst using the CSC channel 156Q. The SES will automatically alternate request bursts between the two allocated request channels, both of which are continuously monitored by all CESs and the NCS in the ocean region. The NCS monitors these channels in case there is no response from a CES to a priority 3 (distress) request, in which case a co-located NCS will initiate a call see Fig. 11.8.

The request burst (1) is addressed to the required CES and will contain the type of communication required, the CES identity, the priority of the communications (priority 3 or routine), the type of channel if for telephony (companded or uncompanded) and the identity number of the SES. The addressed CES will then send a 'request for the telephony channel assignment' to the NCS (2). The NCS will assign a channel and

11.2 Details of the Inmarsat-A system

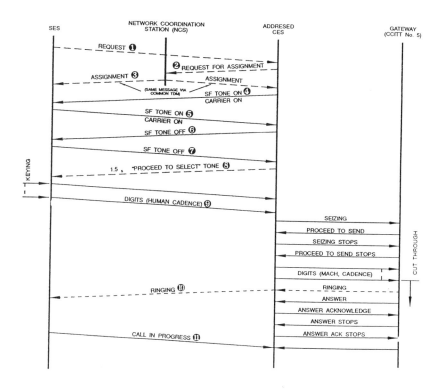

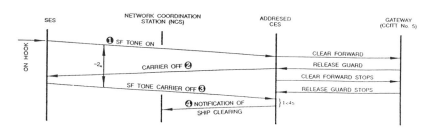

Fig. 11.8 Inmarsat-A telephone channel call set-up and clearing (ship originated) (courtesy Inmarsat)

announce the assignment to both the CES and the SES over the common signalling channel (3). On receipt of the assignment from the NCS the CES will tune one of its modems to the assigned channel and switch on both an FM carrier and a single frequency tone (SF) (2600 Hz) (4). On receipt of the assignment from the NCS the SES will tune to the assigned channel and wait for a signal from the CES. On receipt of the carrier and the SF tone from the CES the SES will turn on its carrier and modulate with an SF tone (5). When the CES receives the SF tone from the SES it removes the SF tone from its own transmission (6). The SES recognizes the removal of the SF tone from the CES frequency and will turn off its own SF tone (7). This completes the 'handshaking'

248 The Inmarsat-A system

and the CES will now send a 'Proceed to select' (PTS) signal to the SES. This tone of 400 Hz lasting for 1.5 s informs the SES operator to input the dialling information in order to contact the called party (8).

Dialling information comprises:
- the telephone service code—a two-digit code depending upon the service required. 00 for automatic calls. Followed by,
- the destination code—either a country access code or maritime access code for another SES. Country code for the United Kingdom is 44. Followed by,
- the called subscriber's number—either a land-based subscriber or another SES. Followed by,
- the end of number selection code—the # character to signify the end of the calling sequence.

The call then proceeds through the International Telephone network to the required country and subscriber. The ringing signal is returned to the SES by the CES (10). When the called subscriber answers, the chargeable time starts (11).

Call clearing When the caller replaces the telephone handset the SES will place the SF tone on the channel (1). The CES recognizes this tone and removes the carrier from the channel (2). This is recognized by the SES which turns off its SF tone and removes its carrier from the channel (3). The SES then returns to the idle state and monitors the CSC. The removal of the SES tone and carrier is recognized by the CES which notifies the NCS that the occupied telephone channel is now available for re-assignment (4).

Call initialization and clearing for telex working For telex traffic each CES has one or more exclusive TDM channels. Each TDM channel contains 22 time slots. When an SES receives an assignment from the CES over the common signalling channel for telex traffic it retunes its terminal to the assigned TDM and listens to its assigned time slot within that TDM channel. The transmission on each TDM is split into 22 separate time slots each containing traffic for different SESs. The TDM/TDMA formats are illustrated in Fig. 11.3. Each CES TDM transmit channel has a corresponding receive channel, which again is divided into 22 time slots. These time slots are for the CES receive channel. The SES, when handshaking or transmitting to the CES, must only transmit in its own assigned TDMA time slot. The telex channel call set-up/clearing is shown in Fig. 11.9.

The request burst (1) is addressed to the required CES and will contain: the type of communication required (telex), the priority and the identity number of the SES. The CES will assign one of the time slots on the TDM to the SES. This assignment is passed to the NCS (2) for retransmission to the SES over the common signalling channel (3). On receipt of this assignment the SES will retune to the assigned TDM and, while listening to that time slot and transmitting into a corresponding time slot, will carry out satellite call transition (handshaking) with the CES (5 & 6).

Satellite Call Transition is the exchanging of a series of 'marks' and 'spaces' with the CES in the assigned time slots. After passing the assignment message to the NCS the CES will transmit the date, UTC time and CES identity and request the SES answerback by transmitting WRU (Who Are You?) (7). The SES will respond with its answerback (8) and the CES will transmit GA+ (Go Ahead) (9). The SES operator will now key in the telex numbers for connection to be made (10).

11.2 *Details of the Inmarsat-A system* 249

Telex Channel Call Set-up (Ship Originated)

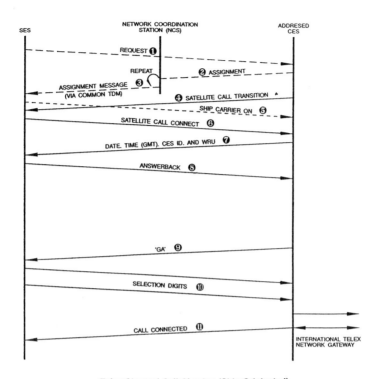

Telex Channel Call Clearing (Ship Originated)

Fig. 11.9 Inmarsat-A Telex Channel Call Set-up and Clearing (ship originated) (courtesy Inmarsat)

Dialling information comprises:
- telex service code—a two-digit code depending upon the service required. 00 signifies an automatic connection. Followed by,
- destination code—either a country access code or maritime access code for another SES. Country code for the UK is 51. Followed by,
- called subscriber's number—either a landbased subscriber or another SES. Followed by,
- end of number selection—the '+' character.

The call is then connected and the chargeable time starts from receipt of the called subscriber's answerback.

Telex call clearing When the operator on board the ship finishes the telex call and switches the telex machine 'Off Line', the SES will exchange a series of 'marks' and 'spaces' with the CES (1).

The CES will then send to the SES confirmation of satellite clearing (2). On receipt of this confirmation, the SES will turn off its carrier in the assigned time slot and return to the idle state monitoring the common signalling channel (3). The CES will then inform the NCS that the SES has cleared (4) so that the NCS can remove the SES identity from its 'Ship Busy' list.

The two sequences describing call initialization and clearing should be studied in conjunction with the SES manufacturer's operators manual. Reference may also be made to the operations documentation issued by some CESs for mobile users.

11.3 Inmarsat-A SES equipment

Whilst equipment is no longer being designed for use on the Inmarsat-A system there are tens of thousands of Inmarsat-A SESs in use in the world and consequently the system will be in use for many years to come.

The council of Inmarsat has adopted the following policy in relation to Inmarsat-A services:

- the services will be supported until at least the years 2005/2006,
- Inmarsat will continue to support new commissionings of Inmarsat-A SESs for a minimum of three years after the introduction of the Inmarsat-B and Inmarsat-M systems,
- Inmarsat will continue to support the re-commissioning of Inmarsat-A SESs after the introduction of the Inmarsat-B and M systems.

Indeed in July 1993 the Inmarsat-A Network Control Stations were extensively redesigned in order to support a major change in system access and improve the allocation of communication channels. The original Inmarsat-A NCSs were commissioned some 15 years earlier and, as a consequence their electronic processing capability, had become slow and limited when compared with modern systems. In addition to improving electronic processing the new NCSs are capable of channel interleaving/inserting in order to improve greatly access time and channel efficiency.

Originally, all Inmarsat-A SESs communicated with the appropriate ocean region NCS on the common signalling channel (CSC) TDM0 in order to request a communications channel. This situation caused queues to be formed during busy periods, resulting in unnecessary waiting periods. From July 1993 there are two common signalling channels available for use by SESs in an ocean region, known as TDM0 and TDM1. All SESs commissioned by that date were allocated alternately one of the two CSCs. This in effect means that the access time for an SES may be reduced by half under those conditions where two ships require access simultaneously. Assuming, of course, that the two SESs are each allocated one of the two different CSCs, TDM0 and TDM1.

In order to ensure that an access clash was unlikely, approximately half of the world's Inmarsat-A SESs were re-tuned to the CSC TDM1, whilst the other half remained tuned to TDM0. The criteria used to determine which SES was re-tuned centred on the unique Inmarsat Mobile Number (IMN) allocated to an SES. Those SESs which possess an even digit as the fourth digit of their IMN number remain tuned to TDM0. Those with an odd digit for the fourth digit re-tuned to TDM1. This situation continues with new commissionings.

Because Inmarsat-A SESs are monitoring different CSCs, distress alerting and message information is broadcast to the SESs on both CSCs. Whilst the modifications to the system are fully transparent to the user, some SESs will revert to the original CSC after being switched off. Re-tuning is fairly straightforward at switch on. Reference should be made to the manufacturer's manual.

A further Inmarsat-A enhancement is to customize the uncompanded channel 02 for data and facsimile use. Customizing this channel required significant alignments to be made at each CES and SES. The circuit alignments involve the limiting of the audio signal level to the frequency modulator in order to limit frequency deviation.

There are currently 13 manufacturers of Inmarsat-A fixed equipment units, some of whom will manufacture the Inmarsat-B, C and Inmarsat-M systems. Because of the extensive design, development and type approval costs, Inmarsat-A equipment manufacturers tend to be large international companies. Such companies are more able to provide the full world-wide maintenance support which an SES may need in ports throughout the world.

Equipment type-approval and commissioning

Inmarsat operates an extensive type-approval service through which a proposed Inmarsat terminal must pass before being allowed access to the space segment. The document 'Technical Requirements For Inmarsat Ship Earth Stations' sets out rigid technical standards for the proposed equipment. Once the equipment has passed through the design and development stage, a pre-production unit is sent for standards testing. If all is well and the equipment passes, the manufacturer goes into production and hopefully sells sufficient units to cover the extensive development costs. Each MES sold is professionally installed, commissioned by Inmarsat and issued with a unique Inmarsat identification number (IMN) normally known as the ID.

Some SESs are provided with two IDs which enable users to receive and place link calls on two separate telephone numbers. Because of their design, dual ID SESs are not able to connect two calls simultaneously. To achieve this a multi-channel SES, consisting virtually of two systems in the one unit, is available from some companies.

SES system description

Essentially Inmarsat-A SES equipment consists of two parts, the above decks equipment (ADE) and the below decks equipment (BDE). The ADE is easily recognizable by the large opaque radome which weather-proofs the unit see Fig. 11.10.

In general, an ADE consists of the following:

- The parabolic antenna—On older equipment this may be in excess of 1.2 m in diameter, whereas on newer equipment it is likely to be approximately 0.8 m. Ever since the introduction of Inmarsat-A SES equipment the practice has been to reduce the size of the antenna and consequently the size and weight of the ADE. As a large proportion of the receiver signal gain and transmitter EIRP is produced by the antenna, the area of the dish can only be reduced if the transmitting power from the satellite transponder is increased and/or the gain of the receive pre-amplifiers can be increased without an appreciable increase in noise.
- The stable platform—This is the mounting bed which remains stable when the ship is pitching or rolling in extreme weather conditions. It is essential that the stable platform holds the dish antenna in its correct Azimuth elevation AZ/EL angular positions despite movements of the ship. The platform usually consists of a solid bed mounted in such a way that four gyro compasses are able to sense movement and correct any errors which are detected. It is a form of electronic gimbals.

252 *The Inmarsat-A system*

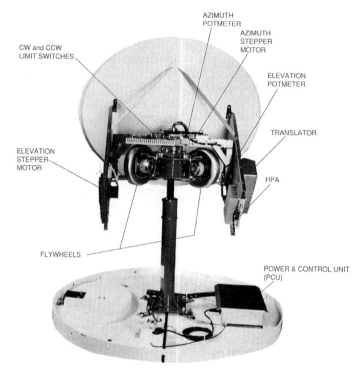

Fig. 11.10 An ADE stabilized antenna assembly minus the radome (courtesy ABB NERA AS)

- The tracking system—The dish antenna is controlled in AZ/EL by stepping motors, which in turn are electronically controlled in a simple feedback system. This electro-mechanical arrangement enables the dish to maintain lock on a satellite despite navigation course changes. As the ship changes course both AZ/EL control corrections must be made.
- The computer—The ADE processor controls all ADE functions, which include satellite tracking and electronic control.
- The RF electronics—Contains the transmitter high-power amplifier (HPA) and the receiver RF 'front end' low-noise pre-amplifier stages (LNA), plus all the critical bandpass filtering stages.
- The multiplexer—In modern equipment it is common practice to reduce the number of cables between ADE and BDE. This is achieved by multiplexing up/down signals or commands onto the one coaxial feeder.

The BDE houses the following units.

- The system computer—This processor controls the whole system operation. Some of its functions may be distributed to second-level dedicated processors.
- Modulator/demodulator—This forms the nucleus of any communications system. In this case it modulates the baseband signals onto the intermediate frequency (IF) carrier and extracts the baseband signals from the IF carrier of the received signal.
- The synthesizer—This produces the highly stable frequencies required for modulation/demodulation and signal switching.
- Audio processing—This performs baseband processing of the audio signals from the microphone on transmit, and from the demodulator output on receive.

- Data processing—This supervises data processing of both transmit and receive telex signals. Monitors out-of-band signals.
- Interfacing—This performs all the tasks of signal conditioning between peripherals and signal processing.
- Operator display and control—System functional control may be achieved by a dedicated control panel as was often the case in older units, or more likely by an IBM PC clone on the latest equipment. Increasingly, manufacturers are using versatile off-the-shelf computer terminals for control instead of designing a dedicated control panel.

The preceding section broadly outlines the systems included in an Inmarsat-A SES terminal. Although each manufacturer inevitably produces a different design from his/her competitors, and although SESs tend to very look different, they do in fact perform the same functions. The following sections look at the basic design of SESs from two major manufacturers—ABB NERA AS based in Norway and The Japan Radio Company, JRC.

SES ADE antenna specifications and siting

The communications antenna of an Inmarsat-A SES is of necessity relatively large and heavy. Over the past decade the ADE, which comprises the mechanical antenna assembly, the antenna control electronics and gyroscopes, the microwave electronic package and the dish antenna, has reduced considerably in both physical size and weight. This reduction, brought about by greater EIRP from satellite transponders coupled with GaAs-FET technology at the front end of the receiver leading to higher gain low-noise RF amplifiers, has made the fitting of Inmarsat-A SES terminals on small craft a reality. Some of the very early ADEs weighed several hundred kilograms which, when mounted 6 m above the top deck of a vessel led to stability problems. Currently, an ADE for a SES is likely to weigh less than 100 kg and include a parabolic antenna of approximately 0.8 m in diameter.

The technical principles of parabolic antennae are detailed in Chapter 8. Inmarsat does not specify the physical size of antenna or ADE it but does specify EIRP and G/T characteristics to be met. Inmarsat specifications for an SES Inmarsat-A antenna system are as follows.

- Gain: antenna gain at both transmit and receive frequencies must be such that the specified transmit EIRP of 36 dBW and the receive G/T of -4 dBK are satisfied.
- Polarization: right-hand circular (RHC) for both receive and transmit in accordance with CCIR recommendation 573.
- Sidelobes: the peaks of the sidelobes must not exceed an envelope described by the following expressions
  $G = 8$ dBi $\qquad 16° \leq \theta \leq 21°$
  $G = 41 - 25\log\theta$ dBi $\quad 21° \leq \theta \leq 57°$
  $G = -3$ dBi $\qquad \theta > 57°$
  where: G is the gain of the sidelobe envelope relative to an isotropic antenna. θ is the angle in degrees between the main beam and the direction considered.
- Steerability and tracking: the antenna beam must be capable of being steered in the direction of any geostationary satellite whose orbital inclination does not exceed 3° and whose longitudinal excursions do not exceed $\pm 0.5°$. Means must be provided to point the antenna beam automatically towards the satellite with sufficient accuracy to ensure that the G/T and EIRP requirements are satisfied continuously under operational conditions.

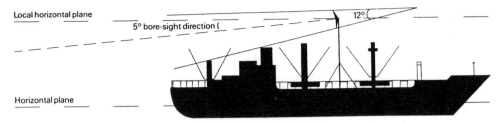

Fig. 11.11 Ship Earth Station Design and Installation Guidelines (courtesy Inmarsat)

Careful consideration should be given to the siting of an Inmarsat-A radome. Essentially, the focal point of the parabolic antenna must be pointing directly at the geostationary satellite being tracked without any interruption of the microwave beam caused by obstructions on the ship. Inmarsat specify that there should be no obstacle which is likely to downgrade the performance of the equipment in any angle of azimuth down to an elevation of $-5°$. This is not easy to achieve. Figure 11.11 illustrates an antenna satisfying these guidelines but with the disadvantage that it is very high above the vessel's deck and would be impossible to install. This antenna would probably be adversely affected by vibration and it would be difficult to gain access for maintenance purposes.

If structures do interrupt the communications beam, blind sectors will be caused leading to degraded communications over some arc of azimuth travel. However, it should be remembered that the communications RF beam of energy possesses a width of usually 12°. Consequently, objects within 10 m of the radome which cause a shadow sector greater than 6° are not likely to degrade significantly the performance of the equipment. It is preferable that there should be no objects within 3 m of the antenna. Beyond this distance, small objects measuring less than 15 cm can be ignored.

If, as is often the case, it is impossible to find a mounting position free from all obstruction, any blind sectors identified should be recorded. It may be possible for the operator, when in an area served by two satellites, to select the satellite whose azimuth and elevation angles, with respect to the ship, are outside the blind sector.

Electromagnetic RF signals at high radiation levels are known to be hazardous to health. It is inadvisable to permit human beings to stand close to the radome of an SES when the system is communicating with a satellite at a low elevation angle. It is not possible to quote a safe distance because international regulations concerning this subject vary. Inmarsat recommends that the radiation levels in the vicinity of the antenna should be measured. The distances from the antenna which result in radiation levels of 100 W/m, 25 W/m and 10 W/m should all be recorded. Radiation plan diagrams may be produced and located near the antenna as a warning, or distances from the antenna may be physically labelled.

Satellite location and tracking

An Inmarsat-A B or M mobile antenna must be capable of locating and tracking the geostationary satellite selected for communications. Locating and tracking may be done automatically, as in the case of a SES installation, or manually only, as with a portable system.

It is common practice to believe that the satellite is 'fixed' and that once the link has been established it will remain so as long as the mobile does not move. It should be remembered that the satellite is not stationary, it is in fact travelling at some 3.073 km/s

Direction (azimuth angle)

The azimuth angle is the angle between north and the horizontal satellite direction as seen from the MES.

The actual azimuth angle for the various satellites due to the Saturn terminal can be found on the map on page 261.

Example:
259° azimuth

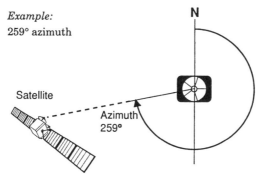

Height (elevation angle)

The elevation angle is the satellite height above the horizon as seen from the MES.

The actual elevation angle for the various satellites due to the Saturn terminal can be found on the map on page 262.

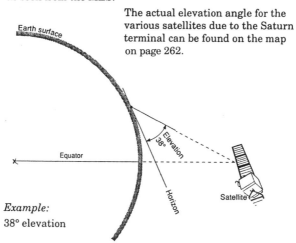

Example:
38° elevation

Fig. 11.12 AZ/EL angle explanations (courtesy ABB NERA AS)

and is geostationary when viewed from the surface of the earth. A satellite is under the influence of a number of variable astrophysical parameters which cause it to move around its station by up to several degrees. A mobile tracking system must therefore counteract this by repositioning the MES antenna at regular intervals. The carrier signal is monitored, and if a reduction in its amplitude is detected a close programmed search is initiated until the carrier strength is again at maximum. No loss of signal occurs during this process which is automatically initiated.

Obviously, the greatest tracking problem will arise when the mobile station is moving at speed with respect to the satellite. An Inmarsat SES antenna may be moved through any angle in azimuth (AZ) or elevation (EL) as the vessel moves along its course. See Fig. 11.12.

256 *The Inmarsat-A system*

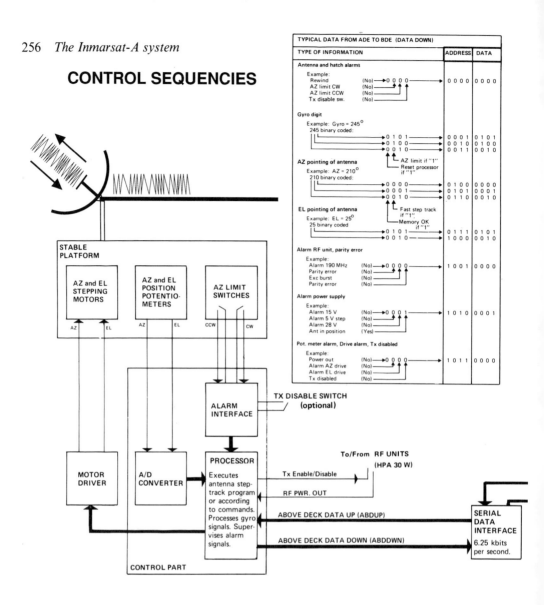

Fig. 11.13 Saturn 3S.90 Inmarsat-A SES. Antenna Control Sequences (courtesy ABB NERA AS)

11.3 Inmarsat-A SES equipment

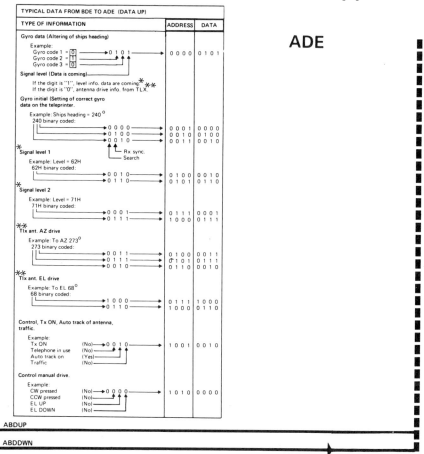

258 The Inmarsat-A system

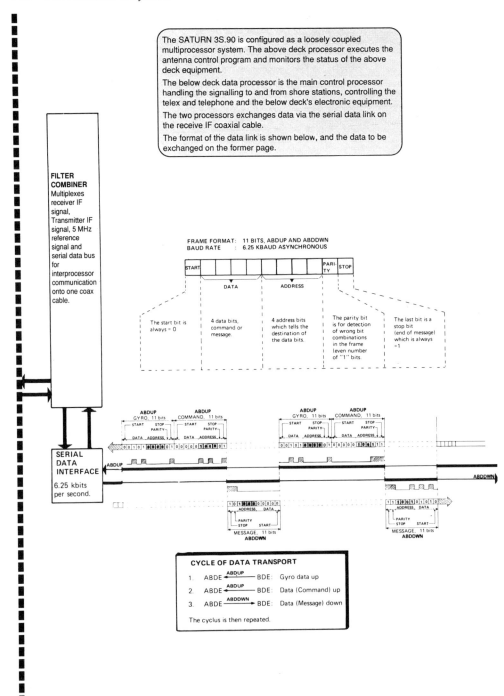

Fig. 11.13 (cont.)

11.3 Inmarsat-A SES equipment

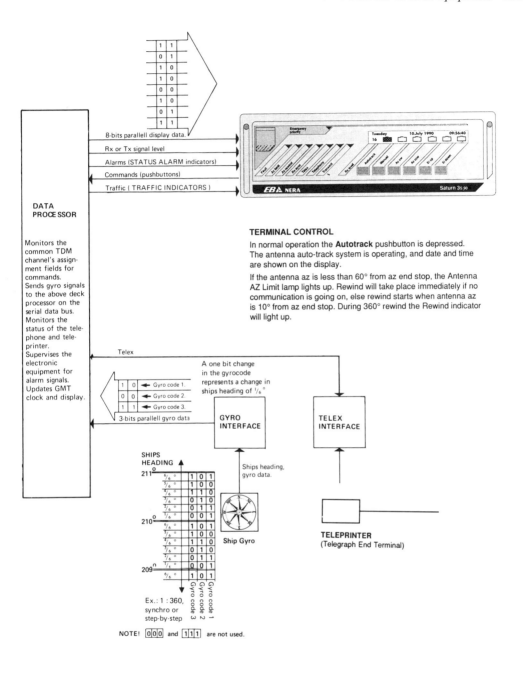

TERMINAL CONTROL

In normal operation the **Autotrack** pushbutton is depressed. The antenna auto-track system is operating, and date and time are shown on the display.

If the antenna az is less than 60° from az end stop, the Antenna AZ Limit lamp lights up. Rewind will take place immediately if no communication is going on, else rewind starts when antenna az is 10° from az end stop. During 360° rewind the Rewind indicator will light up.

CONTROL SEQUENCIES
Saturn 3S.90
System Description

It is essential therefore that electronic control of the antenna is provided. In practice, a simple electronic feedback control system is used. The major Inmarsat SES manufacturer, ABB NERA AS, approaches the problem with the Inmarsat-A SES SATURN 3S.90 terminal as described below, with reference to Fig. 11.13.

Antenna positioning in both AZ and EL is controlled by two stepping motors respectively. Each motor is controlled by a 4 bit digital code applied to four connections in the two phase system.

Each change of one bit in the logic combination represents an antenna change in AZ or EL of $(\frac{1}{6})°$. The step track board in the processor of the ADE executes the antenna step track program according to commands input by the operator or automatically initiated by a change in the ship's heading or a rewind sequence.

- Manual commands: when the operator has selected manual control, elevation is commanded by Up and Down keys whereas azimuth positioning is controlled by Clockwise (CW) and Counter Clockwise (CCW) keys. This command would be used when the vessel's position is known and the satellite position is known. AZ/EL angles can be derived and input to the equipment by using the two AZ/EL charts provided by ABB NERA. See Fig. 11.14.

  Once the antenna starts to pick up the satellite signal the operator display indicates signal strength. Fine positioning can now be achieved by moving the antenna in AZ/EL in $(\frac{1}{6})°$ steps until maximum field strength is achieved.
- Automatic control: once satellite lock has been achieved the system will automatically monitor signal strength in order to maintain this lock as the vessel changes course.
- Automatic search: an automatic search routine commences 1.5 minutes after switching on the equipment or may be initiated by the operator. The EL motor is caused to search between 5° and 85°, whereas the AZ motor is stepped through 10° segments. See Fig. 11.16. If the common signalling channel signal is identified during search the step tracking systems take over to swing the antenna above/below and each side of the signal searching for the maximum signal strength. Fig. 11.15 shows the ordinary step track and the fast step tracking sequences where the antenna is oscillated either side and above/below the beam until maximum signal strength is identified.
- Gyro control: once lock has been achieved it is maintained irrespective of changes to the vessel's course by the ship's gyro repeater circuitry. Gyro repeater pulses are converted into a 3 bit logic Gray Code. Changes in the ship's course produce changes in the Gray Code corresponding to $(\frac{1}{6})°$ variations in azimuth. The AZ stepper motor is commanded accordingly. Satellite signal strength is monitored and, if necessary, the EL stepper motor is commanded to search for maximum signal strength.
- Antenna rewind: the antenna in the ADE is pivoted on a central mast and is coupled by various control and signal cables to the stationary stable platform. If the antenna was permitted to rotate continuously in the same direction the feeder cables would eventually become so tightly wrapped around the central support that they would either prevent the antenna from moving or they would fracture. To prevent this happening a sequence known as antenna rewind is necessary. Fig. 11.17 illustrates the process.

In fact, the antenna is permitted to rotate through 540° in azimuth, although the antenna would normally operate within the segment 60° to 480°. If the antenna moves beyond its operational area into one of the designated rewind areas, 10° to 60° or 480° to 530°, and if no traffic is in progress, the antenna will automatically rewind through 360°.

11.3 Inmarsat-A SES equipment 261

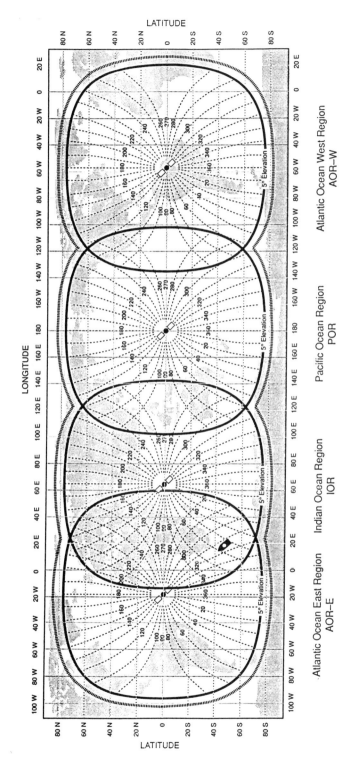

Example:

Azimuth angle for the plotted position 🚢

315° for the AOR-E satellite
55° for the IOR satellite

Be careful not to read the wrong angle in areas where two satellites overlap.

Fig. 11.14 Azimuth angle map. Elevation angle map (courtesy ABB NERA AS)

262 *The Inmarsat-A system*

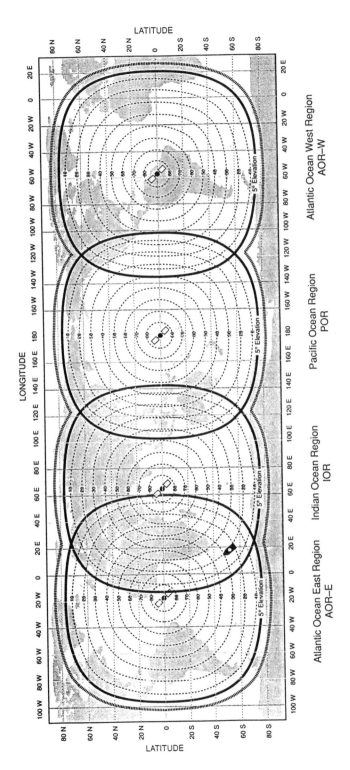

Example:

Elevation angle for the plotted position ⬧

24° for the AOR-E satellite
17° for the IOR satellite

Be careful not to read the wrong angle in areas where two satellites overlap.

Fig. 11.14 (cont.)

11.3 *Inmarsat-A SES equipment* 263

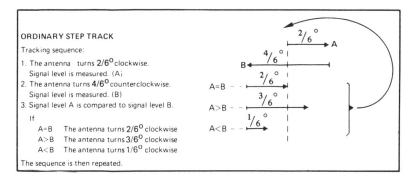

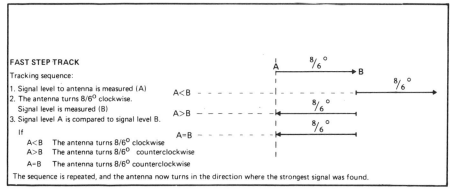

Fig. 11.15 Antenna ordinary step-track sequence and antenna fast step-track sequence (courtesy ABB NERA AS)

In Fig. 11.17 the antenna would automatically move from position 1 to position 2, the same AZ/EL angle with the satellite to maintain lock.

If the SES is processing a call when the antenna runs into the rewind area no rewind will occur unless it continues into the AZ limit regions. If this happens rewind will take place and the call will be lost. The AZ warning indicator on the operator display will light to indicate that antenna rewinding is in progress.

The necessity for antenna rewind is absolute. Using current technology it is not yet possible to have any other method of connection between antenna and cable feeds. Receive signal levels are extremely small and consequently vulnerable to high loss if a system of slip rings or some form of local transmission system was used. The problem of rewind does not apply to transportable Inmarsat-A equipment where the whole unit is manually rotated to point at a satellite. Clearly it would not be possible to fix the antenna of a SES and swing the ship in order to achieve lock! However, it should be noted that the use of an Inmarsat-A transportable land MES in a maritime environment is not permitted as it is not type-approved for a marine environment and is incapable of generating priority 3 distress alerts.

ABB NERA AS—SATURN 3 Series

ABB NERA AS Satcom Marine, based in Billingstadsletta, Norway, is a major manufacturer of all Inmarsat standards of equipment. The company was, in fact, the first European company to design and manufacture SESs. Currently, nearly 3000 SATURN terminals are in service on ships, on land and on off-shore oil platforms.

264 *The Inmarsat-A system*

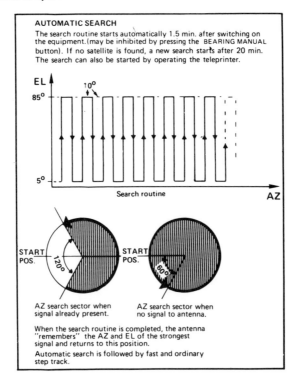

Fig. 11.16 Antenna automatic search (courtesy ABB NERA AS)

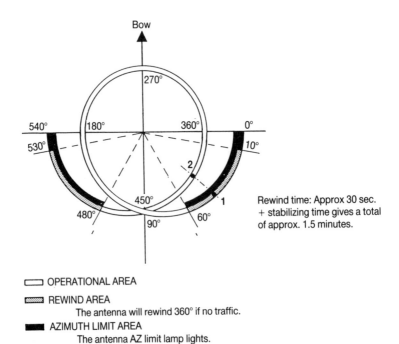

Fig. 11.17 Antenna azimuth limit (courtesy ABB NERA AS)

11.3 *Inmarsat-A SES equipment* 265

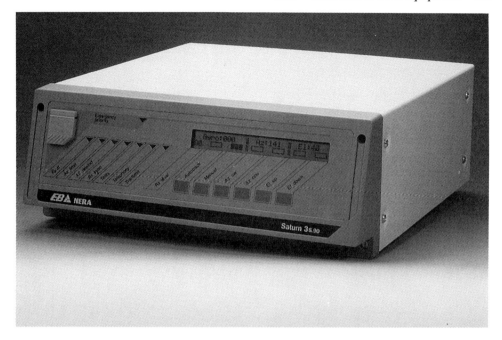

Fig. 11.18 The Saturn 3S.90 SES (courtesy ABB NERA AS)

ABB NERA also manufacture and supply CESs. The CESs at Eik, Thermopylae, Jeddah, Beijing and Perth were all supplied by the company. Aeronautical GESs in the UK, Singapore, Norway, Australia, France, Germany and the USA are currently being supplied by ABB NERA. The Saturn 3S.90. SES, shown in Fig. 11.18, is a good example of a modern satellite communications transceiver using advance technology. A system description follows based on the system block diagram in Fig. 11.19.

When in the idle state, the system continuously receives the assigned CSC, in this case TDM0 on 1537.750 MHz. The duplex filter diverts the received carrier from the output of the transmitter high-power amplifier (HPA) and applies it to the down converter via the low-noise amplifier (LNA). The LNA is particularly important in all satellite receiving systems. The required signal from the dish antenna is likely to be -128 dBm (0.1 μV) which must be amplified before processing can take place. All active devices produce noise. If noise is added to the signal at this stage it will simply be amplified along with the signal throughout the receiver. The LNA contains active devices made from gallium arsenide (GaAs FET) which have been proved to be very efficient low-noise devices.

The down converter is a signal mixer which mixes the amplified TDM signal with a locally generated frequency to produce the nominal Intermediate Fequency (IF) of 210 MHz. The VHF signal thus produced is much easier to handle than the L-band signal. The locally generated frequency 1330 MHz, traditionally known as the local oscillator signal, is derived from the high stability 5 MHz reference oscillator in the BDE. It is essential that this oscillator is highly frequency stable over a considerable range of operating conditions. Any change of frequency here, however slight, will cause major problems with down conversion, up-conversion, data processing and modulation/demodulation. The 5 MHz reference frequency is applied from the timing frequency unit (TFU) to the filter combiner in the BDE where it is multiplexed onto the coaxial cable

266 The Inmarsat-A system

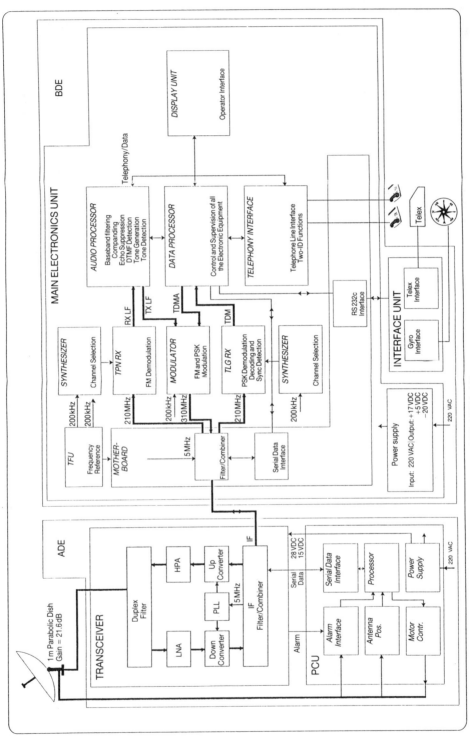

Fig. 11.19 Saturn 3S.90 SES system description (courtesy ABB NERA AS)

carrying the received signal from ADE to BDE. A 190 MHz oscillator is phase locked to this 5 MHz to maintain stability.

The received telex data now travel from ADE to BDE and through the de-multiplexer in the filter combiner to the telegraphy receiver (TLG). This unit performs PSK demodulation and differential decoding to produce the baseband data which is applied to the data processor (DP) for action as required. The TLG also performs synchronizing detection and demodulates the signal amplitude level. Telephony signals are transferred to the telephony receiver (TPN) where FM demodulation takes place before the baseband signal is applied to the audio processor board.

Data processor actions:

- Signal level—The DP continuously stores and compares the CSC received carrier amplitude level to a pre-determined level. If the DP registers a drop in this level it assumes that the dish antenna is in danger of losing satellite lock. The DP will now initiate a step track sequence to move the dish back to the maximum signal strength position. Command data from the DP is applied via the serial data bus to the ADE processor to command the operation.
- Synchronization—PSK demodulation requires accurate synchronization between the receive signal and a locally generated frequency. If synchronization is not accurate it is possible that demodulation will fail completely. The DP monitors this process and provides control to the synthesizer circuits as required.
- Baseband TDM data—The DP continuously monitors the Inmarsat SES IMN identification numbers transmitted in the data frame and the message type in order to determine what action to take. If the DP detects a broadcast message it will activate the telex unit via the interface and cause the message to be printed. Equally, if the DP detects its own IMN number it will activate the appropriate peripheral.

If the DP detects, by monitoring the common signalling channel, the announcement of traffic, duplex telex for instance, destined for itself it will automatically activate the required receive and transmit circuits. A 'control' output line from the DP to the synthesizer commands the correct channel selection in the TLG RX in order to match the assigned TDM channel identified by the NCS. In addition, the TLG RX is told which assigned time slot to search within the channel. A second 'control' line output from the DP is also applied to the MOD, which selects the corresponding duplex transmit paired channel and the assigned time slot. The DP activates the telex interface and the corresponding telex machine to enable operator access. Characters from the telex machine via the interface are encoded in TDMA format before being applied to the MOD for PSK modulation onto the nominal 310 MHz TX IF. This signal is coaxially coupled between BDE and ADE where it is up-converted to the correct channel frequency before being applied to the HPA. The 25 W of signal amplitude from the HPA is duplex filtered and applied to the dish for upward transmission to the satellite.

When telephony communication is required, different duplex channels will be selected by the DP as commanded by the CES. The process of selection is similar to that describing duplex telex channel selection. The primary difference between telex and telephony communications lies in the modulator and signal processing. The telephony modulator comprises a standard FM system with bandwidth strictly controlled by amplitude, and frequency limiting the baseband signal.

Audio processor This performs the following baseband signal processing functions on telephony receive and transmit.

268 The Inmarsat-A system

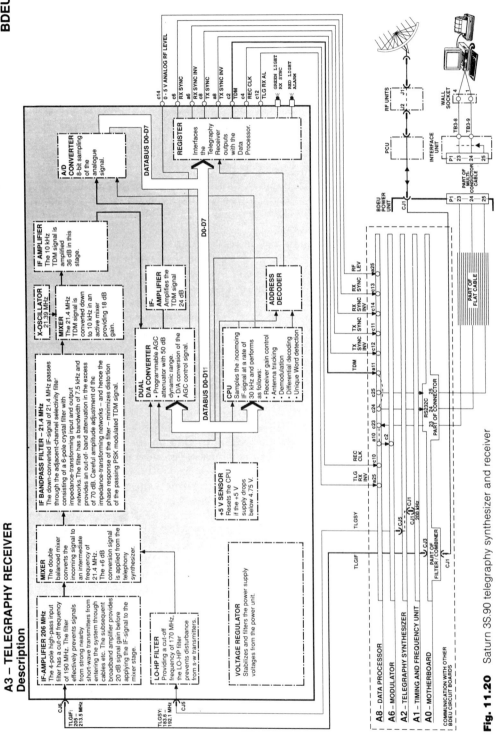

Fig. 11.20 Saturn 3S.90 telegraphy synthesizer and receiver

11.3 Inmarsat-A SES equipment 269

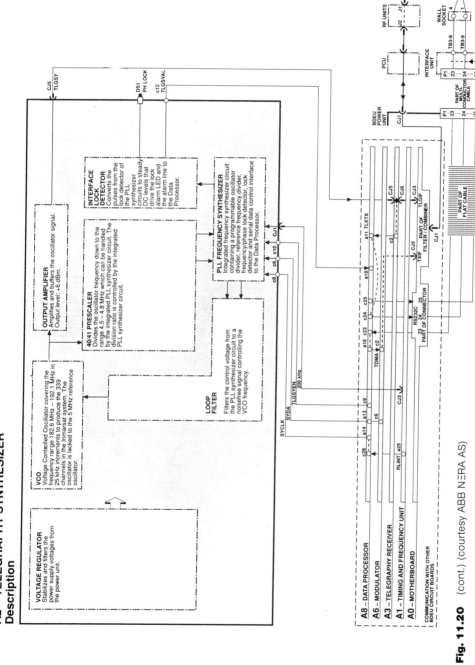

Fig. 11.20 (cont.) (courtesy ABB NERA AS)

- Signalling tone generation. This circuit generates and detects the 2600 Hz SF tone which confirms channel selection with the CES by handshaking.
- Baseband filtering. This filters the audio signals from the telephone handset to the band 300 to 3000 Hz.
- Companding. Emphasis circuitry is not used for baseband signal processing. When a voice channel is selected the baseband audio signal to be transmitted is compressed prior to modulation and the received signal is expanded after demodulation. The process of compression and expansion is known as companding.
- Echo suppression. In view of the great distances over which the RF carrier is transmitted in satellite systems, an echo will be produced. In order not to aggravate the situation, echo suppression circuits are used in the interface between the voice baseband signals and the telephone handset. The circuitry suppresses echoes which may be present when the handset is connected via a two-wire or a four-wire system.

The telegraphy receiver and synthesizer boards are described in order to illustrate the complexity of modern communications equipment. See Fig. 11.20.

In common with virtually all modern frequency synthesizers this equipment generates and maintains the stability of specific frequencies by using a phased locked loop (PLL). The PLL is standard and is essentially a voltage controlled oscillator (VCO) with a prescaler, a data input circuit and a loop filter forming the feedback circuits. In this instance, the VCO generates highly stable frequencies in the range 183.6 MHz to 192.1 MHz in 25 kHz steps in order to produce the 339 channels required. The VCO possesses a standard 5 MHz reference oscillator the frequency of which is divided down by the prescaler to the range 4.5 MHz to 4.8 MHz, the division ratio of which is controlled by the PLL synthesizer circuit. This circuit contains a programmable oscillator divider and other systems which produce an error frequency and hence a dc level from the difference between the frequency output of the prescaler and the inputs from the data processor board. The error voltage thus produced is filtered by the loop filter circuit before being applied to the VCO to control the output frequency, which is further amplified and coupled to the telegraphy receiver board as signal TLGSY.

The down-converted PSK TDM telegraphy signal from the ADU, in the range 205 MHz to 213.5 MHz, is coupled to the telegraphy receiver board as signal input TLGIF. In common with all superhet receivers, maximum signal gain is provided by the intermediate amplifiers (IF). Signal level changes are provided in decibel notation in order for the reader to relate the action of the receiver stages to the overall link-budget parameters explained in Chapter 2, Section 1.

Signal amplification of 20 dB is achieved at approximately 206 MHz in the first IF amplifier before the signal is further down-converted to 21.4 MHz. A received channel frequency will be in the range 205 MHz to 213.5 MHz which, when mixed with the output of the telegraphy synthesizer, signal TLGSY, will produce a further IF of 21.4 MHz. This frequency is always the same as it represents the difference between the two input signals to the mixer. Bandpass filtering, with a bandwidth of 7.5 kHz, provides full attenuation of out-of-band signals and thus limits interference. The PSK signal at 21.4 MHz is again down-converted to the standard telegraphy frequency of 10 kHz before receiving 36 dB of amplification in the amplifier following mixer two.

A D/A converter follows in which an analogue signal is produced for AGC purposes. Another IF amplifier provides 24 dB of gain before the TDM signal is A/D converted and applied as an 8-bit data signal, via the register, to the data processor and thence to the VDU.

The Saturn 3S.90 BDE electronics unit control panel is a good example of the

11.3 Inmarsat-A SES equipment

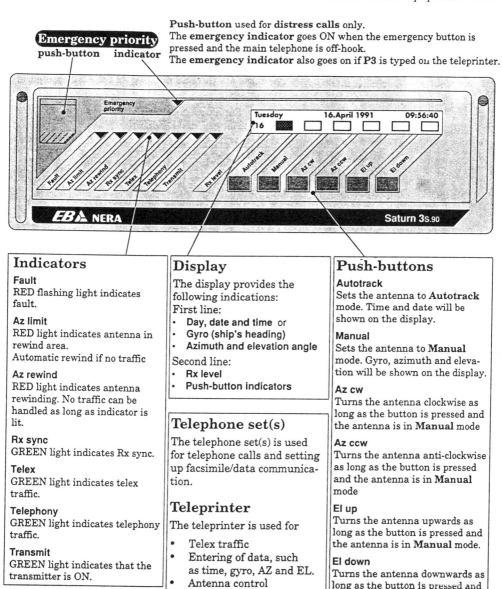

Fig. 11.21 Saturn 3S.90 operator control panel

versatility and simplicity of control which can be achieved with computer controlled equipment. The seven light emitting diodes, on the display, warn the operator of actions taking place in the unit as follows:

- a fault condition;
- the antenna has reached its AZ limit and a rewind sequence is occurring;
- the receiver has achieved synchronization with the satellite signal;
- telex or telephony traffic is in the process of being passed; and:

272 *The Inmarsat-A system*

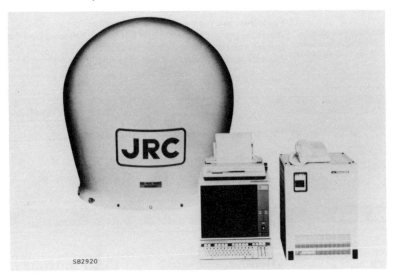

Fig. 11.22   Japan Radio Corporation JUE-45A Inmarsat-A SES)

- the transmitter is on.

The liquid crystal display provides a very accurate day/date lock indication, the ship's heading and the AZ/EL angles of the antenna.

Six push buttons enable the antenna to be automatically or manually controlled.

In the automatic mode the antenna will search for and maintain lock on a satellite.

When manual is selected the antenna may be moved clockwise (AZ CW) or anticlockwise (AZ CCW) in azimuth or upwards (EL UP) or downwards (EL DOWN) in elevation whilst the operator looks for the satellite using the tables shown in Fig. 11.14. The signal strength meter will indicate when a satellite has been found.

JRC—JUE-45A SES Series

JRC has been manufacturing the JUE series of Inmarsat-A SESs since 1977 and has become the world's largest supplier of ship earth stations. The company manufactures all types of satellite and radio communications equipment and has a proven record for reliability. The JUE-45A mark 2, fully compatible with the GMDSS is described below. See Figs 11.22 and 11.23.

The ADE comprises the antenna assembly and the antenna control unit.

- The antenna assembly. The antenna is a parabolic dish of 0.89 m diameter with cross dipole feed producing right-hand circular (RHC) polarization. Antenna gain exceeds 21 dBi with a radome fitted. The twin bandpass filter diplexer enables duplex communication to be effected. The received signal at approximately -125 dBm is amplified by about 40 dB in the LNA before being applied to the frequency translation unit (FTU). The FTU downconverts the received signal to 366.25 MHz (CSC channel reference) before being applied to the Triplexer which multiplexes three signals onto the common coaxial line between ADE and BDE. The three signals are: the receiver IF, the transmitter IF and the upline synthesizer signal from the BDE.

11.3 Inmarsat-A SES equipment

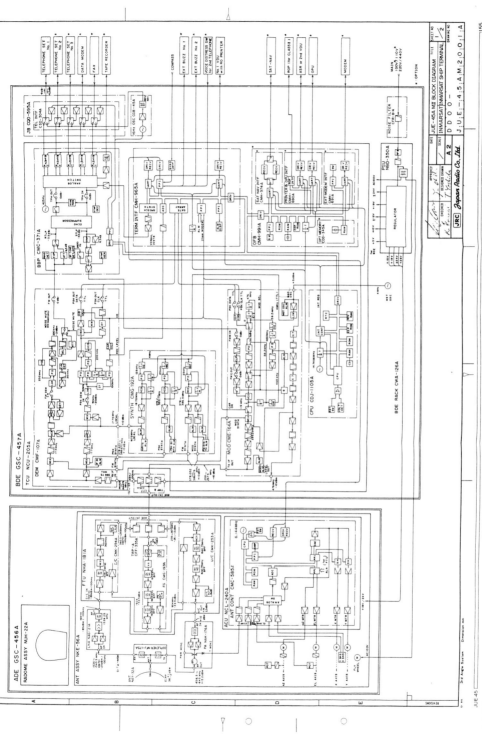

Fig. 11.23 JRC JUE-45A SES system diagram

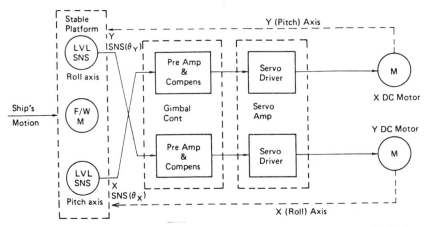

Fig. 11.24 X/Y axis control of an Inmarsat-A antenna assembly (courtesy Japan Radio Corporation)

The transmitter IF, in the range 259.0 MHz to 267.5 MHz, at this point is applied to the up-converter (U/C) where the frequency is increased to that required for transmission. The PA stage provides a gain of 0.1 dB/MHz which gives an overall output power of approximately 15.2 dBW. Output from the dish antenna has an effective isotropic radiated power (EIRP) of approximately 36 dBW.

- The antenna control. Antenna control is achieved on four axes X, Y, EL and AZ. The X and Y axes control the stable platform and compensate for the ship's rolling and pitching. The AZ axis compensates for the ship's turning and yawing. Both the AZ/EL axes perform satellite tracking in much the same way as previously described.

The X and Y axes components form the gimbals. Both axes are controlled to compensate for the ship's motion by combining passive and active controls. Passive control is achieved by the gyro effect of flywheels fitted to the four gyro's mounted on the stable platform. Active control is performed by the servo system which consists of a two-axes inclinometer, servo amplifiers and dc motors.

The level sensor LVL SNS mounted on the roll axis detects the inclination of this axis. The detected signal is fed to the Y servo line and drives the Y motor to compensate for roll axis inclination changes. The pitch axis is stabilized by a similar action performed by the X servo line.

Automatic satellite tracking is commanded by the ANT CONT computer which receives TDM signal level information from the BDE.

The BDE comprises the central processing unit (CPU), the synthesizer (SYNTH), the modulator (MOD), the demodulator (DEMOD), the baseband processor (BBP), the interface boards (TEL INTF) (TERM INTF), the optional interface board (OFB) and the display. The two signal processing boards are now considered. See Fig. 11.25.

- The modulator board (MOD). Two forms of modulation are required, PSK for telegraphy and FM for voice telephony. Since both modulators will not be required at the same time a common phase locked loop (PLL) circuit is used to achieve modulation with the correct system selected by IC11.

In the PSK mode PSK DATA and PSK CLOCK signals are fed to IC28 and IC30

11.3 Inmarsat-A SES equipment

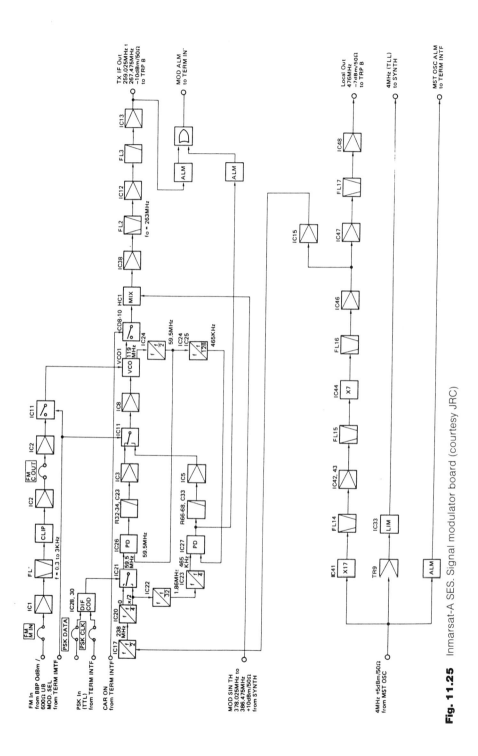

Fig. 11.25 Inmarsat-A SES. Signal modulator board (courtesy JRC)

276 The Inmarsat-A system

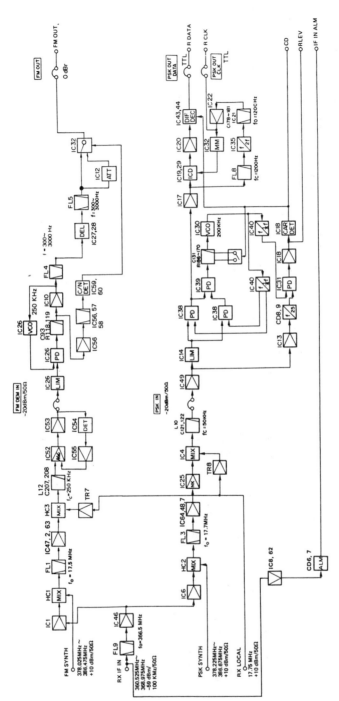

Fig. 11.26 Inmarsat-A SES. Signal demodulator board (courtesy JRC)

11.3 Inmarsat-A SES equipment 277

where they are differentially coded and then modulated. Two 90° phase shifted 59.5 MHz signals are fed to gate switch IC21. One of the signals is selected by the differentially coded data. The phase difference between two input signals is 90° and the PLL operates as a times two multiplier. The phase difference at the output is therefore 180°. The phase detector (PD) IC26 compares the phase of the signal from IC21 with that of the signal from IC24. The output of the PD is fed to VCO1 through the loop filter. One of the two outputs of the VCO (121 MHz) is fed to IC24, thus forming the loop. Another output is taken to the mixer HC1 through the carrier on/off switch (CD8-CD10). This signal (119 MHz) is mixed with the local oscillator signal in the band (378.025 MHz to 386.475 MHz) from the SYNTH and is converted to the IF band signal (259.025 MHz to 267.475 MHz) by HC1. The output is then coupled to the triplexer TRP B and up to the ADE.

In the FM mode the audio signal from the BBP is fed to the bandpass filter FL1 (300 Hz to 3000 Hz) through the U link. The output signal from FL1 is coupled to the amplitude peak clipper and then to the modulation input of VCO1. Two signals are fed to PD IC27. One is the reference frequency (464.84 kHz) which is obtained by the divider IC17, IC20, IC22 and IC23 from the 476 MHz signal. The other is the feedback signal (464.84 kHz) which is obtained from the divider IC24 and IC25 from the 119 MHz signal. The output of the PD is fed to IC8 through the loop filter. The output of IC8 is applied to VCO1 to form the loop. The FM signal thus obtained is converted to the IF band signal and applied to TRP B.

The primary function of the DEM board is PSK and FM demodulation as required. There are four outputs from the PSK demodulator circuitry (see Fig. 11.26):

- Received data (R DATA): 1.2 kbit/s.
- Received clock (R CLK): 1.2 kHz.
- Received level (R LEV): +5V dc.
- Carrier detection (CD) TTL.

Carrier recovery and demodulation Demodulation occurs in a PLL arrangement. IC38/39 are phase detectors the output of which is amplified and fed to the VCO IC30. IC30 output frequency is divided by four and applied to the phase detector IC31 with final carrier detection occurring in IC18.

Carrier detection The input signal is amplified by IC13, peak rectified by CD8 and CD9 and applied to IC31. When the two signals are applied to IC31 the input signal is rectified to the dc level (R LEV). R LEV is used for antenna tracking purposes.

Clock recovery The output of IC38 is amplified by IC17 and then squared in IC35. IC21 extracts the clock frequency from the output of IC35, which is then raised to the TTL level by IC22 and used as the recovered clock signal (R CLK).

Data recovery The output of IC17 is integrated in IC19 and applied to a comparator IC20. The output of IC20 is differentially decoded in IC43/44 to produce the R DATA signal. R DATA is controlled by the carrier detecting signal CD.

The FM demodulator uses a PLL. After limiting, the signal is fed to the phase detector IC26 which compares this input with that provided by the VCO IC26. The output of IC26 is filtered and applied to the dc amplifier IC10. The IC10 output feeds the VCO to form the loop. This output is also filtered to provide the baseband signal which is passed to the baseband processor board (BBP).

11.4 Inmarsat-A transportable equipment

A number of companies are manufacturing Inmarsat-A transportable equipment, which in effect is a take-anywhere 'satellite phone'. The major news gathering services of the world are the premier users of this system. Instantaneous on-scene reporting of major news items directly into the news wire services has now become commonplace using transportable equipment. Weighing between 20 and 50 kg, the equipment is self-contained in one or two suitcases, which contain a fold-away parabolic antenna, the communications electronics and the telephone or computer terminal. Such equipment must conform to Inmarsat-A specifications and be able to provide all Inmarsat-A services except priority 3 distress alerting.

The two main features which make this type of equipment different from an Inmarsat-A SES installation are the power supply and the antenna arrangement.

Power supply When used without the aid of a generator the transportable unit will operate from 24 V lead acid batteries. Power consumption on receive is approximately 75 W whereas on transmit it doubles to approximately 150 W, leading to a current drain of approximately 6.25 A. However, it should be remembered that the transmitter is not on the air continuously so the continuous current drain will not be that high. Also, a number of power saving systems are used such as auto power-down.

Antenna As with all Inmarsat-A land earth stations (LESs) the equipment is dependent upon antenna gain for both transmit and receive, and consequently uses a parabolic dish antenna. The antenna, between 0.8 and 1 m in diameter, is sectionalized or folded to fit in the carrying case. It must, therefore, be assembled and manually aligned with the satellite. This is done using charts like those produced by ABB NERA AS Communications (Fig. 11.15) or by following the simple routine produced by Inmarsat.

Mobile antenna alignment

The AZ/EL values to be used when aligning the antenna may be derived as follows.

- Determine the LES position relative to the satellite.
- Find the difference in longitude between the LES and satellite (e.g. if the satellite longitude is 15.5° west (AORE) and the LES longitude is 15.5° east, the difference is 31°).
- Round off the longitude to the nearest 5°.
- Round off the latitude of the LES to the nearest 5°.
- From the LES position, calculate in which quadrant the LES lies relative to the orbital location of the satellite.
- Read-off the AZ and EL figures from the relevant quadrant table. There are four published tables for when the LES is located NW, SW, NE or SE of the satellite. The NW table is shown in Fig. 11.27.

Example 11.1.
LES latitude   23° 45' North (25° North)
   longitude   139° 11' East (139° East)
Satellite longitude 180° East (POR)
Difference in longitude   180° − 139° = 41° (40° to nearest 5°)
   The LES latitude is 25° North (to the nearest 5°)
   The LES is NorthWest of the satellite, use the table shown in Fig. 11.27.
   At latitude 25° North and longitude 40° West, the AZ angle is 117° and the EL angle is 37°.

11.4 Inmarsat-A transportable equipment

LATITUDE (DEGREES NORTH)	DIFFERENCE IN DEGREES LONGITUDE BETWEEN SHIP AND SATELLITE																
	0	5	10	15	20	25	30	35	40	45	50	55	60	65	70	75	80
0	180	090	090	090	090	090	090	090	090	090	090	090	090	090	090	090	090
	90	84	78	72	66	61	55	49	44	38	33	28	22	17	12	07	02
5	180	135	116	108	103	101	099	097	096	095	094	093	093	092	092	091	091
	85	82	77	71	66	60	55	49	43	38	32	27	22	17	11	06	01
10	180	153	135	123	116	110	107	104	102	100	098	097	096	095	094	093	092
	78	77	73	69	64	59	53	48	43	37	32	27	21	16	11	06	01
15	180	161	146	134	125	119	114	110	107	105	102	100	098	097	095	094	093
	72	71	69	65	61	56	51	46	41	36	31	26	21	16	11	06	01
20	180	166	153	142	133	126	121	116	112	109	106	103	101	099	097	095	093
	66	66	64	61	57	53	49	44	39	34	30	25	20	15	10	05	01
25	180	168	157	148	139	132	126	121	117	113	110	106	104	101	099	096	094
	61	60	59	56	53	50	46	41	37	33	28	23	19	14	09	05	00
30	180	170	161	152	144	137	131	126	121	117	113	109	106	103	100	098	095
	55	54	53	51	49	46	42	38	34	30	26	22	17	13	09	04	00
35	180	171	163	155	148	141	135	129	124	120	116	112	108	105	102	099	-
	49	49	46	46	44	41	38	35	31	28	24	20	16	12	08	04	-
40	180	172	165	157	150	144	138	133	127	123	118	114	110	107	103	100	-
	44	43	43	41	38	37	34	31	28	25	21	18	14	10	07	03	-
45	180	173	166	159	153	147	140	135	130	125	121	116	112	108	104	101	-
	38	38	37	36	34	33	30	28	25	22	19	16	12	09	05	02	-
50	180	173	167	161	155	149	143	138	132	127	123	118	114	110	106	102	-
	33	32	32	31	30	28	26	24	21	19	16	13	10	07	04	01	-
55	180	174	168	162	156	150	145	139	134	129	125	120	115	111	107	102	-
	27	27	27	26	25	23	22	20	18	16	13	11	08	05	03	00	-
60	180	174	168	163	157	152	146	141	136	131	126	121	117	112	107	-	-
	22	22	21	21	20	19	17	16	14	12	10	08	06	04	01	-	-
65	180	174	169	164	158	153	148	142	137	132	127	122	118	113	-	-	-
	17	17	16	16	15	14	13	12	10	09	07	05	04	02	-	-	-
70	180	175	169	164	159	154	148	143	138	133	128	123	118	114	-	-	-
	11	11	11	11	10	09	09	07	07	05	04	03	01	00	-	-	-
75	180	175	169	164	159	154	149	144	139	134	129	124	-	-	-	-	-
	06	06	06	06	05	05	04	04	03	02	01	00	-	-	-	-	-
80	180	175	170	165	160	155	150	-	-	-	-	-	-	-	-	-	-
	01	01	01	01	01	00	00	-	-	-	-	-	-	-	-	-	-

For any calculation the table is arranged with the Azimuth reading above the Elevation reading e.g. **AZM**
ELN

Fig. 11.27 AZ/EL data for a vessel situated North and West of the satellite (courtesy Inmarsat)

280 The Inmarsat-A system

Forced clear down
Used if normal release for telephone (on-hook) or telex (off-line) does not clear the station.

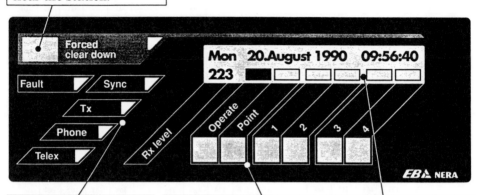

Indicators

Fault
RED flashing light indicates fault.

Sync
GREEN light indicates Rx synch.

Telex
GREEN light indicates telex traffic.

Phone
GREEN light indicates telephony traffic.

Tx
GREEN light indicates that the transmitter is ON.

Pushbuttons

Operate
Sets the CompacT to operating mode. Date and time is shown on the display.

Point
The antenna can be repositioned. Adjust to obtain maximum Rx level on the display.

1, 2, 3, 4
For testing only.

Display
The following may be displayed:
First line:
- **Day, date and time**
- **Rx level indicator**
- **Status/alarms indications**

Second line:
- **Rx level (3 digits)**
- **Pushbutton indicators**

Telephone set(s)
The telephone set(s) is used for telephone calls and facsimile/data communication.

Teleprinter
The teleprinter is used for
- Telex traffic
- Equipment tests

Fig. 11.28 Saturn Compact-T transportable LES operator control panel (courtesy EB NERA AS)

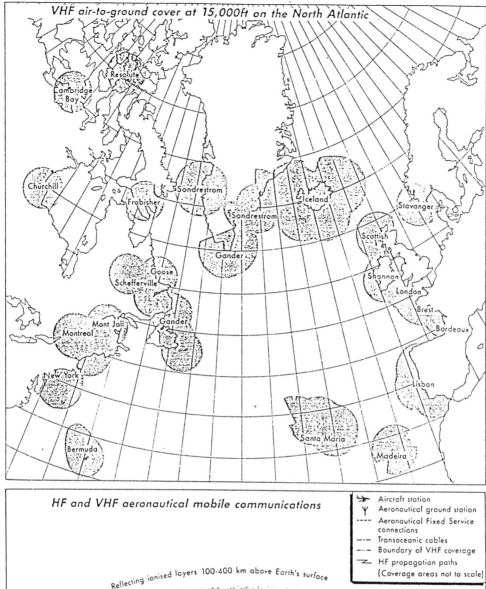
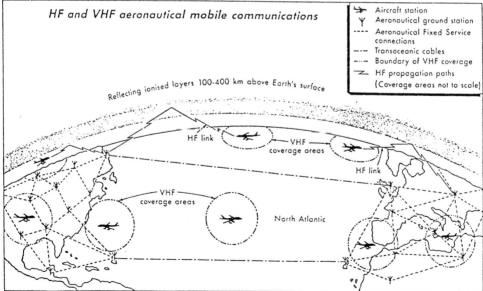

Fig. 11.29 VHF air-to-ground cover at 15,000 ft on the North Atlantic. HF and VHF mobile communications (courtesy *Ocean Voice*)

282 The Inmarsat-A system

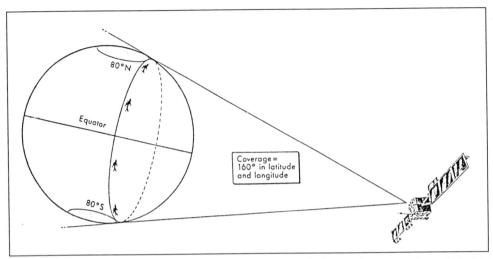

Fig. 11.30 Aero mobile communications coverage from a single geostationary satellite (courtesy *Ocean Voice*)

If the transportable equipment is mounted in a vehicle it is possible to use dynamically driven antennae which automatically track the satellite once aligned.

The SATURN Inmarsat-A Compact-T

After the antenna is assembled and the power connected, the unit runs through a diagnostic check as part of the 'system startup' procedure. Any error which is diagnosed will be indicated on the front panel by the red fault LED and a corresponding indication on the display. When the system is successfully initialized, it enters the operate mode. See Fig. 11.28.

The antenna is now aligned by pressing the 'point' command and using AZ/EL figures derived from the charts shown in Fig. 11.27. The signal strength is indicated on the display. The unit may now be used for telephone/telex traffic using the standard Inmarsat-A calling sequences.

11.5 Inmarsat-A aeronautical service

The impact of satellite communications on aeronautical communications has been dramatic. It is currently possible for a member of the flight crew or a passenger to have duplex communications with a mainland subscriber whilst in flight. Aeronautical communications in the past have relied upon the use of VHF and HF as part of the 'flight following' system.

Aircraft operators are required to file a flight-plan, indicating aircraft type, intended route and communications requirements, before an aircraft takes off. The plan is transmitted to radio stations via the point-to-point Aeronautical Fixed Telecommunications Network (AFTN). The Captain of the aircraft follows the flight-plan reporting to each air traffic control station in progression. Because of this previously planned programme of communications between aircraft and ground, if an expected communication does not materialize, an 'alert' is triggered. If the missing report cannot be attributed to equipment malfunction or natural causes an 'emergency situation' is

11.5 *Inmarsat-A aeronautical service* 283

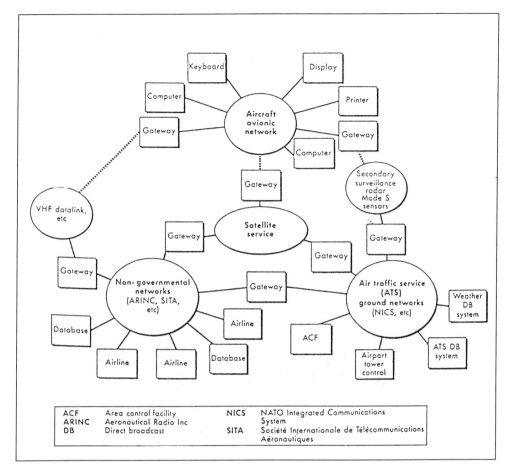

Fig. 11.31 FANS proposal for a future aeronautical communications system (courtesy *Ocean Voice*)

initiated. In this way communications via the fixed AFTN precede the aircraft, and ground stations are able to 'follow' the flight.

Flight control mobile communications take place on VHF within the band 117.975 – 136 MHz. As has been previously described communications using the VHF band are essentially line-of-sight. Consequently, a large number of VHF stations are required for flight paths anywhere over the surface of the earth. Increasingly, the HF band is being used for communications between aircraft and ground, producing problems because of the band's inherent fading and noise problems.

The VHF air-to-ground cover for the North Atlantic routes is shown in Fig. 11.29, along with VHF/HF aeromobile communications.

The use of satellites alleviates many of the existing aeronautical communications difficulties whilst introducing others. The use of geostationary satellites provides coverage of 160° of latitude and longitude, as shown in Fig. 10.3, but provides no coverage in the polar regions. As many of the earth's main aeronautical flight paths pass over the north pole it is essential that good communications be made available in the polar regions. This is achieved by the use of both geostationary satellites and the proposed highly elliptical inclined orbiting satellites. A satellite in such an orbit with a very high

284 *The Inmarsat-A system*

apogee is able to keep a broad area of the earth's surface in view and provide polar coverage. See Fig. 1.3.

The International Civil Aviation Organization (ICAO) has proposed the use of space technology to create a single integrated aeronautical system meeting the diverse needs of air traffic control, airlines and passengers. The system, known as the Future Air Navigation System (FANS), is shown in Fig. 11.31.

12
The Inmarsat-B system

12.1 Introduction

The Inmarsat-B system is to be the ultimate successor to the ageing Inmarsat-A system for the provision of mainstream professional mobile communication services into the next century. Inmarsat-A system technology is now ageing, although it has been expanded to the effective limit of its analogue technology. The number of Inmarsat-A users has grown dramatically as have their needs. Inmarsat-A is supporting far more traffic than was intended in the original design and, consequently, a change to new digital technology, as utilized in the Inmarsat-B system is necessary. Continuing system development is driven by consumer demand, and better system access, better quality and more sophisticated services, amongst other things, are required. These factors in no way detract from the capabilities of the Inmarsat-A system, which has been the workhorse of mobile satellite communications for nearly two decades. Indeed, Inmarsat-A will continue to provide communications services, in parallel with Inmarsat-B, until at least the year 2005 when it will be totally replaced by the new system.

The new fully digital Inmarsat-B system became operational in 1993 to provide global mainstream satellite communications between fixed stations (LES) and mobile terminals (MES).

Inmarsat-B coverage, performance, availability and SES environmental conditions are identical to Inmarsat-A and are fully compatible with IMO requirements for distress operation within the GMDSS radionet. It should be noted however, that the technology of Inmarsat-B is far superior to that used in the Inmarsat-A system and, consequently, the two systems are not technologically compatible.

12.2 Outline system specifications

Introduction

Inmarsat-B system design is based on the very latest digital systems technology, which has been used extensively in order to reduce satellite channel requirements (when compared with the Inmarsat-A system) by approximately 50%. The use of digital coding and modulation techniques, voice-activated carrier switching and forward power control on the SCPC forward communications carriers has led to both bandwidth requirements and power reductions of this order.

SES power efficiency is further improved by providing for automatic adjustment of EIRP settings over a range of values depending upon satellite transponder gain and receive G/T values. The extensive use of phase shift keying, offset-quadrature PSK modulation on the SCPC channels, bi-phase BPSK modulation on the TDM channels and FEC particularly enables bandwidth savings to be made.

As an example, the voice channel bandwidth has been reduced from 50 kHz to 20 kHz and the typical transmit power output from a MES has been reduced from 40 W to 25 W when compared with the Inmarsat-A system.

Inmarsat-A voice baseband processing using FM has been replaced with digital voice coding at 16 kbit/s information rate with adaptive predictive coding with maximum likelihood quantization (APC-MLQ). The use of digital speech processing enables large bandwidth and power savings to be made whilst maintaining the excellent quality produced by Inmarsat-A MESs. A combination of all these factors plus the use of unpaired transmit and receive channel frequencies leads to a considerable improvement in the use of the frequency spectrum available.

Inmarsat-B has available for use a much greater radio frequency spectrum band, which at 20 MHz is almost three times the space available to Inmarsat-A operations at 7.5 MHz. RF communications channels used in the Inmarsat-B system are demand assigned (unpaired) in order to provide flexible channel assignment. This flexible system considerably improves spectrum management over the ageing Inmarsat-A system where channels were paired with a fixed separation of 101.5 MHz.

Spectrum efficiency will be further improved with the use of the spot beam communication facilities to be used on the new Inmarsat INM3 satellites. Instead of using the one-third earth coverage transponder, two ships in the same ocean region may be able to use different spot beam transponders with a much narrower beamwidth enabling simultaneous channel re-use to be made in the same ocean region.

Figure 12.1 compares the technical specifications of the Inmarsat-B system with those of the Inmarsat-A, C and M systems.

The return request channel and the telex message channel details refer to 1/2 FEC. Forward error correction (FEC) is used in order to reduce satellite and SES EIRP requirements without compromising the bandwidth economy previously achieved. In addition to improving efficiency, the use of FEC provides an efficient error correction and detection protocol in the transmission path. In common with all error detection and correction systems FEC requires that a number of data bits are added to the information bits to form an appropriate bit 'pattern' which may be checked at the receiver. Adding redundant bits in this way greatly increases the number of bits to be transmitted leading to greater inefficiency in the communications system. FEC used in a satellite link uses a procedure which operates continuously on the bits in the data stream using a technique called Viterbi Convolutional Encoding. It is only necessary to appreciate that using this system different levels of correction can be achieved.

The notation FEC 3/4, 1/2 etc indicates the level of encoding and signifies the ratio of output bits to input bits. FEC 1/2 encoding, as used in the Inmarsat-B channel, indicates that the number of output bits from the encoder will be approximately twice the number of input bits. FEC 3/4 indicates four bits out of the encoder for three bits input and so on. FEC 1/2 consequently provides a greater error detection and correction method than FEC 3/4.

The space segment

Inmarsat-B uses the same space segment as all the other Inmarsat systems, as described for the Inmarsat-A system. The new INM2 satellites carry a number of transponders, and part of one of these is dedicated to Inmarsat-B operation. The next generation of satellites, INM3, currently under development, will also carry spot beam facilities for use with Inmarsat-B.

The use of spot beams is another significant difference between the Inmarsat-A and B systems. Using spot beams will lead to greater economy of operation in regions of high traffic density and make channel re-use in an ocean region possible.

12.2 Outline system specifications

PARAMETER	A	B	M	C	AERO
Digital Data Services	–	9.6/16 kbit/s	2.4 kbit/s	600 bit/s	9.6 kbit/s
Channel Assignment SCPC	NCS	NCS	NCS	NCS/CES	GES
TDM/TDMA	CES	CES	–	–	GES
SCPC Frequency Assignments	Paired	Unpaired	Unpaired	Unpaired	Unpaired
Spot-beam Identification	–	Yes	Yes	Yes	Yes
Bulletin Board	(Service Announcement)	Yes (Forward & return signalling channels)	Yes (Forward & return signalling channels)	Yes (Log-in)	Yes (Log-in)
Forward Signalling/Assignment Channel	As Telex	TDM, BPSK	TDM, BPSK	TDM, BPSK	TDM, BPSK
Return Request Channel	Aloha BPSK (BCH) 4800 bit/sec	Aloha O-QPSK (1/2 – FEC) 24 kbit/s	Slotted Aloha BPSK (1/2 – FEC) 3 kbit/s	Aloha BPSK (1/2 – FEC) 600 bit/s	Aloha BPSK (1/2 – FEC) 600 bit/s
Signal Unit Size	–	96 bits	96 bits	Variable	96 bits
MES G/T	–4 dBK	–4 dBK	–10 or –12 dBK	–23 dBK	–13 dBK
MES Antenna diameter (typical)	0.9 to 1.2 m	0.9 m	0.3 to 0.5 m	0.1 m	0.5 m
MES EIRP	36 dBW	33–25 dBW	19–27 dBW	11–16 dBW	13.5–25.5 dBW
MES HPA (typical)	Class – C 30–40 W	Class – C 25 W	Class – C 20 W	Class – C 10 W	Class – C 60 W
Transmit Chains	2	1	1	1	1
Receive Chains	2	1	1	1	1 (min)
Synthesizer Step Size	25 kHz	10 kHz	5 kHz	5 kHz	17.5 kHz
Voice Channel Coding/Modulation	NBFM, 2:1 Companding	16 kbit/s APC, O-QPSK	4.8 kbit/s IMBE coding O-QPSK	–	9.6 kbit/s APC O-QPSK
Voice Channel FEC	–	3/4	3/4	–	1/2
Voice Channel Rate	–	24 kbit/s	8.0 kbit/s	–	21 kbit/s
Sub-band signalling rate	–	2 kbit/s	400 bit/s	–	864 bit/s
Voice Channel interleaving	–	–	–	–	Yes
Voice Channel Bandwidth	50 kHz	20 kHz	10 kHz	–	17.5 kHz
Forward Carrier Vox	Yes	Yes	Yes	–	Yes
Forward Power Control	–	Yes	Yes	No	Yes
Satellite L-band EIRP	17.5 dBW	16 dBW	19 dBW	–	23 dBW
Voice-band Data Rate	up to 9.6 kbit/s	up to 2.4 kbit/s	–	–	up to 300 bit/s
Tlx Msg channel modulation/coding: Forward Return	TDM BPSK TDMA BPSK	TDM BPSK (1/2-FEC) TDMA O-QPSK (1/2-FEC)	–	BPSK(1/2-FEC) BPSK(1/2-FEC)	–
Tlx/Msg channel rate: Forward Return	1200 bit/s 4800 bit/s	6 kbit/s 24 kbit/s	–	600 bit/s 600 bit/s	–
Tlx/Msg channel bandwidth: Forward Return	25 kHz 25 kHz	10 kHz 20 kHz	–	2.5 kHz 2.5 kHz	–
Tlx Capacity per carrier	22	56	–	variable	–

Fig. 12.1 Inmarsat-B comparison with other Inmarsat standards

Inmarsat-B is able to support the new Inmarsat High Speed Data Services and, consequently, is fully compatible with the CCITT integrated services digital network (ISDN).

The ground segment

The system utilizes the same ground control, access and communications station network as previously described in Chapter 10, Section 2. Most of the existing NCSs and LESs currently providing Inmarsat-A services have refitted signalling, access and control equipment in order to operate within the new technology used in Inmarsat-B. Enhanced user services are offered at the discretion of the telecommunications company owning the LES, and thus will differ.

A number of Inmarsat-B LESs exist in each ocean region to provide the fixed end of the communications link. It is the responsibility of these LESs to assign the TDM/TDMA channel (data communications).

As with other Inmarsat systems, the NCS in each ocean region plays a vital role in the operation of the system by co-ordinating access to the communications channels. With reference to Fig. 10.8 it can be seen that the NCS is provided with signalling links with both the MES and the LES.

The NCS is responsible for the assignment of communication channels for SCPC (voice) communications. In addition, the NCS monitors all access, control and signalling channels and maintains a database on MES status etc.

Distress requests and replies are monitored by the NCS, which is able to take executive action should no reply to a distress call from an SES be forthcoming. Full details of distress procedures can be found in the companion volume *Understanding GMDSS*.

Ship earth stations

Inmarsat-B SESs are similar in design to those currently in use in the Inmarsat-A system. The equipment operates in the transmit band 1626.500 to 1646.500 MHz and in the receive band 1525.000 to 1545.000 MHz. This wider band of frequencies coupled with reduced bandwidth requirements and the introduction of new technology essentially leads to more channels being available. Also, channels are unpaired and are assigned on a demand basis from a pool of frequencies, leading to very efficient management of the limited spectrum available.

Typical ADE equipment is very similar in appearance and size to that used in the Inmarsat-A system. This includes a parabolic antenna of approximately 0.9 m in diameter, inside a protective radome, with full stabilization and satellite tracking assemblies. The antenna produces a gain of approximately 20 dB which easily satisfies the Inmarsat signal specification of -4 dBk receive G/T at 5° satellite elevation and a nominal transmit EIRP of 33 dBW per carrier, typically 25 W. Transmit EIRP levels are capable of being automatically reduced in two steps of 4 dB each for use in the future with INM3 satellite spot beams and those with higher transponder receive gains.

As with the Inmarsat-A system the antenna assembly is fully stabilized in order to satisfy the IMO regulations for compliance with the SOLAS carriage requirements for vessels operating within the GMDSS radionet.

Vessel motion specifications include, $\pm 30°$ roll, $\pm 10°$ pitch, 6° turning rate and 1°/second^2 angular acceleration rate.

The use of the same antenna and stabilization specifications of the Inmarsat-A system enable Inmarsat-B equipment manufacturers to reduce development costs significantly by the re-use of major portions of existing designs.

12.2 Outline system specifications

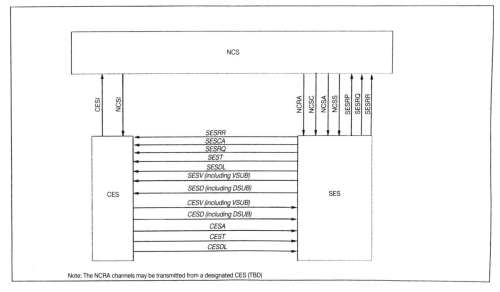

Fig. 12.2 Inmarsat-B system functional channel configuration

With the exception of voice communication, the Inmarsat-B BDE equipment closely resembles the design of an Inmarsat-C terminal based on a personal computer.

There are two classes of Inmarsat-B SES as follows:

- Class 1 SES, telephone and telex services,
- Class 2 SES, telephony only (plus the capability to receive Inmarsat service announcements).

Access control and signalling

To achieve and maintain good control of system access and communication channels the NCS, CES and SES utilize various signalling channels as detailed below.

Assignment channels CES to SES and SES to CES
SESRQ (SES Request channel). A burst mode channel using the ALOHA random access protocol to carry SES signalling information to the selected CES. Specifically used for the access of main request and acknowledgement messages. Monitored by the NCS for distress purposes.
CESA (CES Assignment channel). A TDM continuous channel used to carry CES signalling messages to SESs. Used specifically for channel assignments for calls which use TDM/TDMA communications channels (Telex).
NCSA (NCS Assignment channel). A TDM channel used to carry channel assignment messages to SESs for calls which use SCPC communications channels (voice).

Information and monitoring channels
CESI (CES Inter Station Channel). A TDM continuous channel used by each CES to carry signalling information to the NCS.
NCSI (NCS Inter Station Channel). A TDM continuous channel used from the NCS to each CES in the network to carry signalling information.
NCSC (NCS Common Signalling Channel). A TDM channel used to carry NCS

290 *The Inmarsat-B system*

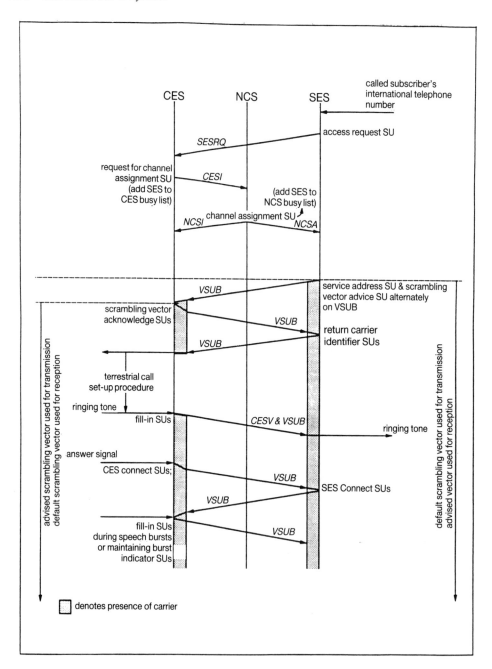

Fig. 12.3 Ship-originated duplex telephone call set-up sequence

signalling messages from shore to SESs for call announcement broadcasts, selective clearing, network status information and Bulletin Board.
SESRP (SES Response Channel). A burst mode TDM channel used to carry SES signalling information to the NCS. Specifically, the response information required for

shore-originated calls, and for the acknowledgement of SES group ID downloading messages.
NCSS (NCS Spot Beam Channel). A TDM continuous channel to be used in the future, in the C-L direction (one frequency per spot beam) to enable SESs to identify the satellite spot beam in which they are located. INM3 satellites will carry spot beam facilities.

SCPC communication channels. (Access controlled by the NCS)
SESV (SES Voice) and CESV (CES Voice). Single-channel-per-carrier (SCPC) digital voice channels supporting a voice coding rate of 16 kbit/s with Adaptive Predictive Coding (APC), CCITT Group 3 facsimile and 9.6 kbit/s data services. Used in both the forward (to ship) and reverse (from ship) directions.
   The channels contain sub-band signalling information (VSUB) for channel control during the communications link.
SESD (SES Data) and CESD (CES Data). SCPC data channel with similar usage and sub-band signalling as with the SCPC channels. The forward data channel is labelled CESD and the return channel SESD. Sub-band signalling is labelled DSUB.

TDMA/TDM Communications Channels. (Access controlled by the CES)
SEST (SES Telex channel). A TDM channel used to carry telex to a CES.
CEST (CES Telex channel). A TDM channel used to carry telex to the SES.
SESDL (SES Low Speed Data Channel). A TDMA burst mode channel used to carry low-speed data to a CES in an asynchronous mode at information rates up to 300 bit/s.
CESDL (CES Low Speed Data Channel). A TDM continuous channel used to carry low-speed data to a SES. Same rate as for SESDL.

Call set-up procedures

Duplex radiotelephony from the mobile
   The set-up sequence is initiated by a call, which includes the following information, from the SES to the selected CES on the SESRQ channel. The access-request message, Fig. 12.5b includes the following information:
CES identification;
SES identification;
Request service type. (Telephony, telex or data);
Priority. (Normal or priority 3, maritime SES only);
SES antenna azimuth zone (for frequency assignment purposes);
SES antenna azimuth angle (distress priority only);
SES antenna elevation angle (distress priority only);
Satellite spot beam identification (for future use).
   The information is contained in the access-request message O3H (S3) (or message 04H (S4) if distress is selected) which includes azimuth and elevation angle zones to facilitate location of the casualty in future. See Fig. 12.3 which illustrates the action sequence between the signalling units (SU).
   Assuming the access request carries normal priority, the CES returns a request-for-channel-assignment message using the CESI channel to the NCS. The NCS will respond to the SES over the NCSA channel by sending a SCPC-channel-assignment message 06H (S6) and, using the NCSI, informing the CES of the channel assignment. The SES identity is then included in the network SES busy list at the NCS and the local SES busy list at the CES.
   Note: when the new generation of INM3 satellites are in use, the NCS will attempt to

292 The Inmarsat-B system

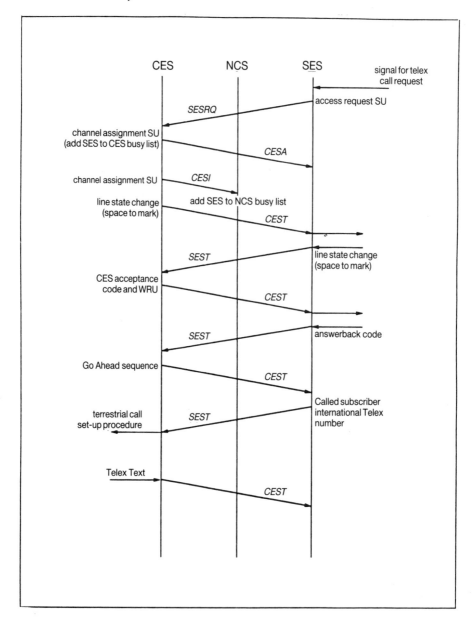

Fig. 12.4 Ship-originated duplex telex call set-up sequence

assign a spot-beam channel frequency. If one is not available, or until the time when INM3 satellites are in use, a global-beam channel is assigned. This sequence establishes the communication channel link.

The SES now transmits the called party identification to the CES via the forward sub-band channel (VSUB) using the service-address message along with a scrambling-vector message to ensure call security is maintained. The messages are sent continuously until acknowledged by the CES using the scrambling-vector-acknowledgement message on the forward sub-band channel VSUB. The duplex voice channel now starts to carry traffic.

Duplex radiotelephony from shore
A CES receives, via landline from the Maritime Satellite Switching Centre, the number of the SES to be called. A 'T' digit is used to indicate that the Inmarsat-B system is to be used for the call. The CES now sends a request-for-call-announcement message via the CESI channel to the NCS. If the SES identification is not included on the busy list a call-announcement message 01H (S1) is sent on the NCSC to the SES.

The SES replies using a response-message on the SESRP channel to the NCS. The procedure then continues as before.

Duplex radiotelegraphy from the mobile (telex)
The telex call set-up initially follows the same lines as those outlined for voice communications with the transmission by the SES of an access-request message 03H (S3) to the CES on the SESRQ channel.

Telex communication uses TDMA/TDM channels which are assigned by the CES. Consequently the CES now transmits a channel-assignment message 07H (S7) to the SES using the CESA channel. This message includes the communications channel to be used plus the time-slot assignment information. The SES identity is now added to the CES local busy list. The CES informs the NCS by means of the CESI channel of the channel-assignment and the SES is added to the network busy list.

Figure 12.5 shows a sample of the access signalling units mentioned above.

INMARSAT-B TDM/TDMA CHANNEL ASSIGNMENT Frame S7									
BIT number								OCTET No
8	7	6	5	4	3	2	1	
Message type								1
SES ID								2
^								3
^								4
Spare			SES transmit time slot					5
SES EIRP		SES receive time slot						6
SES receive channel number								7
^								8
SES transmit channel number								9
^								10
CCITT CRC								11
^								12

Fig. 12.5 Inmarsat-B signalling units

294 The Inmarsat-B system

8	7	6	5	4	3	2	1	OCTET No
INMARSAT-B ACCESS REQUEST (NON-DISTRESS) Frame S3								
BIT number								OCTET No
Message type								1
SES ID								2
SES ID								3
SES ID								4
CES ID								5
Azimuth angle zone				Priority			Elevation angle zone	6
Service nature			Service type					7
Channel parameters								8
Terrestrial network ID								9
init repeat	spare			Spot beam ID				10
CCITT CRC								11
CCITT CRC								12

(Courtesy Inmarsat)

Fig. 12.5 (cont.) (b)

13
The Inmarsat-C system

13.1 Introduction

Inmarsat-C is an all-digital text/data system which operates between the mobile earth station (MES) and a 'gateway' or land earth station (LES). The principal advantages of the system, when compared with Inmarsat-A, are that it is low cost, small and uses a very small omni-directional antenna. Its main disadvantage is of course that voice communication is not possible. It should be remembered however, that information transfer by voice communication is very slow and generally requires more bandwidth per carrier than data communications.

This go-anywhere system became fully operational, after a long period of pre-operational trials, in January 1991 and has been further developed since that date. Inmarsat-C mobile terminals generally use a laptop computer as the operator interface and this becomes the data terminal equipment (DTE), whereas the interface between this unit and the satellite is known as the data circuit terminal equipment (DCE).

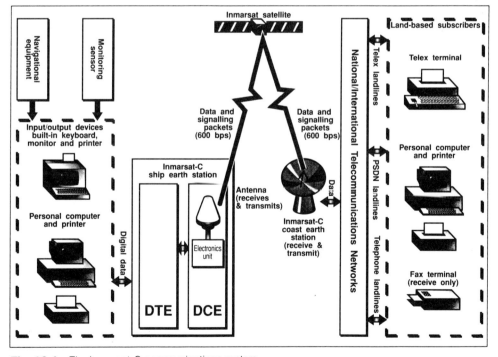

Fig. 13.1 The Inmarsat-C communications system

13.2 Outline system specifications

Inmarsat-C terminal's store-and-forward feature enables the system to interface with any terrestrial message or data networks including telex, X.25, X.400 and voice band data networks via the public service telephone network (PSTN) or the integrated services digital network (ISDN). Data is transferred between MES and LES at the rate of 600 bit/s, for second generation satellites and 300 bit/s for first generation satellites, in a 'packetized' stream which is formed of interleaved 8.64 second frames providing

		MARECS	INTELSAT-V MCS	INMARSAT SECOND GENERATION
DC Power (end of life) (W)		723	297	1200
Eclipse capability		Full	Full	Full
C-to-L repeater				
Receive band	(MHz)	6420–6425	6417.5–6425	6425–6441(1)
Transmit band	(MHz)	1537.5–1542.5	1535–1542.5	1530–1546(1)
Receive G/T(2)	(dBK)	−15	−12.1	−14
L-band EIRP(2)	(dBW)	34.5	33.0	39
C-band antenna Receive polarization		RHC(3)	RHC	RHC
L-band antenna Transmit polarization		RHC	RHC	RHC
Capacity (Standard-A voice channels)		60 (80)5	35 (50)5	125 (250)5
L-to-C repeater				
Receive band	(MHz)	1638.5–1644	1636.5–1644	1626.5–1647.5(1)
Transmit band	(MHz)	4194.5–4200	4192.5–4200	3600–3621(1)
Receive G/T	(dBK)	−11.2	−13.0	−12.5
C-band EIRP(2)	(dBW)	16.5	20.0	24
L-band antenna Receive polarization		RHC	RHC	RHC
C-band antenna Transmit polarization		LHC(4)	LHC	LHC
Capacity (Standard-A voice channels)		90	120	250

Notes:

(1) Includes 1 MHz at upper end for aeronautical communications
(2) Minimum value at edge of coverage
(3) Right-hand circular polarization
(4) Left-hand circular polarization
(5) Capacity with voice activated carrier suppression

Fig. 13.2 Inmarsat satellite transponder characteristics

13.2 Outline system specifications 297

10,000 frames per day. The relatively slow rate of data communications coupled with an efficient error checking system ensures that even when signal fading exceeds one second, an exceedingly long duration in data communication systems, there will be no data packet loss.

The space segment

The Inmarsat-C service was developed to use a mixture of satellites, each with varying characteristics. First generation satellites, MARECS and INTELSAT-V MCS, provide lower transponder gains than the second generation INM2 satellites. Consequently, an overall reduction in carrier-to-noise power spectral density (C/N_o) exists of approximately 3 dB, or half power. This, in turn, leads to the need for a reduction in data transmission speeds, from MES to LES, from 600 bit/s to 300 bit/s when using first generation satellites. The change is fully automatic and is based on information, received on the NCS common channel, via the NCS bulletin board. First and second generation satellite transponder characteristics are shown in Fig. 13.2.

Each of the four satellite ocean regions are supported in the same way as those described for Inmarsat-A operation, which of course uses the same satellites.

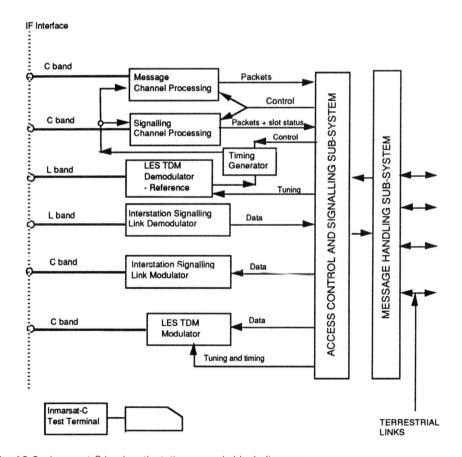

Fig. 13.3 Inmarsat-C land earth station example block diagram

298   *The Inmarsat-C system*

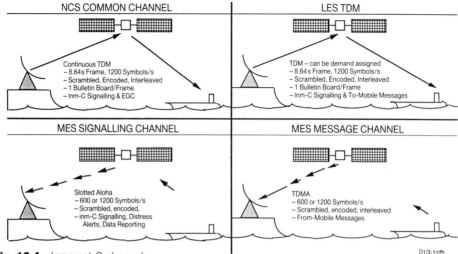

Fig. 13.4 Inmarsat-C channels

The ground segment

The network control centre NCC at the Inmarsat headquarters in London, supervises the four interactive NCSs using interstation signalling links (ISL). The four NCSs are as follows; the AORE and AORW NCS is Goonhilly Down in the UK, the IOR NCS is Thermopylae in Greece and the POR NCS is at Singapore. Unlike Inmarsat-A NCSs, the stations communicate using ISLs and consequently construct an updated scenario of the system and its active status. NCSs communicate with both MESs and each LES operating in an ocean region. An example is shown in Fig. 13.3.

LESs provide the gateway through which traffic flows between the satellite and land-based communications networks.

The signalling processing stages of an Inmarsat-C LES shown in Fig. 13.4 are typical of the structure used to provide the interface between the RF stages and the landline network.

Currently there are 12 LESs operating in the system with a further ten planned for the near future.

Access control and signalling

A number of different types of channels are used in the Inmarsat-C system. Using the channels, information is transferred in data packets each of which contains a checksum permitting ARQ error correction to be made.

All channels use a common standard as follows.

Modulation: BPSK 1200 symbols/s (second generation satellites)
Information rate: 600 bit/s (300 bit/s for the return channel on first generation satellites)
FEC coding: forward direction—1/2 rate convolutional coding with interleaving
                  reverse direction—1/2 rate convolutional coding with interleaving for the MES message channel.
Frequency bands: TX—1626.500 to 1646.500 MHz
                  RX—1530.000 to 1545.000 MHz
                  both in 5 kHz steps.

13.2 *Outline system specifications* 299

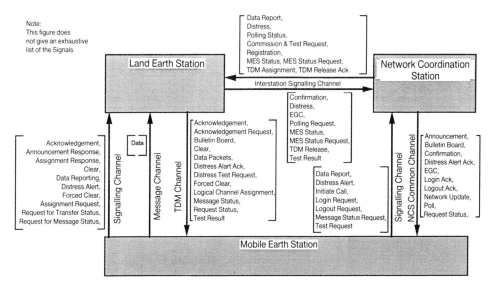

Fig. 13.5 Inmarsat-C channels and signals

Types of channel
Refer to Fig. 13.5.

Signalling channels
MES signalling channels. Used during the call set-up phase. These channels are used by the MES, in TDM format, to transmit signalling packets of information to both the NCS and the LES. Each LES has one or more signalling channels assigned to it and, in addition, there is a signalling channel associated with each NCS common channel. The channels are used, amongst other things, for requesting a message channel assignment from the LES and to the NCS for login purposes.
  Access is achieved by using a combination of slotted ALOHA with reservation, the operation of which is described in Chapter 3, Section 7.

Fixed station signalling channels
NCS – NCS signalling channel: Interstation signalling channels which carry system information between ocean regions.
NCS – LES signalling channel. Another channel carrying system information between the NCS in an ocean region and all its attendant LESs.

TDM channels
NCS common channel. A continuously transmitted TDM channel to which all MESs, in a particular ocean region, are tuned when in the idle state. The channel carries the important bulletin board packet which contains the information required by an MES to gain access to the system. The bulletin board is transmitted as the first packet in each TDM field. It contains information on the static operational parameters of both the NCS and LES which are transmitting the TDM channel. Access to the channel is on a first-come first-served basis for packets of the same priority. There are three levels of priority:

- Inmarsat-C call announcement, polling, EGC distress priority messages, distress alert acknowledgement;

300 The Inmarsat-C system

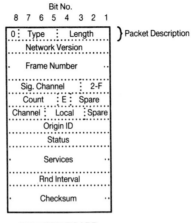

Fig. 13.6 Bulletin board signal channel descriptor frames (courtesy Inmarsat)

- Inmarsat-C signalling and
- other EGC messages.

LES TDM channel. This channel has the same structure as the NCS common channel and contains, amongst other information, call set-up signalling. Access to the channel is on a first-come first-served priority basis. There are four levels of priority:

- distress priority packets,
- logical channel assignments,
- other protocol packets, and
- messages.

TDM channel information can be displayed on the VDU of an Inmarsat-C SES. This information may be as follows.

- Frame number. The number of the current frame in use. Each 24 hour period is divided into 10,000 frames (8.64 seconds per frame) which are reset to 0000 each midnight UTC and advance in number throughout the day.
- TDM channel number. The channel number represents the frequency of the TDM channel.
- Bulletin board. This is a unique packet of information which indicates the operational status of the NCS and CES being used. The slot also contains the current frame number. Reception of consecutive bulletin board packets is used by the receiver to measure the quality of the reception. This is displayed as the 'bulletin board error rate' which, suprisingly, should be a low value. Zero is the ideal state. See Fig. 13.6.

Communication channel
Message channel. Used for the transfer of traffic. LES message channels are used by the MES to send, store and forward messages to an LES. Access to the channel is allocated on a TDMA basis. The access protocol used is a form of slotted ALOHA with explicit reservations. Because more than one MES may be sharing the same message channel, the destination LES allocates a transmission time to each MES. Once assigned a start time, the MES transmits all of its message without interruption. Message channels are allocated by the NCS to an LES depending upon the amount of traffic to be handled. Consequently, each LES may have more than one message channel to use.

Channel timing relationships

In the example shown in Fig. 13.7, the TDM Channel frame (8.64 s) is received after a propagation delay of between 239 and 277 ms depending upon the LES to MES range via the satellite. There is a further variable time delay dependent upon the number of

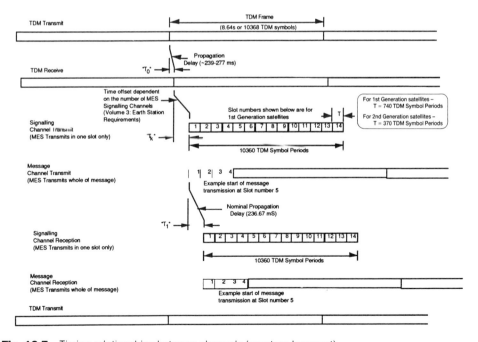

Fig. 13.7 Timing relationships between channels (courtesy Inmarsat)

302 *The Inmarsat-C system*

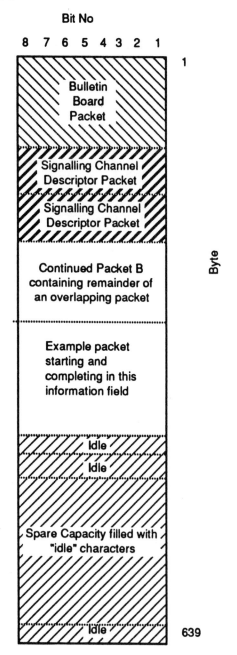

Fig. 13.8 TDM frame information field (courtesy Inmarsat)

MES signalling channels. The greater the number of channels available for use the longer will be the possible delay.

The TDM frame of 8.64 s containing 10368 symbols is divided into 14 slots (first generation satellites) each slot containing 740 TDM Symbol periods. The MES transmits signalling information in one slot only and receives in the corresponding slot on the signalling channel (reception), shown after a further propagation delay of 236.67 ms.

Assuming the MES is told to use slot 5 of the allocated message channel it uses this slot to transmit the whole message to the LES.

To receive information the MES will use the corresponding slot 5 on the message channel (reception).

The use of channels for call set-up and message transfer is fully described later.

Frame structure

The NCS common channel and the LES TDM channels share a common structure.

The structure is based on fixed length frames of 10,368 symbols transmitted at 1200 symbols/s (600 bit/s)—for INM2 satellites—prior to convolutional coding, producing a frame duration of 8.64 s.

Each frame carries a 639 byte frame information field, an example of which is shown in Fig. 13.8.

The 639 bytes of information plus a flush byte are scrambled before being passed to the encoder. Thus, the frame is now a total of $640 \times 8 \times 2 = 10,240$ symbols. Two unique words, each of 64 symbols are added to complete the 10,368 per frame which is transmitted.

ALOHA and the return link signalling channel

A number of return link frequencies associated with each TDM channel are assigned to the MES as signalling channels. The channel is organized to allow random access using a form of the slotted ALOHA protocol with explicit reservations. This method of access provides the MES with a greater likelihood of gaining access without collisions occurring, it also improves the throughput of the channel.

If more than one MES tries to gain access at precisely the same instant, a collision will occur as the two burst packets are transmitted in the same signalling channel slot. Neither MES will gain access to the system. MESs are not able to monitor their own transmission via the spacecraft and consequently the NCS or LES will detect the collision and will tell the MESs to try again after a random time interval. The random time interval often simply occurs because each of the MESs will be at different transmission distances from the satellite and are very unlikely to transmit together a second time.

In order to reduce further the likelihood of collisions and improve channel access, the Inmarsat-C uses a system whereby transmissions may only commence at the start of time slots and not part way through. Additionally, there are reserved slots provided for transmission which are indicated in the LES bulletin board.

Call set-up procedures

From-mobile call
Refer to Fig. 13.9.

The message to be sent is input to the equipment and stored in the DTE, for store-and-forward transmission, until the system indicates that it is ready to receive the traffic.

The DCE receiver checks its memory to find the number corresponding to the frequency of the selected LES TDM channel. The DCE knows this because it has been continuously monitoring the NCS common signalling channel since login.

The DCE now changes its tuning from the NCS common signalling channel to the LES TDM channel. (Depending upon the make of terminal, the operator may see this

304 The Inmarsat-C system

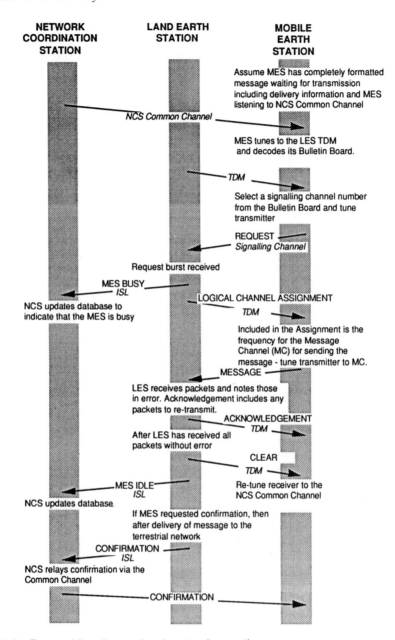

Fig. 13.9 From-mobile call procedure (courtesy Inmarsat)

act on the VDU display as the channel number changes and the signal levels vary at changeover.)

When the DCE receiver is synchronized to the LES TDM channel, it decodes the bulletin board information to find the number corresponding to the frequency of the SES signalling channel. This is the LES calling channel to be used.

The DCE tunes to this channel and transmits an assignment request to the LES to request a message channel—the communications channel.

13.2 *Outline system specifications* 305

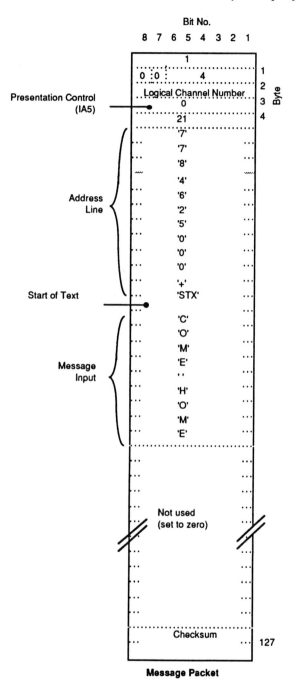

Fig. 13.10 Message transfer frame (courtesy Inmarsat)

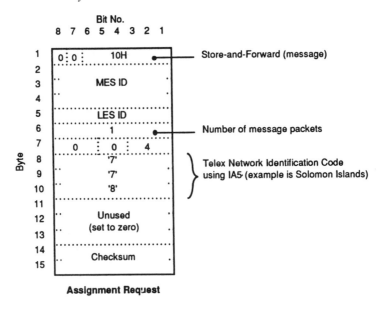

Fig. 13.10 (cont.)

The LES receives the assignment request as the initial indication that the MES wants to send a message and, if a channel is available, transmits a channel assignment packet in the LES TDM channel.

The DCE receiver decodes the channel assignment packet and identifies the frequency of the channel to be used for transmission.

The LES now sends an MES busy packet over the interstation link to the NCS indicating that the MES is busy.

The NCS uses the MES busy packet to update its database concerning this MES.

The MES DTE now passes the message to the DCE transmitter which then sends the message in data packets on the message channel to the LES. See Fig. 13.10.

The LES receives, decodes and stores the incoming data packets, and performs error checking and correction processes. If any errors are detected the following occur.

- The LES returns an acknowledgement packet to the SES with details of the data packets received in error.
- The DCE receiver decodes this information and instructs the DCE transmitter to re-transmit the packets received in error.
- This sequence may be repeated.

When the LES is satisfied that the message is correct, it returns a channel clear packet on the LES TDM channel to the MES, instructing it to retune to the NCS common signalling channel. The channel clear packet also contains details of the confirmation of reception of the message, which is printed for information.

The LES sends an MES idle packet to the NCS announcing that the MES is now free.

The NCS updates its database accordingly.

To-mobile call

The procedure is illustrated in Fig. 13.11.

The idle MES is tuned to the NCS common signalling channel and updates its memory accordingly.

13.2 Outline system specifications 307

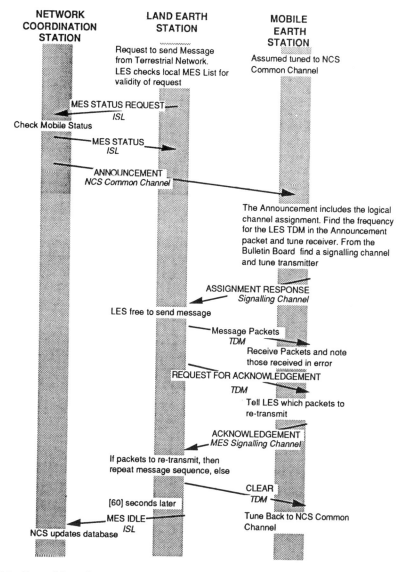

Fig. 13.11 To-mobile call procedure (courtesy Inmarsat)

The message, from the terrestrial network, appears at the LES which, upon receipt, uses the ISL to check the status of the designated MES. If the MES is free, an announcement is made on the NCS common channel advising the MES of the traffic and telling it to tune to the LES TDM channel for information.

Once the LES bulletin board has been read and a signalling channel has been found the MES tunes to that channel and sends an assignment response message to the LES. The LES now sends the traffic as message packets using the TDM channel and requests an acknowledgement. If any packets are received in error the MES tells the LES to re-transmit those packages. If no repeat is required an acknowledgement is sent and the LES responds by sending a clear message to the MES on the TDM channel.

The MES now re-tunes to the NCS common signalling channel. The LES advises the NCS that the MES is idle and the NCS updates its database accordingly.

Data reporting and polling

This Inmarsat-C service is provided to enable land-based companies or government agencies to receive data directly from mobiles. As an example, shipping companies may require positional, or other data on a daily basis, leading to greater operating efficiency. An important application of data required by governments is to assist ships in distress, and to prevent marine pollution. For these reasons, governments, through various international organizations (SOLAS convention etc) have developed requirements for ships to make reports to shore authorities. Government bodies have, for some years, been involved in receiving information from ships primarily for weather reporting. Position reporting is not new. International ship position reporting data have in the past been sent by morse code and used by the US Coast Guard service as part of their AMVER organization.

Ship data may be interfaced directly with an Inmarsat-C terminal from the various navigational, engine or weather sensors carried on board. There are two ways in which these data may be sent, either automatically or manually at pre-determined times from the MES, or alternatively, the MES may be interrogated (polled) from shore. When a polling command is sent to an MES, the unit will respond by automatically returning data in a pre-determined format.

The poll command contains information about how and when the MES should respond. The types of polling available are as follows.

- Individual polling—An explicit poll command to one MES. If the MES is busy the poll is queued until the terminal is idle
- Group polling—A single poll command to a group of mobiles. Typically a fleet call.
- Area polling—A single poll to a number of mobiles located in a specific geographical area.
- Enhanced Group Calling (EGC)—This is a message broadcast service to Inmarsat-C users from land-based information services.

There are two types of EGC calls:

- FleetNET which is for commercial use by individual companies to their mobiles, and
- SafetyNET which is provided exclusively for maritime users. See *Understanding GMDSS*.

It is possible that future polling commands may include remote monitoring and control instructions whereby data polled from an SES is used by an office ashore for remote control of some of the operating parameters of a vessel.

Data reporting and polling formats use standardized short binary encoded data messages for the transfer of information. Data reports are limited to three data packets which, with header information, provide 32 bytes of data. In addition, Macro-encoded Messages (MEMs) may be used. These are pre-defined text messages which are compressed to provide more efficient use of reporting formats. In the Inmarsat standard maritime report, a total of 128 MEM codes are available, of which about half are defined by Inmarsat.

13.3 Mobile earth stations

Inmarsat-C MESs are classed as follows (the term EGC refers to the Inmarsat Enhanced Group Call used extensively in a distress scenario).

13.3 Mobile earth stations   309

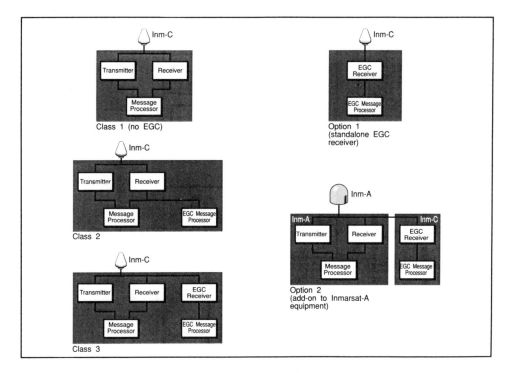

Fig. 13.12 Different classes of Inmarsat-C ship earth station

- A class 0 MES: EGC receivers may be fitted in the following ways.
  (i) Option 1: as a stand-alone EGC receiver, without communications options; or,
  (ii) Option 2: as a stand-alone EGC receiver added to an Inmarsat-A terminal.
- A class 1 MES is used for two-way communications to/from the mobile but is not able to receive EGC messages.
- A class 2 MES is capable of two modes of operation;
  (i) as in class 1 and also for receiving EGC messages;
  (ii) exclusively for EGC reception.
- A class 3 MES has two independent receivers, one for EGC reception and one for communications, thus making simultaneous operation in both modes possible. See Fig. 13.12.

There are two parts to an Inmarsat-C SES. The data terminal equipment (DTE) which forms the below decks electronics unit and the data circuit terminating equipment (DCE) which is the antenna, transmitter and receiver.

The DCE part of an Inmarsat-C SES normally comprises a right-hand circular (RHC) polarized omnidirectional antenna fitted inside a protective radome. Also contained in this unit are the RF electronics, details of which are shown in Fig. 13.13.

The main characteristics of the antenna unit are:

- minimum G/T at 5° elevation: -23 dB/K,
- minimum EIRP at 5° elevation: 12 dBW, and
- the antenna gain pattern: the minimum EIRP and G/T figures must be matched down to $-15°$ elevation in order to accommodate ship motion. See Fig. 13.14.

310 The Inmarsat-C system

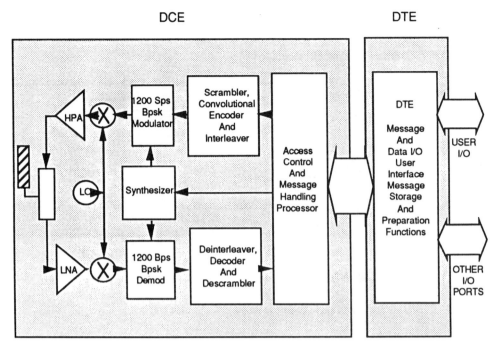

Fig. 13.13 Inmarsat-C mobile earth station example diagram

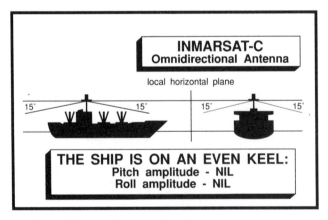

Fig. 13.14 Inmarsat-C omnidirectional antenna location

In practice, the antenna should be mounted in such a position that it possesses an unrestricted view of the sky from $-15°$ elevation, and through $360°$ in azimuth. There should be no metal objects within 3 m of the antenna which would cause blind arcs. Downlink signals from the antenna are coupled, via a low noise amplifier, to the mixer where they are downconverted to an intermediate frequency. BPSK demodulation produces the raw data which is descrambled before being passed down to the DTE for processing. The DTE comprises a laptop computer with NBDP and possibly other peripherals. Decoded messages are either stored, with a message displayed to warn the operator, or are printed directly upon receipt.

13.3 Mobile earth stations 311

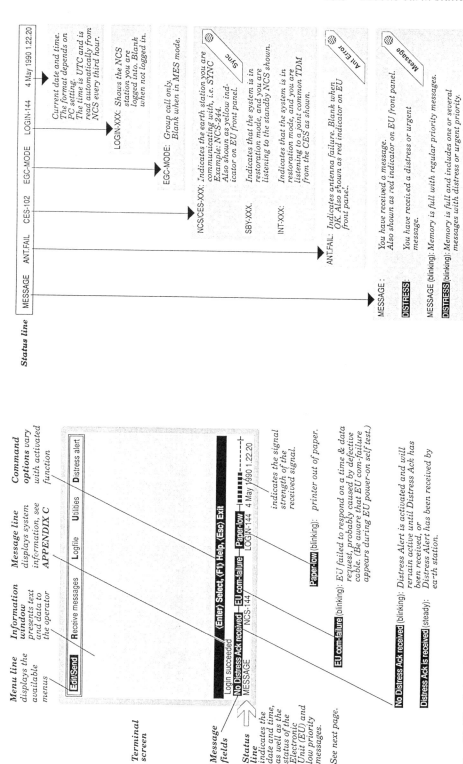

Fig. 13.15 Saturn-C screen display and status line explanation (courtesy ABB NERA AS)

312 *The Inmarsat-C system*

Messages to be sent are input via the operator keyboard and prepared for transmission using word processing, after which they are stored. Using the Inmarsat-C 'store and forward' system the MES automatically gains access to the satellite at a convenient time and sends the message to a selected LES for onforwarding.

SES VDU display

The video display unit (VDU) provides the operator with an extremely versatile information display. The Saturn-C uses a laptop PC computer which provides full editing facilities, communications control and information display. See Fig. 13.15.

The Saturn-C screen display carries all the vital information for system operation. The top menu line has five pull-down menus which may be demanded by the operator. The communications message display (information window) is shown blank. The two

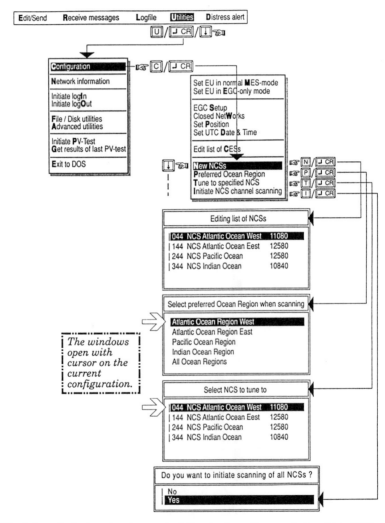

Fig. 13.16 Saturn-C display showing part of the configuration mode in which NCSs and Ocean Regions are selected (courtesy ABB NERA AS)

13.3 Mobile earth stations 313

illustrations here demonstrate how versatile a computer screen is when used in this context. Many readers will be computer literate and consequently will feel happy interpreting the displayed information.

As an example of the use of pull-down menus, part of the system configuration display is shown in Fig. 13.16.

The ship is currently logged into the AORW. By using one or more of the commands shown it is possible to change ocean regions and NCSs providing, of course, that the ship is within the footprint of another satellite ocean region. Figure 13.17 shows part of the message display which announces one of 43 different system messages requiring action from the operator.

The messages 1-15 listed below appear in the message line on the screen display.

The messages contain system information and are displayed for a period of 1 minute.

Ref	Message/Description	Action
1	New message from shore has arrived A message has arrived.	Retrieve message: *Receive messages / Read new message.*
2	Distress test is initiated automatically The operator failed to manually initiate a distress test within 2 minutes during a distress test sequence.	Do nothing, wait until PVT is finished.
3	Position not updated, done by NMEA device Position update attempt failed because EU is set to accept position from an NMEA device only.	If NMEA device is not connected, enable manual setting: Utility /Configuration/Position.
4	Result of test has arrived Denotes arrival of test results to EU from NCS/CES during a PV-test.	Read the results: *Utilities / Get results of last PV-test*
5	Aborting transmission succeded: [...reason...] Denotes that CES has accepted the abort call from Saturn C. "Abort Call" has been initiated by the operator or automatically from EU (due to protocol error).	
6	Sending the message succeeded, message ref. number is: XXXXX CES reports the message as completely received and also reports the message reference number	
7	Login succeeded NCS confirms login after such request	

Fig. 13.17 Saturn-C display showing the first seven of over 40 systems messages

14
The Inmarsat-M system

14.1 Introduction

The Inmarsat-M system offers mobile satellite communications of the highest quality whilst keeping equipment costs and physical size to a minimum. Inmarsat-M MESs are significantly smaller than either of their two cousins, Inmarsat-A or B, enabling their use in areas where physical size is a major consideration factor. The system, which bears much in common with Inmarsat-B became operational at the same time in order to provide an alternative voice-grade, data and facsimile service for smaller vessels or land earth stations, either mobile or transportable. Voice quality is as good as a current cellular telephone service and enables MES connection to a PSTN subscriber. The data service is relatively low speed at 2.4 kbit/s. However, this service, along with facsimile, makes the system attractive to the smaller user.

Two types of Inmarsat-M MES are available:

- a land mobile MES, which may be vehicular mounted or transportable and is expected to form the bulk of MES sales; and
- a maritime mobile MES, intended for use on smaller vessels where physical size and weight are of particular importance. It should be noted that currently this MES does

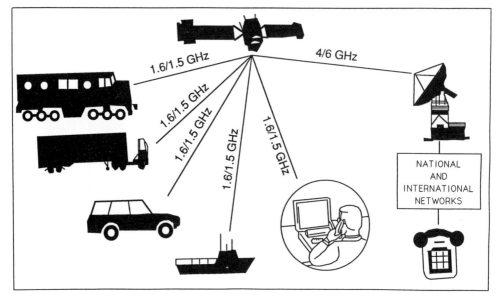

Fig. 14.1 Inmarsat-M network configuration

not meet the rigid criteria set by the IMO for distress and safety operation within the GMDSS radionet.

Intentionally, the Inmarsat-B and M systems share much in common, thus enabling design costs to be kept to a minimum. Access control and sub-systems are identical as is the use of digital technology and signal processing to improve the efficiency of use of satellite power, bandwidth and channels. Inmarsat-M uses the same space and ground sub-systems as previously described and thus is transparent to the user. Third-generation satellites, (INM-3) spot beam facilities will also be used by the system.

Due to its low cost and physical size, it is expected that Inmarsat-M will stimulate the land MES market. Indeed, Inmarsat market research has predicted that approximately 600,000 Inmarsat-M MESs will be in use by the year 2005. Clearly, if this forecast is anywhere near correct the Inmarsat-M system will be a major provider of satellite mobile communications well into the next century.

14.2 Outline system technical specifications

The use of digital technology throughout enables significant savings to be made in terms of power and occupied bandwidth.

Digital speech processing is used which, although the coding rate is slower than that used in Inmarsat-B at 6.4 kbit/s (including error correction coding) or 4.8 kbit/s (prior to error coding), produces speech of the quality to be found in most cellular mobile radio systems.

Offset-quadrature phase-shift keying (O-QPSK) voice coding ensures bandwidth efficiency, whilst forward error correction (FEC) on single channel per carrier (SCPC) communication channels improves power efficiency over the Inmarsat-A system operation.

MES antennae are provided with a transmit beamwidth which is wide in elevation and narrow in azimuth in order to maintain satellite lock and simplify tracking requirements. Thus, the overall size, complexity, power consumption and cost may be reduced.

Unlike the relatively inefficient fixed channel allocation used by Inmarsat-A, this system, in common with Inmarsat-B, uses unpaired transmit and receive RF channel frequencies allocated on demand. This fact greatly improves management of the available spectrum and subsequently achieves better efficiency and is more flexible.

Optional facilities include:

- data communications at a rate of 2.4 kbit/s in both duplex and fixed–originated simplex modes;
- group call facilities (fixed–originated mode); and
- facsimile (group 3) at 2.4 kbit/s in both modes.

The space segment

The space segment is transparent to the user and utilizes the same satellites as the other Inmarsat systems.

The ground segment

Inmarsat-M uses the same ground operational organization of LES, NCS, OCC, SCC and TT&C stations, although the designated NCS and some LESs for each ocean region may be different from the other Inmarsat systems. Inmarsat-M became a fully global system in Autumn 1993 when the CES at Yamaguchi in Japan was inaugurated to operate in both the IOR and POR ocean regions, and the CES at Perth, Western Australia, was commissioned to operate in the IOR. These stations along with the CESs at Santa Paula, ocean region POR, Southbury, ocean Region IOR and Goonhilly UK serving the regions AORE/AORW provide full global coverage. Further CESs serving one or more of the four ocean regions will come on stream in future.

Mobile earth stations

An Inmarsat-M MESs, maritime or land based, is more cost effective and compact than MESs used for Inmarsat-A and B communications. The very much more compact externally mounted equipment (EME), previously known as the above decks equipment (ADE), unit includes a steerable antenna of approximately 0.3 to 0.5 m in diameter, which, because of its modest mass, requires a much smaller and less complex satellite acquisition and tracking system, RF head and signal translation unit.

Maritime MESs
The use of a more compact EME makes the system more attractive to companies or individuals operating smaller vessels. Although smaller and cheaper than SESs in the Inmarsat-B system, Inmarsat-M MESs offer full duplex voice communications, although of a lesser quality, but sufficiently good for connection to a national PSTN. Additionally, the MES has the capacity for communications using digital data services at 2.4 kbit/s, which is slower than the Inmarsat-B system but faster than the Inmarsat-C system. While the maritime MES is provided with a distress priority system, the reduced power and operating parameters of the MES leads to the slight possibility of a link being unavailable at high latitudes – low satellite elevation angles. Consequently, the Inmarsat-M maritime MES *does not* comply with the IMO requirements for operation within the GMDSS radionet.

Land mobile MES
Land mobile MESs are still in the early stages of development. The EME design introduces a new set of design challenges. In most cases the MES will be identical to the maritime MES but satellite acquisition and tracking techniques from a rapidly moving vehicle require electro-mechanical systems which will rapidly respond to changes in azimuth and elevation. The antenna is unlikely to be a parabolic type, but is more likely to be a low mass phased array or a crossed dipole array. The transmit beam is wide in elevation and narrow in azimuth to maintain satellite lock and simplify tracking needs.

MES specifications
Receive G/T figures for communication links with satellites in the range 5° to 90° elevation are:
maritime MES -10 dB/K
land mobile MES -12 dB/K
Transmit EIRP figures:
maritime MES 27 dBW
land mobile MES 25 dBW

Both figures are capable of being reduced in steps to 21 dBW and 19 dBW respectively to work into satellites with spot-beam coverage facilities.
Frequency band—transmit:

maritime MES 1631.5 – 1646.5 MHz
land mobile MES 1656.5 – 1660.5 MHz.
Frequency band—receive:
maritime MES 1530 – 1545 MHz
land mobile MES 1555 – 1559 MHz

Transmit and receive channels are spaced at 10 kHz and are demand assigned from a pool and not paired.

Access control and signalling

Access control and signalling methods are similar to those used in the Inmarsat-B system in that communications channels are demand assigned in response to requests from a MES. Frequencies are assigned from a common pool by the NCS. The system uses the following access control and signalling channels in common with Inmarsat-B: NCSC, NCSA, NCSI, and NCSS

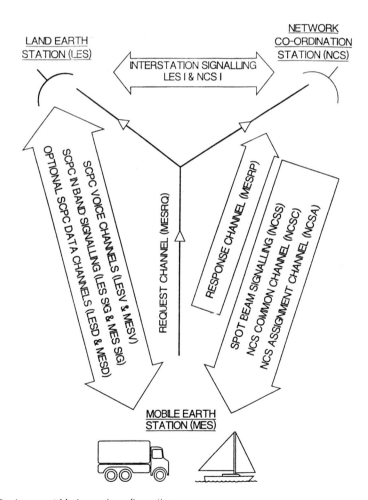

Fig. 14.2 Inmarsat-M channel configuration

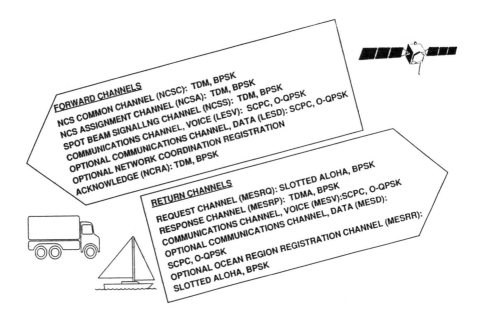

Fig. 14.3 Inmarsat-M channel description, mobile satellite links

LESA and LESI
MESRQ and MESRP.
   See Fig. 14.2.
   SCPC (voice) communication channels are titled LESV and MESV, in the forward and reverse directions respectively, and optional TDMA/TDM (data) communication channels are labelled LESD and MESD.
   Additional optional network coordination registration acknowledgement (NCRA) and ocean region registration (MESRR) channels are included for future use.

Call set-up procedures from the MES

As with the Inmarsat-B system which is very similar in its operation, the sequence is initiated by the MES which sends an access-request message on the MESRQ channel to the LES. The message contains the following information:
LES identification;
MES identification;
Request service type (telephony or data);
Priority (for use by Maritime MES only);
MES antenna AZ angle zone (for frequency assignment purposes);
MES antenna AZ angle (for maritime distress purposes only);
MES antenna EL angle (for maritime distress purposes only);
Satellite spot beam identification (for future use);
Received signal strength report (for land mobile MES forward link power control.
   The process then continues as described for the Inmarsat-B system.
   Figure 14.3 shows the full forward and return channels along with the access/modulation methods used in the system.

15
Satellite mobile frequency bands

In order for any communications system to expand to keep pace with ever increasing demands for channels it is critically important that more frequency spectrum is made available. The World Administrative Radio Conference met in 1992 (WARC92) in Torremolinos in Spain to consider the problems encountered by mobile satellite services (MSS). As a result, the frequency bands available for MSS have been increased.

The original frequency bands allocated to mobile satellite communications were 1535.000 – 1542.500 MHz (receive) and 1636.500 – 1644.500 MHz (transmit). A small band of frequencies indeed in which to offer a service on a global scale. Because of the increase in the number of SESs in use and other systems being introduced the band available has been increased by previous meetings of WARC to 63 MHz, 1530.000 – 1548.000 MHz (receive) and 1626.500 – 1649.500 MHz (transmit).

After WARC92 a further 33 MHz of global frequency spectrum has immediately become available for use by the MSS, thus increasing the spectrum available by over 50%. The bands, to be used by the new Inmarsat INM3 satellites are 1524.000 – 1559.000 MHz receive and 1626.500 – 1660.500 MHz transmit. Further huge increases of 80 MHz near 2 GHz, and 70 MHz near 2.5 GHz, will become available after the year 2005 ensuring that MSS will expand and develop to meet future needs as new technology becomes available. See Fig. 15.1.

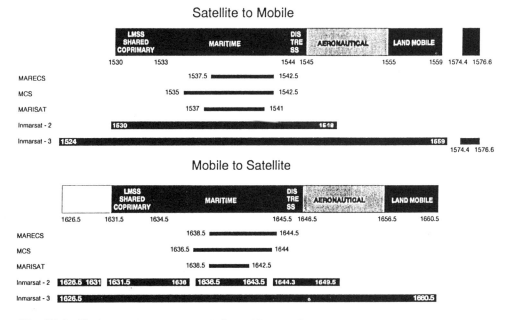

Fig. 15.1 The Inmarsat space segment channel frequencies

Section Three
COMMUNICATIONS WITH FIXED SATELLITE SERVICE

Introduction

The fixed satellite service (FSS) covers those services available from major operators such as Intelsat, Eutelsat, Arabsat etc. The FSS involves satellites in geostationary orbit providing 24 hour per day service with minimal tracking constraints on the earth station. The space segment for these systems consists of operational satellites, with an in-orbit spare satellite, together with a Telemetry, Tracking Command and Monitoring (TTC&M) system. The function of TTC&M is to monitor the satellite performance and provide the necessary control functions to keep the satellites operational. The ground segment consists of all earth stations that have access to the satellite. Types of earth station and their operational requirements are discussed in the chapters that follow.

A complete telecommunications link can be established via the FSS with a variety of configurations, including:

- direct links such as business-to-business;
- connections via terrestrial links;
- connections to very small aperture terminals (VSATs) with a central station acting as a switching centre.

The systems to be considered in this section are restricted to Intelsat and Eutelsat since they are seen to be representative of international and regional systems respectively. Operational variations between these systems and others will differ only in detail.

16
The Intelsat Organization

16.1 Introduction

An organization known as COMSAT was set up in the US in 1962 and two years later placed a contract for a geostationary satellite to be known as Early Bird. In August of 1964 the International Telecommunications Satellite Organization (Intelsat) was formed, consisting of 11 member nations. Two international agreements providing interim arrangements for the international co-operative were signed in Washington DC by representatives of governments and operating entities. Early Bird, launched by the National Aeronautics and Space Administration (NASA) in 1965 became Intelsat 1. As the viability of international communications via satellites became increasingly proven, many more countries joined the organization. In 1971, on the anniversary of the formation of Intelsat, agreements for definitive arrangements for Intelsat opened for signature in Washington DC. These definitive arrangements came into force in February 1973, superseding the interim arrangements of 1964. The agreements that came into force in 1973 provided Intelsat with a four tier organizational structure as follows:

- Assembly of Parties
- Meeting of Signatories
- Board of Governors
- Executive Organ.

The Assembly of Parties is composed of representatives of governments of the Intelsat member countries (Parties to the agreement) and normally meets every two years to consider resolutions and recommendations, on general policy and long term objectives, as proposed by the Meeting of Signatories or the Board of Governors. Each Party has one vote. Extraordinary meetings may be convened, as necessary, for consideration of specific issues of interest to the Parties. The 17th extraordinary meeting was held in 1992 to confirm the appointment of a new Director-General, and Chief Executive Officer, for Intelsat, while the 18th meeting of the Assembly was held later in the same year.

The Meeting of Signatories is composed of representatives of signatories to the Operating Agreement (member governments or their designated telecommunications entities) and normally meets once a year to consider issues related to the financial, technical and operational aspects of the system as proposed by the Assembly of Parties or the Board of Governors. Each signatory has one vote. The 22nd Meeting of Signatories occurred in 1992.

The Board of Governors oversees the design, development, construction, procurement, establishment, operation and maintenance of the Intelsat space segment, as well as approving many other activities undertaken by Intelsat. The Board considers all inputs from the Assembly of Parties or the Meeting of Signatories. The Board is assisted by committees such as:

- advisory committee on technical matters and planning
- budget and accounts review committee.

As of February 1992, the Board consisted of 28 governors representing 103 of Intelsat's then 121 signatories. Governors represent an individual signatory or a group of signatories which meet a minimum investment share for board membership, determined each year by the Meeting of Signatories. A governor may represent a group of five or more signatories from within an administrative region of the International Telecommunication Union (ITU), regardless of total investment share. The board normally meets four times a year.

The Executive Organ consists of a Director General and four Vice Presidents, reporting to the Director General, with duties defined as follows.

- Executive Vice President, Operations and Services: to oversee three activities that allow Intelsat to respond to its customers, namely operations, marketing and external relations.
- Vice President and Chief Financial Officer: to direct the related activities of finance, accounting, procurement and headquarters building management.
- Vice President of Engineering and Research: to administer the areas of engineering and research and the future planning of Intelsat's engineering and research programmes.
- Vice President of Information Systems and Administration: to oversee information management which includes computer and electronic data processing services, conference services and language services.

As of March 1992, Intelsat management consisted of 778 staff employees at Headquarters, Intelsat Spacecraft Program Offices in sites in the US and France, and at various TTC&M sites around the world.

Other Intelsat meetings include:

- The Global Traffic Meetings, held annually where representatives of Intelsat user organizations, earth station owners and operators meet to forecast and plan services and system requirements
- The Ocean Region Meetings, where representatives of signatories and user organizations review and plan the region's operational needs.

Additionally, Intelsat participates in meetings organized by bodies such as:

- the ITU, where a recent involvement studied the allocation and improved use of the radio frequency spectrum and simplification of the radio regulations;
- the International Radio Consultative Committee (CCIR) where a contribution is made to that body's technical work for the benefit of signatories, members and users;
- the International Organization for Standardization (ISO) where Intelsat has been involved in the development of protocol standards so that satellites remain a viable transmission medium for computer communications.

Also, Intelsat has made recommendations to the International Telephone and Telegraph Consultative Committee (CCITT) to ensure the compatibility of satellite transmissions with emerging services such as the broadband integrated switched digital network (B-ISDN) and the synchronous digital hierarchy (SDH), which are considered the systems of the future.

Intelsat's 122 signatories make capital contributions based on their investment (i.e.

ownership) shares and pay space segment charges to finance Intelsat operations. Non-member users only pay space segment charges.

Each signatory investment share is proportional to its percentage of all utilization of the Intelsat space segment. There is a minimum investment share of 0.05%. The relative utilization percentages, and investment shares, are based on signatory's utilization charges for the 180 day period preceding each investment share determination, which is done annually or when a signatory joins or withdraws. There are currently 122 signatories with the major investment shares belonging to the United States of America, at over 21.86%, and the United Kingdom at just over 12%. Other signatories with significant investment shares are France, Germany and Japan.

Intelsat space segment charges are designed to cover:

- operating, maintenance and administration costs;
- provision of operating funds, as determined by the Board of Governors;
- amortization of signatories investment;
- compensation for the use of capital.

The sum of the above costs forms Intelsat's revenue requirement.

The ground segment provision is the responsibility of the signatories, who must also provide the interconnection facilities to link users to a terrestrial network.

16.2 Development of the Intelsat space segment provision

Intelsat I

Intelsat I, launched in 1965, has already been mentioned. This satellite had two transponders, each of which had a bandwidth of 25 MHz. The frequency used was C-band with 6 GHz for the up-link and 4 GHz for the down-link. The lower frequency is typically used for the down-link because attenuation increases with higher frequencies and only a limited power is available on the satellite. The total traffic capacity was 240 voice circuits (or 1 television channel). The satellite completed a link between two North American and four European earth stations but, because multiple access was unavailable at the time, stations at each end of the link took it in turns to use the space segment, linking with each other via terrestrial networks. Intelsat I was the first commercial communications satellite and paved the way for the development of a further generation of Intelsat satellites.

It had been intended that there would be two satellites in the Intelsat I series but, although a second satellite was assembled it was never launched, its design being overtaken by the next generation. The design life of Intelsat I was 18 months but it gave three and a half years of satisfactory operational service.

Intelsat II

1967 saw the successful launch of three satellites in the Intelsat II series. The first and third of these provided service in the Pacific Ocean Region, the latter in the capacity of a spare, while the second provided additional capacity in the Atlantic Ocean Region. A single transponder was incorporated with a bandwidth of 130 MHz to provide traffic capacity of 240 voice circuits (or 1 television channel). Frequencies used were again 6/4 GHz. Multiple access was incorporated allowing multipoint communications capacity between earth stations. Intelsat II operated using MCPC/FM/FDMA.

There was provision for five satellites of the Intelsat II series but only three were launched successfully. One satellite failed to reach its orbit while another was assembled but not launched.

The design life of this series was three years with the last of the series being retired after five years' service.

Intelsat III

The period from late 1968 to early 1970 saw the successful launch of the next generation of satellites, designated Intelsat III. The provision of these satellites established global coverage, with satellites servicing the Atlantic Ocean Region (AOR), Pacific Ocean Region (POR) and Indian Ocean Region (IOR). Each satellite had two transponders, with a total bandwidth of 300 MHz and with a capacity of 1500 telephony circuits and four television channels. Like their predecessors, the Intelsat III satellites received transmissions at 6 GHz and relayed transmissions to earth at 4 GHz. The biggest advance on earlier satellites provided by Intelsat III satellites was the provision of a global antenna. Previous satellites had used spin stabilization with radiation patterns symmetrical about the axis of spin. Thus, most of the satellite RF power was radiated into space and wasted. The Intelsat III satellites were spin stabilized but with a de-spun antenna which was kept pointing at the earth by spinning at the same speed as the satellite but in the opposite direction of rotation. The de-spin motors that controlled the spin of the antenna were operated using infra-red sensors which detected the earth's horizon.

The Intelsat III series was designed to comprise eight satellites but only five were successfully launched while three did not reach geostationary orbit because of the failure of the launch vehicle. Of the remaining five, two failed in orbit leaving just three satellites to complete their design life of five years.

The global beam antenna of the Intelsat III series was effective in providing a directional beam towards the earth but the level of power flux density (PFD) reaching the earth's surface was barely adequate to operate the service. In order to increase capacity it would be necessary to increase the available PFD at the surface of the earth; this could be done by making the antenna more directive or by increasing the transmitter power, or a combination of both. Increasing the antenna gain makes the beam more directive and reduces the area covered by the beam. Thus, to cover the area beneath the satellite effectively, more antennae would be required with interconnections between the receive and transmit antennae according to traffic flow. Alternatively, the smaller beams could be made steerable to cover the areas where traffic density demanded it.

Intelsat IV

The eight satellites in the Intelsat IV series were launched in the period 1971 to 1975 (only seven were operational since one failed to reach orbit). The satellites had increased capacity with 12 transponders, each with a bandwidth of 36 MHz, with 4 MHz guard bands between transponders. The capacity was 4000 telephony circuits and two television channels. Frequencies used were again 6 GHz for the up-link and 4 GHz for the down-link. The satellites adopted the antenna configuration of a global beam horn antenna and two spot beam parabolic reflector antennae. All the transponders had inputs from the global beam antenna and four of the transponders had outputs connected directly to the global beam antenna. The remaining eight transponders could be connected via a switch, operated under ground control, to the global beam antenna or one of the spot beam antennae. The direction of the spot beams was also controllable from the earth, thus allowing the beam to be steered to the area of greatest traffic density.

Because of the rapid growth in traffic in the early 1970s Intelsat utilized a series of six satellites developed from the Intelsat IV series. Only five were successfully launched, with one failing to reach its orbit. The design life of this series was estimated to be seven to ten years. The new satellites were improved versions of the original Intelsat IV design giving increased capacity. The launching period of the Intelsat IV-A satellites commenced in 1975 and continued until 1978. The major difference between the IV and IV-A series lay in the simultaneous use of the same frequencies through the use of directional antennae for both reception and transmission. This effectively gave a traffic capacity of nearly twice the number of transponders. Spot beams serving the western and eastern hemispheres were connected to transponders such that twofold frequency re-use existed. Frequency re-use allowed the transmission of signals at 6 GHz from the west to a satellite transponder via a directional antenna. After processing, the signal was transmitted at 4 GHz via a directional antenna to the eastern hemisphere. Similarly, a second transponder was fed via a directional antenna with signals at 6 GHz from the east and after processing the signals were transmitted to their destination in the west at 4 GHz. The beams were shaped by using offset parabolic reflectors fed via a horn array. Switching matrices allowed the selection of the required horns to give the required beam directivity. The beam diversity allowed by this arrangement enabled the transponder output power to be concentrated in specific areas of the eastern or western hemisphere according to the traffic demands. Each satellite had 16 transponders, arranged in pairs, connected to the directional antennae and four transponders connected to global antennae. Total bandwidth was 800 MHz, an increase of 300 MHz on the Intelsat IV satellites. Total capacity was 6000 telephony circuits and two television channels. Stabilization of the satellite was again achieved using de-spun antennae.

The last of the Intelsat IV satellites was withdrawn from service in 1989.

Intelsat V

The first of the Intelsat V series of satellites was launched in 1980 and showed a radical departure from the previous series.

Differences in shape (box-like rather than cylindrical) meant that spin stabilization could no longer be used and, instead, the satellite is stabilized along each of its three axes using an automatic control system; the stabilization is known as three-axis stabilization.

Intelsat V is shown in Fig. 16.1 in its deployed state.

The design life of the Intelsat V series was estimated at seven years.

Intelsat V incorporated directional antennae just as on Intelsat IV and, in addition, provided orthogonal polarization to give a fourfold increase in frequency re-use. The use of orthogonal polarization is discussed in detail in Chapter 9, Section 2. It is sufficient to state here that transmissions from an antenna can be polarised (using, say, horizontal and vertical polarization) to allow two transmissions on the same frequency without interference. In practice, the type of polarization chosen was left-hand and right-hand circular polarization. More details on the polarization of Intelsat V transmissions is shown in Table 16.9 (page 349).

Because of the allocation of increased frequency bands in 1979, Intelsat V also incorporated transmissions in Ku-band at 14 GHz/11 GHz. This too is shown in Table 16.9.

Intelsat V uses global beams, hemi, zone and spot beams to achieve its coverage. The global beams are left-hand circularly polarized (LHCP) for the up-link and right-hand circularly polarized (RHCP) for the down-link. The hemi beams have the same polarization as the global beams but utilize a different part of the frequency spectrum. The zone beams for the most part lie in the hemi beam area, and to avoid interference the

328 *The Intelsat Organization*

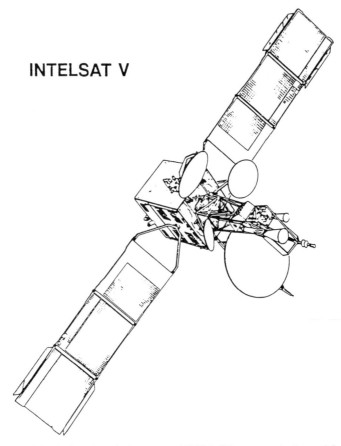

INTELSAT V

Fig. 16.1 Intelsat V in deployed mode (courtesy of British Telecommunications plc)

zone beams are polarized orthogonal to the hemi beams. The transmit beam coverage for the Intelsat V Atlantic 319.5° East satellite is shown in Fig. 16.2.

The spot beams are operated at 14 GHz for the up-link and 11 GHz for the down-link. Figure 16.2 shows the separation between the east spot and west spot beams which none the less are linearly polarized with vertical polarization for the up-link and horizontal polarization for the down-link. Transmission of global beams is via single antennae which can handle only one polarization. Transmission of hemi and zone beams is via a single antenna which has a single reflector with a horn array. Reception of transmissions is also via a single antenna and both the transmit and receive antennae are able to operate with LHCP and RHCP beams at the same time. The horn array can be reconfigured, using switches that can be controlled from the ground, to alter the zone beam coverage should circumstances demand. Spot beam antennae are offset paraboloids with reflectors capable of a limited amount of movement, via ground control, to vary the spot coverage.

Intelsat V has 27 transponders (21 in C-band, six in Ku-band) and operates these transponders via seven receivers. The receivers are connected according to the beam type being transmitted, i.e global beam; hemi beam east and west; zone beam east and west; spot beam east and west. Transmissions from each received beam are passed through an array of bandpass filters, which selects the required transponder for amplifi-

16.2 Development of the Intelsat space segment provision

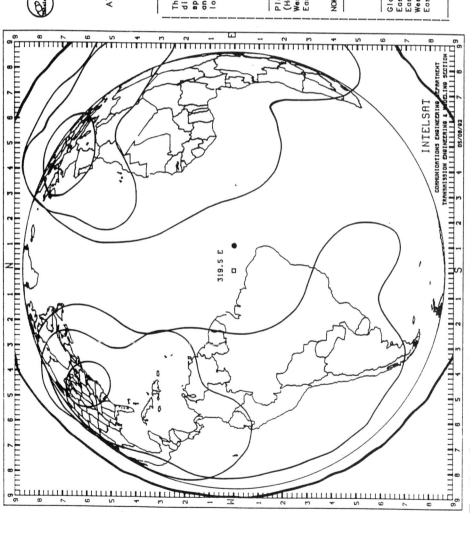

Fig. 16.2 Transmit beam coverage for Intelsat V (courtesy of Intelsat)

330 The Intelsat Organization

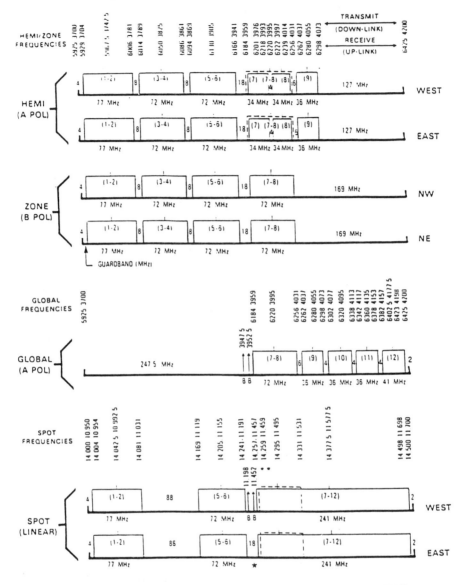

Fig. 16.3 Transponder layout of Intelsat V satellite (courtesy of British Telecommunications plc)

cation and frequency down-conversion before selecting one of the seven beams for down-link transmission. The path required for a link is capable of being selected using a switching matrix which can be controlled from the ground and altered as traffic demands vary.

The transponder layout of an Intelsat V satellite is shown in Fig. 16.3.

The bandwidth of the Intelsat V satellite is 2144 MHz and the total capacity is 12,000 telephony circuits and two television channels.

16.2 Development of the Intelsat space segment provision

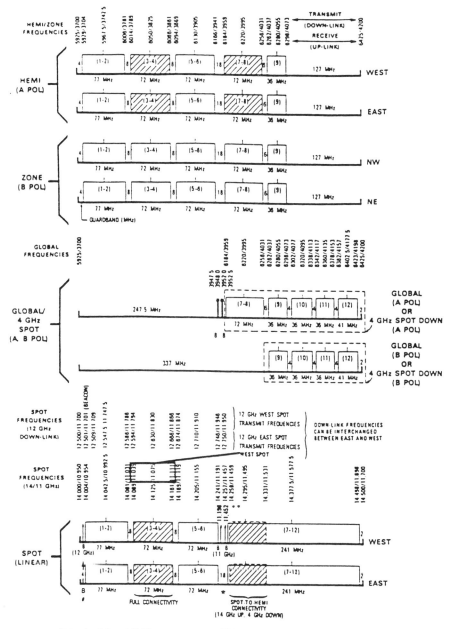

Fig. 16.4 Transponder layout of Intelsat VA satellite (courtesy of British Telecommunications plc)

Various Intelsat V satellites were used, commencing in 1982, to provide a service to Inmarsat, for communications with ships, using the maritime communications system (MCS) package. As stated in Section 2, three of these satellites are currently used by Inmarsat as spares.

332 The Intelsat Organization

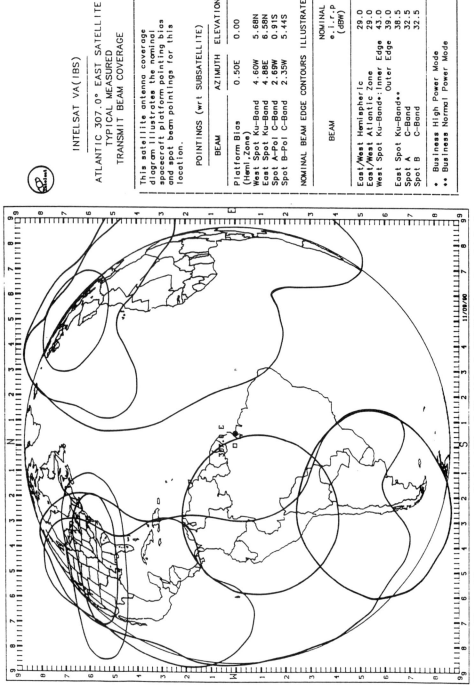

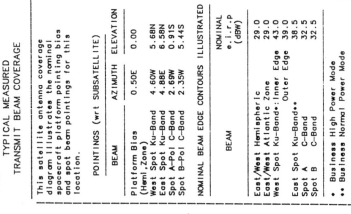

Fig. 16.5 Transmit beam coverage of Intelsat VA satellite (Atlantic 307.0° E) (courtesy Intelsat)

16.2 Development of the Intelsat space segment provision

Intelsat VA satellites came into use starting in 1985. The main difference between these satellites and their predecessors was the provision of down-link spot beams at 4 GHz using the same reflectors as the 14/11 GHz spot beam antennae. The 6/4 GHz spot beams were designed for domestic leased services. With an increased number of C-band transponders (32 as against 21 for the Intelsat V satellites), the bandwidth of the Intelsat VAs was increased to 2250 MHz while the traffic capacity increased to 15,000 telephony circuits and two television channels.

Later versions of the Intelsat VA satellites were modified to increase their power and coverage and to provide the facilities for a business service. These satellites operate in the Ku-band at both 14/11 GHz and 14/12 GHz, with the choice of operating frequency being made from the ground. The reason for using 12 GHz is that it can take advantage of a frequency band which is allocated exclusively to satellite communication without the danger of terrestrial interference. The frequency band 12.5 GHz to 12.75 GHz is for satellite use only in ITU region 1 (which includes Europe, Africa and some areas in the Middle East).

The Intelsat VA transponder layout for the business service use is shown in Fig. 16.4.

This satellite has greater flexibility in the connections that can be made between input and output beams so that an incoming signal can be retransmitted using any of the zone, hemi or spot beams.

The transmit beam coverage for the Intelsat VA (IBS) Atlantic 307.0° East satellite is shown in Fig. 16.5.

The order for 15 V and VA satellites was among the most successful for Intelsat, with only two launch failures. This series of satellites, with a design life of nine years, is still operational.

Intelsat VI

The next generation of satellites is the Intelsat VI series, the first of which was launched in 1989. These five satellites, two of which are deployed in the Indian Ocean Region (IOR) with the others in the Atlantic Ocean Region (AOR), show improved frequency re-use compared with their predecessors. Sixfold frequency re-use is achieved in the C-band while, additionally, satellite switched time division multiple access (SS/TDMA) is used. SS/TDMA (see Chapter 3, Section 5) allows interconnections between the receive and transmit hemi and zone beams to be changed several times during a TDMA frame according to traffic requirements.

The Intelsat VI satellite is spin-stabilized and is shown in Chapter 9, Fig. 9.1.

The satellite consists of four basic modules:

- the antenna farm;
- the de-spun compartment containing the repeater;
- the spun structure containing power, propulsion and other housekeeping systems;
- a pair of cylindrical solar arrays.

In orbit, Intelsat VI is a dual spin satellite. The lower portion, which includes the solar arrays, spins while the remainder, which includes the communications payload and antennae, is de-spun keeping the antennae pointing at earth.

Transponder centre frequencies, bandwidths and polarizations are shown in Fig. 16.6.

The main differences in communications facilities between the Intelsat V and Intelsat VI satellites are shown in Table 16.1 while other details regarding the communications payload are given in Tables 16.2 to 16.4.

334 The Intelsat Organization

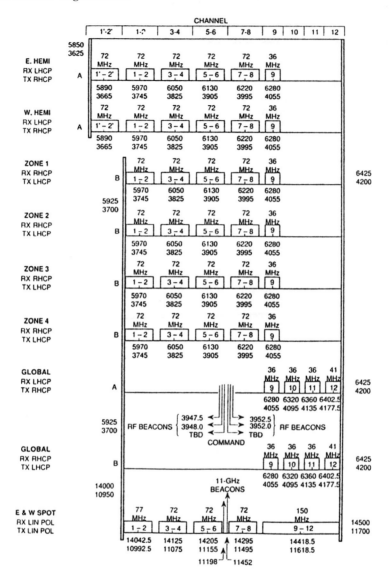

Fig. 16.6 Transponder frequency plan for Intelsat VI (reproduced by permission of *COMSAT Technical Review*, CTR, Vol.21, No.1, Spring 1991, p 108)

The G/T figures quoted in Table 16.2 are derived using assumed values for: antenna gain at the edge of beam coverage; low noise figures and expected losses.

The saturation EIRP figures of Table 16.3 are derived using amplifier output levels, beam edge antenna gains and expected losses.

Improved G/T figures allowed a decrease in flux density needed to achieve saturation without increasing the up-link thermal noise. The figures are shown in Table 16.4. The flux density at the high gain setting is based on the performance of the hemi-to-hemi links operating with TDMA. The flux density at the low gain setting is determined by the performance of the hemi-to-hemi links operating with multi-carrier FDMA.

A block diagram of the communications payload is shown in Fig. 16.7.

16.2 *Development of the Intelsat space segment provision* 335

Table 16.1 Comparison of Intelsat V and VI (courtesy of British Telecommunications plc)

Description	V	VI
Hemi coverages (6/4 GHz)	2	2
No. of transponders	10	12
EIRP, dBW	29	31
Zone coverages (6/4 GHz)	2	4
No. of transponders	8	20
EIRP, dBW	29	31
Global coverages (6/4 GHz)	1	2
No. of transponders	3	6
EIRP, dBW	23.5	26.5
Spot coverages (14/11 GHz)	2	2
No. of transponders	6	10
EIRP, dBW (east spot/west spot)	41.4/44.4	41.1/44.4

Table 16.2 Intelsat VI G/T performance (reproduced by permission of *COMSAT Technical Review*, CTR, Vol. 20, No. 2, Fall 1990, p 292).

Coverage	Antenna gain (dBi)	Losses (dB)	Est. noise figure of receiver and transponder (dB)	Noise temp. (dB/K)	Spec. G/T^* (dB/K)	Margin (dB)
Global A	16.0	1.5	3.0	27.6	−15.0 (−14.0)	1.9
Global B	16.0	1.5	3.0	27.6	−15.0 (−14.0)	1.9
E. Hemi	21.9	1.5	3.0	27.6	−8.5 (−9.2)	1.3
W. Hemi	21.3	1.5	3.0	27.6	−8.5 (−9.2)	0.7
Zone 2	23.3	1.5	3.0	27.6	−7.0	1.2
Zone 4	23.5	1.5	3.0	27.6	−7.0	1.4
Zone 1	32.8	1.5	3.0	27.6	−1.0 (−2.0)	4.7
Zone 3	29.2	1.5	3.0	27.6	−1.0 (−2.0)	1.1
E. Spot	33.6	2.0	5.0	29.6	1.0	1.0
W. Spot	37.5	2.0	5.0	29.6	4.3 (1.7)	1.6

* Numbers in parentheses are the revised specification values agreed to during the course of the implementation program.

The key elements of the system are as follows.

Antenna system
Large antenna reflectors are used for the 4 GHz transmit and 6 GHz receive hemi and zone beams. The size is determined by the need to generate four co-frequency and co-polarization zone beams with a minimum interbeam spacing of 2.15° and an isolation better than 27 dB. The C-band beams use circular polarization. A double-offset feed reflector is used with a flat feed array of 146 feed horns. The diameter is 3.2 m for the transmit and 2.0 m for the receive antenna. Groups of feeds are excited together in order to produce the required beam patterns. The coverage of the four zones varies according to the ocean region and depends on a group of earth stations

Table 16.3 Intelsat VI EIRP for hemi/zone beam (reproduced by permission of *COMSAT Technical Review*, CTR, Vol. 20, No.2, Fall 1990, p 293)

Coverage	Amplifier output (W)	Expected losses* (dB)	Antenna gain, inc. feed** (dBi)	Spec. EIRP (dBW)	Margin (dB)
W. Hemi					
80 MHz	16.0	1.7	21.3	31.0	0.6
40 MHz	8.5	1.7	21.3	28.0	0.9
E. Hemi					
80 MHz	16.0	1.7	21.9	31.0	1.2
40 MHz	8.5	1.7	21.9	28.0	1.5
Zone 1					
80 MHz	2.5	1.7	32.8	31.0	4.1
40 MHz	1.3	1.7	32.8	28.0	4.2
Zone 2					
80 MHz	10.0	1.7	23.3	31.0	0.6
40 MHz	5.0	1.7	23.3	28.0	0.6
Zone 3					
80 MHz	2.5	1.7	29.2	31.0	0.5
40 MHz	1.3	1.7	29.2	28.0	0.6
Zone 4					
80 MHz	10.0	1.7	23.5	31.0	0.8
40 MHz	5.0	1.7	23.5	28.0	0.8

* Expected losses due to switches, output multiplex filters, and waveguide.
** Antenna gains include losses due to the beam-forming network and are based on lowest gain for the beam in any ocean region.

Table 16.4 Intelsat VI RF flux density for saturation (courtesy of British Telecommunications plc)

Up-path beam	Bandwidth MHz	Saturation flux density required (dBW/m^2)		
		Low gain	High gain	Extra high gain
14 GHz east spot	72, 77, 150	-73.0 ± 2	-78.0 ± 2	-84.0 ± 2
14 GHz west spot	72, 77, 150	-76.3 ± 2	-81.3 ± 2	-87.3 ± 2
6 GHz	36, 41	-70.1 ± 2	-77.6 ± 2	N/A
6 GHz	72	-67.1 ± 2	-77.6 ± 2	N/A

in each location. The excitation of four zone beams in the AOR is shown in Chapter 9, Fig. 9.14.

The different ocean sets of zone beam patterns are generated via panels selected by switches connected to the 'B' polarization port of every feed. The zone coverage can be changed by command from the ground.

Typical coverage diagrams for the Atlantic Ocean Region and the Indian Ocean Region are shown in Fig. 16.8.

16.2 *Development of the Intelsat space segment provision* 337

INTELSAT VI

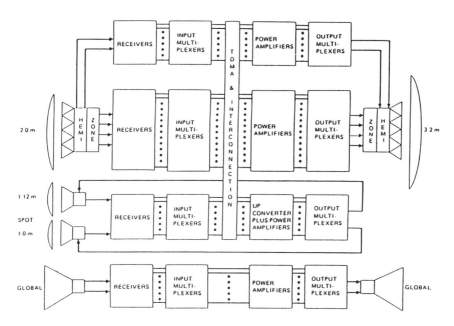

Fig. 16.7 Intelsat VI simplified communications sub-system (courtesy of British Telecommunications plc)

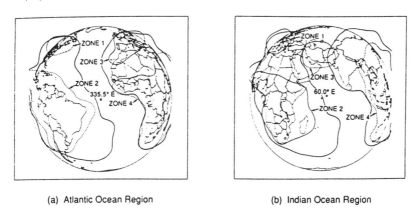

(a) Atlantic Ocean Region (b) Indian Ocean Region

Fig. 16.8 AOR and IOR coverage diagrams (reproduced by permission of *COMSAT Technical Review*, CTR, Vol.20, No.2, Fall 1990, p 243)

For the two hemi beams, groups of 79 and 64 feeds are activated using 'A' polarization feed connectors. The hemi beams are much broader in their coverage than the zone beams and the coverage does not vary with ocean region. A tracking beacon sited at an earth station generates the RF source for pointing the transmit and receive hemi/zone reflectors.

Two 14/11 GHz spot beams are produced using offset-fed circular reflectors of 1.0 m diameter with both frequencies being carried simultaneously using opposite linear polarizations. The reflectors are quasi-parabolic, i.e. shaped, to give optimum performance. The spot beams operate in high traffic density areas, mostly in the northern hemisphere, and are fully steerable anywhere in the visible part of the earth. The beams cover elliptical areas with the east spot beam oriented 23° from the horizontal and the west spot beam oriented 37° from the horizontal.

Global beams at 6/4 GHz are produced using simple conical horns with limited pitch steering. These beams are dual circularly polarized and are separated from the hemi/zone beams by frequency

Receivers/downconverters
Sixteen 6 GHz and four 14 GHz receivers operate in four-for-two redundancy groups to provide for each of the ten satellite beams. (Redundancy is the provision of standby units which can be switched into service should an operational unit fail.)

The 6 GHz receivers operate with bandpass filters followed by GaAs FET pre-amplifiers, which have noise figures less than 3 dB. The pre-amplifiers are followed by downconversion to the 4 GHz down-link band using a microwave integrated circuit (MIC) balanced mixer. The mixer local oscillator (LO) frequency is derived from a crystal oscillator. The LO frequency can be varied ± 12 kHz in steps to allow matching at all hemi/zone receiver conversion frequencies to within 9 kHz of each other.

The 14 GHz receivers use advanced GaAs FET pre-amplifiers which produce noise figures better than 5 dB. The local oscillator, which converts to the 4 GHz band, is controlled from the ground to ensure suitable frequency stability.

Input multiplexers
Each broadband 4 GHz receiver output signal is split into two and, via odd–even multiplexers, is divided into transponder channel blocks of 36, 41, 72, 77 or 150 MHz bandwidth depending upon the frequency band and beam connections. There are ten multiplexers each consisting of a hybrid power divider and two pairs of 2 or 3-channel non-contiguous circulator coupled coaxial resonator bandpass filters. The filters are quasi-elliptic and self-equalized, designed to give flat in-band passband and group delay performance with good out of band rejection. Each of the 50 channels has an individual high performance bandpass filter and the physical implementation allows cross-coupling to achieve group delay equalization. Total group delay is defined as the delay measured between the input to, and the output from, the transmission channel and includes the receive and transmit antennae. A switchable gain-step attenuator for each channel follows the input multiplex.

Intelsat VI has the facility to interconnect the 6/4 GHz and 14/11 GHz bands using a process known as 'cross-strapping'. The allocated 500 MHz is sub-divided into 12 nominal 41 MHz segments and a numbering system from 1 to 12 designates frequency slots with a bandwidth not greater than 41 MHz. When a transponder with a bandwidth greater than 41 MHz is used, a multiple number (e.g. 1–2) indicates the fact. The transponder occupying the new up-link band from 5854 MHz to 5926 MHz is designated (1'–2').

Intelsat VI has five banks of 72 MHz transponders at C-band using channels (1–2) to (7–8) and (1'–2'). Transponders (1–2) to (7–8) have sixfold frequency re-use with the combination of four zone beams and two hemi beams while transponder (1'–2') has twofold frequency re-use with two hemi beams. There is thus a total of twenty-six 72 MHz transponders for C-band. The C-band transponder banks (1–2) and (3–4) are designed to carry 120 Mbit/s TDMA while all other transponder banks are designed for FM/FDMA or PSK/FDMA traffic. Global beam transponders can handle a variety of

16.2 Development of the Intelsat space segment provision

traffic, including SCPC telephony and analogue FM video for occasional-use TV. Transponders (10), (11) and (12) are assigned to global beams permanently while transponder (9) can be switched to provide either twofold re-use through the cross-polarized global beams or sixfold frequency re-use through the zone and hemi beams.

There are four 72 MHz transponder banks at Ku-band, each with twofold frequency re-use. These transponders can be cross-strapped as required. There is also one transponder bank at 150 MHz.

Static/dynamic (SS/TDMA) switches
Hemi and zone repeaters may be interconnected using static switch matrices (SSMs) or dynamic microwave switching matrices (MSMs). Because Ku-band up-link signals are downconverted to C-band, the Ku-band spot repeaters can be additionally statically connected to the hemi/zone repeaters. The global transponders, however, cannot be interconnected with other beams.

SSMs are constructed as 6 × 6 matrices (six inputs and six outputs) employing 15 coaxial switches, or 8 × 8 matrices using 22 coaxial switches. A 6 × 6 configuration is shown in Fig. 16.9.

The 8 × 8 matrices are used in channels 1–2 through to 7–8 allowing interconnection between the four zone beams, the two hemi and the two spot beams. Channel 9 uses a 6 × 6 matrix because only hemi and zone interconnections are required. Switches in each matrix can be connected singly or in a group on a channel by channel basis.

In transponder banks (1–2) and (3–4) both static and 6 × 6 dynamic switches are

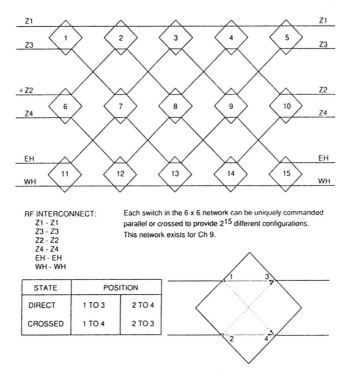

Fig. 16.9 6 × 6 SSM configurations (reproduced by permission of *COMSAT Technical Review*, CTR, Vol.21, No.1, Spring 1991, p 127)

340 *The Intelsat Organization*

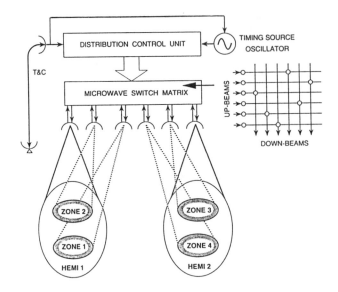

Fig. 16.10 Intelsat VI SS/TDMA system (reproduced by permission of *COMSAT Technical Review*, CTR, Vol.20, No.2, Fall 1990, p 336)

provided. The dynamic switches allow zone and hemi, beam-to-beam interconnections to be changed up to 64 times within each 2 ms TDMA frame period.

Satellite switched time division multiple access (SS/TDMA) utilizes the MSM to connect a TDMA burst on an up-link to a down-link beam. Figure 16.10 shows the arrangement.

The connection may be single point (one connection between a given row and column in the matrix) or multipoint (with many connections between a given row and a number of columns). Each state exists for a period of time that is sufficient for the traffic occurring between the beams.

The Intelsat VI modes of operation also include an Intelsat V/VA compatibility mode and a pseudo-global mode for frequency slot 1'-2'. In the Intelsat V/VA compatibility mode of operation, the Intelsat VI satellite provides two zone beams (West Zone and East Zone) for each ocean region made up of the sum of the Intelsat VI zone beams. As an example, the Intelsat V/VA compatibility zone beam for the West Zone is composed of the Intelsat VI NW and SW zone beams. The Intelsat V/VA compatibility mode may be selected on a transponder bank by transponder bank basis, and is intended to be used during transition from Intelsat V to Intelsat VI. In the pseudo-global mode of operation, the hemispheric frequency slots of 1'-2' are configured to function as a single channel without affecting other frequency slots.

The beam and transponder interconnection capabilites are shown in Table 16.5.

Each beam accessing a given frequency slot is designated by a two or three digit number. The units digit identifies the frequency band while the tens and hundreds digits identify the beam coverage and type. Table 16.6 indicates the Intelsat VI beam connection number together with the frequency slot number and respective centre frequencies.

From Table 16.5 it can be seen that transponder (3–4) can be cross-connected from a west spot beam at 14 GHz to an east hemispheric beam at 4 GHz. From Table 16.6 the connection could be written as 62 up/22 down or simply 62/22.

16.2 Development of the Intelsat space segment provision

Table 16.5 Intelsat VI beam and transponder static interconnection capabilities (courtesy Intelsat)

(6 GHz UP-LINKS)

Satellite reception			Satellite transmission	
Coverage (up-link)	Freq. band (GHz)	Via frequency slot numbers (Transponder)	Coverage (down-link)	Freq. band (GHz)
Global A	6	9, 10, 11, 12	Global A	4
Global B	6	9, 10, 11, 12	Global B	4
East Hemi	6	(1–2),(3–4),(5–6),(7–8),9	East or West Hemi	4
West Hemi	6	(1–2),(3–4),(5–6),(7–8),9	East or West Hemi	4
East Hemi	6	(1–2),(5–6)	East or West Spot	11
West Hemi	6	(1–2),(5–6)	East or West Spot	11
East Hemi	6	(3–4),(5–6)	East or West Zone	4
West Hemi	6	(3–4),(5–6)	East or West Zone	4
East Zone	6	(1–2),(3–4),(5–6),(7–8),9	East or West Zone	4
West Zone	6	(1–2),(3–4),(5–6),(7–8),9	East or West Zone	4
East Zone	6	(3–4),(5–6)	East or West Hemi	4
West Zone	6	(3–4),(5–6)	East or West Hemi	4

(14 GHz UP-LINKS)

Satellite reception			Satellite transmission	
Coverage (up-link)	Freq. band (GHz)	Via frequency slot numbers (Transponder)	Coverage (down-link)	Freq. band (GHz)
East Spot	14	(1–2),(5–6),(7–8),(9–12)	East or West Spot	11
West Spot	14	(1–2),(5–6),(7–8),(9–12)	East or West Spot	11
East Spot	14	(1–2),(5–6)	East or West Hemi	4
West Spot	14	(1–2),(5–6)	East or West Hemi	4
East Spot	14	(5–6)	East or West Zone	4
West Spot	14	(5–6)	East or West Zone	4

The identification of circularly polarized beams is defined as follows:

(a) A pol: Satellite receive (up-link) LHCP
    Satellite transmit (down-link) RHCP
(b) B pol: Satellite receive (up-link) RHCP
    Satellite transmit (down-link) LHCP.

where polarization can be left-hand circular polarization (LHCP) or right-hand circular polarization (RHCP).

Table 16.5 (cont.)

(6 GHz UP-LINKS)

Satellite reception		Via frequency slot numbers (Transponder)	Satellite transmission	
Coverage (up-link)	Freq. band (GHz)		Coverage (down-link)	Freq. band (GHz)
Global A	6	9, 10, 11, 12	Global A	4
Global B	6	9, 10, 11, 12	Global B	4
East Hemi	6	(1'–2'),(1–2),(3–4),(5–6),(7–8),9	East or West Hemi	4
West Hemi	6	(1'–2'),(1–2),(3–4),(5–6),(7–8),9	East or West Hemi	4
East Hemi	6	(1–2),(3–4),(5–6),(7–8)	East or West Spot	11
West Hemi	6	(1–2),(3–4),(5–6),(7–8)	East or West Spot	11
East Hemi	6	(1–2),(3–4),(5–6),(7–8),9	NW,SW,NE or SE Zone	4
West Hemi	6	(1–2),(3–4),(5–6),(7–8),9	NW,SW,NE or SE Zone	4
NW Zone	6	(1–2),(3–4),(5–6),(7–8),9	East or West Hemi	4
NW Zone	6	(1–2),(3–4),(5–6),(7–8),9	NW,SW,NE or SE Zone	4
NW Zone	6	(1–2),(3–4),(5–6),(7–8)	East or West Spot	11
SW Zone	6	(1–2),(3–4),(5–6),(7–8),9	East or West Hemi	4
SW Zone	6	(1–2),(3–4),(5–6),(7–8),9	NW,SW,NE or SE Zone	4
SW Zone	6	(1–2),(3–4),(5–6),(7–8)	East or West Spot	11
NE Zone	6	(1–2),(3–4),(5–6),(7–8),9	East or West Hemi	4
NE Zone	6	(1–2),(3–4),(5–6),(7–8),9	NW,SW,NE or SE Zone	4
NE Zone	6	(1–2),(3–4),(5–6),(7–8)	East or West Spot	11
SE Zone	6	(1–2),(3–4),(5–6),(7–8),9	East or West Hemi	4
SE Zone	6	(1–2),(3–4),(5–6),(7–8),9	NW,SW,NE or SE Zone	4
SE Zone	6	(1–2),(3–4),(5–6),(7–8)	East or West Spot	11

(14 GHz UP-LINKS)

Satellite reception		Via frequency slot numbers (Transponder)	Satellite transmission	
Coverage (up-link)	Freq. band (GHz)		Coverage (down-link)	Freq. band (GHz)
East Spot	14	(1–2),(3–4),(5–6),(7–8),(9–12)	East or West Spot	11
West Spot	14	(1–2),(3–4),(5–6),(7–8),(9–12)	East or West Spot	11
East Spot	14	(1–2),(3–4),(5–6),(7–8)	East or West Hemi	4
West Spot	14	(1–2),(3–4),(5–6),(7–8)	East or West Hemi	4
East Spot	14	(5–6),(7–8)	NW,SW,NE or SE Zone	4
West Spot	14	(5–6),(7–8)	NW,SW,NE or SE Zone	4

16.2 Development of the Intelsat space segment provision   343

Table 16.6 Intelsat VI beam connection and transponder numbering system (courtesy Intelsat)

(6/4 GHz transponders)

Frequency slot number	Beam connection ID number								Centre frequency (MHz)	
	West hemi	East hemi	NW zone (Z1)	NE zone (Z3)	SW zone (Z2)	SE zone (Z4)	Global (A)	Global (B)	Satellite receive (up-link)	Satellite transmit (down-link)
1'–2'	10	20	—	—	—	—	—	—	5890	3665
1–2	11	21	41	51	91	101	—	—	5970	3745
3–4	12	22	42	52	92	102	—	—	6050	3825
5–6	13	23	43	53	93	103	—	—	6130	3905
7–8	14	24	44	54	94	104	—	—	6220	3995
9	15	25	45	55	95	105	35	85	6280	4055
10	—	—	—	—	—	—	36	86	6320	4095
11	—	—	—	—	—	—	37	87	6360	4135
12	—	—	—	—	—	—	38	88	6402.5	4177.5
Polarization ID	A	A	B	B	B	B	A	B	—	—

(14/11 GHz transponders)

Frequency slot number	Beam conn. ID. no.		Centre frequency (MHz)	
	West spot	East spot	Satellite receive (up-link)	Satellite transmit (down-link)
1–2	61	71	14042.5	10992.5
3–4	62	72	14125	11075
5–6	63	73	14205	11155
7–8	64	74	14295	11495
9–12	69	79	14418.5	11618.5
Polarization ID	See text		—	—

The linear polarization of the spot beams is defined as follows:
(a) West Spot Beam Orientation:
  Satellite receive (up-link, 14 GHz)—vertical
  Satellite transmit (down-link, 11 GHz)—horizontal
(b) East Spot Beam Orientation:
  Satellite receive (up-link, 14 GHz)—horizontal
  Satellite transmit (down-link, 11 GHz)—vertical.

where vertical orientation is parallel to the spacecraft's pitch axis (north/south) and horizontal orientation is parallel to the spacecraft's roll axis (east/west).

Power amplifiers

Signals for the global, hemi and zone beams in the 4 GHz band are amplified using transistor amplifiers followed by either travelling-wave tube amplifiers (TWTAs) or solid-state power amplifiers (SSPAs). Since beam gains vary, the type of power amplifier used depends on the beam. For high power, TWTAs in the range 5–20 W are used while 1.8 to 3.2 W SSPAs are used for the smaller zone beams. High gain driver stages are used in channels 1–2 and 3–4 to allow for the loss through the TDMA microwave switches. Three-for-two redundancy is used for all 4 GHz amplifiers.

Signals for the 11 GHz and 12 GHz bands are up-converted, using an image-enhanced double-balanced mixer using a frequency of 7.25 or 7.5 GHz, before being amplified using 20 or 40 W TWTAs. Four-for-two redundancy is used.

Output multiplexers

Contiguous output multiplexers are used at both 4 GHz and 11 GHz. There are two 4-channel global, four 5-channel zone and two 6-channel hemi multiplexers at C-band and two 5-channel Ku-band spot multiplexers.

At 4 GHz triple cavity TE_{111} dual mode cylindrical resonators are mounted on the output waveguide manifold. The RF response, for a 72 MHz channel, is typically -0.6 dB and $+29$ ns worst case delay at ± 36 MHz.

At 11 GHz, similar multiplexers, with triple cavity TE_{113} dual mode cylindrical cavities achieve a performance of -1 dB and $+42$ ns worst case delay at 36 MHz.

Insertion loss is less than 1 dB in all cases.

Each Intelsat VI satellite can carry up to 120,000 telephone calls and three television channels simultaneously.

Intelsat K

In 1989 Intelsat decided to invest in a fully Ku-band satellite to meet its Atlantic Ocean Region television requirements. Intelsat was able to purchase a SATCOM K4 satellite, which was under construction by GE/ASTRO, and rename it Intelsat K. The bus is a standard GE 5000 with the communications payload altered to meet Intelsat requirements, mainly in terms of required antenna coverage and inter-beam switching. This satellite was successfully launched in 1992 and is located at 338.5° E

Intelsat K is a three-axis stabilized satellite with two antenna systems.

- The east-west antenna comprising a parabolic gridded dual-offset reflector fed by four offset feed arrays giving the horizontal (H) and vertical (V) coverages of North America and Europe. The European H and V polarized feed arrays consist of 23 horns. The H polarized horns can operate in the receive and transmit bands while the V polarized horns only operate in the transmit band. There are 14 horns in the North America polarized array and, while the H polarized horns operate in both receive and transmit bands, the V polarized horns operate in the receive band only.
- The South America antenna comprising an offset-fed elliptical parabolic reflector fed by a feed array of four horns using vertical polarization with transmit only facility.

The RF bandwidth of the satellite is divided into 54 MHz segments with 16 transponders providing connections west to west, west to east, east to east and east to west. Transponders allocated to the west may be individually switched to the North America or South America down-link or both combined. Transponders allocated to the east are connected to the European beam. The beam and transponder interconnection capability is shown in Table 16.7.

16.2 Development of the Intelsat space segment provision

Table 16.7 Intelsat K beam and transponder interconnection capabilities (courtesy Intelsat)

(14 GHz UP-LINKS)

			SATELLITE TRANSMISSION (DOWN-LINK)										
	Beam		North America NAH		Europe EUV		EUH	South America SAV		NAH+SAV		NAH+EUV	NAH+EUV+SAV
		DOWN LINK FREQ.	B	A	C	A	A	B	A	B	A	A	A
		SLOT No. (XPDR)	1–4	5–8	1–4	5–8	5–8	1–4	5–8	1–4	5–8	5–8	5–8
S A T. R E C E P T I O N	Europe EUH	1–4	X		X			X		X			
		5–8		X		X	X		X		X	X	X
	North America NAV	1–4	X		X			X		X			
		5–8		X		X	X		X		X	X	X
U P - L I N K	North America NAH	1–4	X		X			X		X			
		5–8		X		X	X		X		X	X	X
	NAH + EUH	1–4	X		X			X		X			
		5–8		X		X	X		X		X	X	X

There is provision for receive/transmit coverage of North America and Europe and transmit coverage of South America. Depending on the transponder connection there is two times frequency re-use by either orthogonal linear polarization or spatial separation.

Figure 16.11 shows the transponder layout while Fig. 16.12 shows the typical measured transmit beam coverage.

Frequency bands used for the coverage are shown in Table 16.8.

346 *The Intelsat Organization*

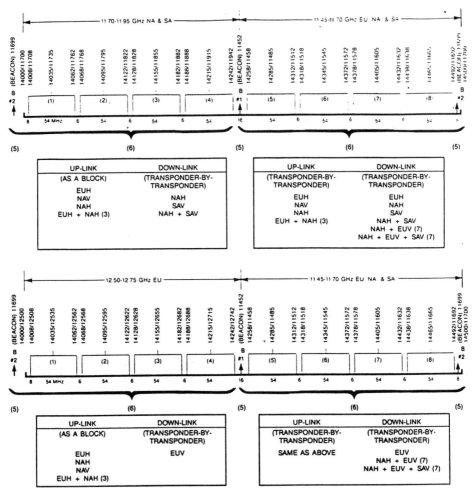

Fig. 16.11 Transponder layout for Intelsat K (courtesy Intelsat)

16.2 *Development of the Intelsat space segment provision* 347

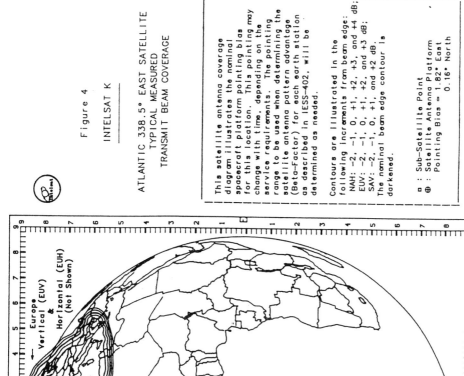

Fig. 16.12 Typical measured transmit beam coverage for Intelsat K (courtesy Intelsat)

Table 16.8 Frequency bands for Intelsat K coverage areas. Up-link frequency band is 14.00–14.50 GHz. (courtesy Intelsat)

Coverage area	Polarization	Down-link frequency band (GHz)
North America	Horizontal	11.45–11.70/11.70–11.95
South America	Vertical	11.45–11.70/11.70–11.95
Europe	Horizontal	11.45–11.70
Europe	Vertical	11.45–11.70/12.50–12.75

Briefly, the communications sub-system comprises the following.

- Receivers/down-converters. There are four low-noise amplifiers (LNAs) of which two are operational using four-for-two redundancy. Reception of signals is at 14.0–14.5 GHz followed by downconversion to three possible down-link frequency bands, namely:

    (1) conversion of 14.25–14.50 GHz band to 11.45–11.70 GHz using a LO of 2.8 GHz with a four-for-two redundancy arrangement;
    (2) conversion of 14.0–14.25 GHz band to 12.50–12.75 GHz using a LO of 1.5 GHz with a two-for-one redundancy scheme;
    (3) conversion of 14.0–14.25 GHz band to 11.70–11.95 GHz using a LO of 2.3 GHz with a two-for-one redundancy scheme.
- Power amplifiers. Driver stages are in two groups and provide amplification in the complete down-link frequency range of 11.45–12.75 GHz. One group operates on the horizontally polarized and South America signals while the other group deals with vertically polarized signals. The high-power amplifiers (HPAs) are 60 W TWTAs arranged in two groups with 11-for-8 redundancy.

Table 16.9 shows the polarization of Intelsat V, K and VI satellites.

Intelsat VII

The new generation of satellites, to replace the ageing Intelsat V series, has been designed mainly to provide for increasing traffic demand in the Pacific Ocean Region (POR) but with the facility to be adapted for use in the other two regions. Among the features regarded as necessary for the design specification were:

- maximum reliability and lifetime;
- satellite parameters optimized for operation to smaller earth stations while maintaining efficient use of bandwidth;
- designed predominantly for digital operation.

In June 1988 the Board of Governors selected Ford Aerospace (now Space Systems/Loral) as the preferred contractor, and signed for five spacecraft in October 1988. Work on the spacecraft has progressed to environmental and performance testing and the first of these satellites was successfully launched in late-1993. The satellite in its fully deployed state is shown in Fig. 16.13.

Intelsat VII is three-axis stabilized in the same way as Intelsat V. The rectangular

16.2 Development of the Intelsat space segment provision

Table 16.9 Polarization of Intelsat transmissions (courtesy British Telecommunications plc)

Satellite	Coverage	Polarization nomenclature	Frequency (GHz)	Up-link	Down-link
INTELSAT V,VA,VA(IBS)	Global	A	6/4	LHC	RHC
INTELSAT VA,VA(IBS)	Global	B	6/4	RHC	LHC
INTELSAT V,VA,VA(IBS)	Hemi	A	6/4	LHC	RHC
INTELSAT V,VA,VA(IBS)	Zone	B	6/4	RHC	LHC
INTELSAT VA,VA(IBS)	Spot*	A	4	LHC*	RHC
INTELSAT VA,VA(IBS)	Spot*	B	4	RHC*	LHC
INTELSAT V,VA,VA(IBS)	West Spot	Linear	14/11/12**	Vertical	Horizontal
INTELSAT V,VA,VA(IBS)	East Spot	Linear	14/11/12**	Horizontal	Vertical
INTELSAT V,VA,VA(IBS),VI	Global (TC&R)	–	6/4	Horizontal or LHC	RHC
INTELSAT V,VA,VA(IBS),VI	Global (Beacons)	–	4	–	RHC
INTELSAT V,VA,VA(IBS),VI	Spot (Beacons)	–	11	–	RHC
INTELSAT VA(IBS)	West Spot (Beacons)	Linear	12	–	Horizontal
INTELSAT VA(IBS)	East Spot (Beacons)	Linear	12	–	Vertical
INTELSAT V(MCS)	Global	B	6/4	RHC	LHC
INTELSAT V(MCS)	Global	–	1.6/1.5	RHC	RHC
INTELSAT VI	Global	A	6/4	LHC	RHC
INTELSAT VI	Global	B	6/4	RHC	LHC
INTELSAT VI	Hemi	A	6/4	LHC	RHC
INTELSAT VI	Zone	B	6/4	RHC	LHC
INTELSAT VI	West Spot	Linear	14/11	Vertical	Horizontal
INTELSAT VI	East Spot	Linear	14/11	Horizontal	Vertical
INTELSAT K	North America	Linear	14/11/12	Horizontal & vertical	Horizontal
INTELSAT K	Europe	Linear	14/11	Horizontal	Horizontal & vertical
INTELSAT K	Europe	Linear	14/12	Horizontal	Vertical
INTELSAT K	South America	Linear	11/12	–	Vertical
INTELSAT K	North America (TC&R)	Linear	14/11/12	Vertical	Horizontal
INTELSAT K	Europe (TC&R)	Linear	14/11/12	Horizontal	Vertical
INTELSAT K	South America (TC&R)	Linear	11/12	–	Vertical
INTELSAT K	Omni (TC&R)	Linear	14	Horizontal	–

LHC: Left-hand circular
RHC: Right-hand circular
* Global up-link
**The 12 GHz band is used only on the INTELSAT VA (IBS)

main body incorporates the electronics equipment internally and supports the communications antennae externally on the east, west and earth faces. The communications repeater is mounted on the north and south panels.

The design specification of Intelsat VII was simplified, compared with that of Intelsat VI, with some modifications as follows:

- no reconfigurable antennae;
- use of fourfold frequency re-use at C-band and not sixfold as per Intelsat VI;
- limit coverage to two locations in POR and four in AOR;
- not use satellite switched time division multiple access (SS/TDMA).

A comparison of some key parameters of Intelsat VI and VII satellites is shown in Table 16.10.

350 *The Intelsat Organization*

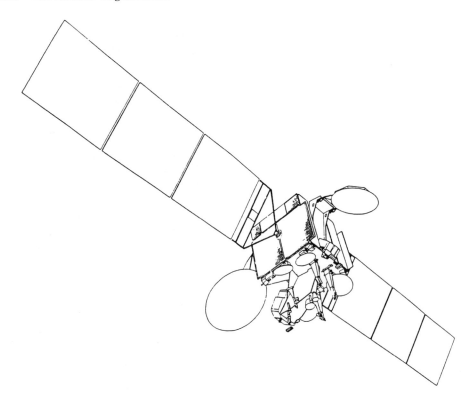

Fig. 16.13 Intelsat VII in deployed state (courtesy of British Telecommunications plc)

Table 16.10 Comparison between Intelsat VI and VII. (courtesy Intelsat)

		VI		VII	
C-band	Coverage	Number of transponders	EIRP (dBW)	Number of transponders	EIRP (dBW)
	hemi	12	31	10	33
	zone	20	31	10	33
	global	6	26.5	6	26.5/29.5
Ku-band	spot	10	44–47	10	45–48

16.2 Development of the Intelsat space segment provision

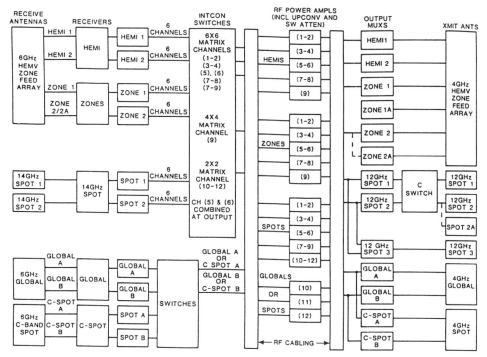

Fig. 16.14 Simplified block diagram of Intelsat VII communications repeater (courtesy of British Telecommunications plc)

Among the improvements incorporated in Intelsat VII compared with earlier satellites are:

at C-band

- the use of solid-state power amplifiers (SSPAs) for all channels;
- improved zone coverage (and Ku-band coverage of South East Australia when operating in the POR);
- independently steerable spot beam;
- channel 9 transponder switchable between hemi, global or C-band spot;

at Ku-band

- linearizers are used on the TWTAs;
- an Intelsat business service (IBS) payload with business band transponders selectable on a per channel basis;
- increase in bandwidth from 72 MHz to 112 MHz on four transponders to facilitate higher bit-rate carriers;
- gain selectable in 1 dB steps over a range of 15 dB on each channel.

In both bands higher power levels are provided to enable the use of smaller earth stations.

A simplified block diagram of the communications repeater is shown in Fig. 16.14.

352 The Intelsat Organization

Antenna system
At C-band the coverage is:

- global receive and transmit beams;
- two hemi beams;
- four zone beams, of which only two are for simultaneous down-link transmission;
- a spot beam.

The coverage areas are divided into groups one and two. Either zone 1 or zone 2 up-link beam coverages can be enhanced by ground command. In this case, signals from diametrically opposite zone beams (1A or 2A) can be combined in the satellite although the total signal is directed to a single down beam.

The Intelsat VII hemi/zone antennae use offset fed reflectors with more than 120 feed horns.

The steerable C-band spot beam antenna shares transponders on a channel-by-channel switchable basis with the global coverage and uses a high performance dual-polarized transmit/receive feed.

At Ku-band the coverage is:

- spots 1,2 and 3 comprising elliptical or circular beams. These beams are steerable over the whole of the visible earth. The spot beam antennae have offset feeds and shaped reflectors and, to maintain beam shape regardless of beam-pointing position, the reflector and feed are steered as one unit. In the POR an extra feed on spot 2 enables illumination of South East Australia

The frequency plan of the Intelsat VII satellite is given in Fig. 16.15 with other important parameters defined by Tables 16.11 and 16.12.

Table 16.11 Intelsat VII RF flux density to produce saturation at beam edge (courtesy of British Telecommunications plc)

Beam	Lowest power flux density (dBW/m^2)	Highest power flux density (dBW/m^2)
C-band beams		
Hemi 1,2	−87	−73
Global		
Zone 1,2 or enhanced		
C-band spot		
K$_u$-band beams		
Spot 1,2,3 inner	−90	−76
Spot 1,2,3 outer	−87	−73

Power amplifiers
For C-band, the signals are amplified to their transmit levels using highly linear solid-state power amplifiers (SSPAs) at 10, 16, 20 and 30 W power levels.

For Ku-band, the high-power amplifiers are travelling wave tube amplifiers (TWTAs) at 35 and 50 W power levels. For the first time linearizers will be used with the TWTAs. The linearizers are of the pre-distortion type giving highly linear characteristics. The use of TWTAs in this configuration should result in enhanced channel performance when operating in the multicarrier mode.

16.2 Development of the Intelsat space segment provision

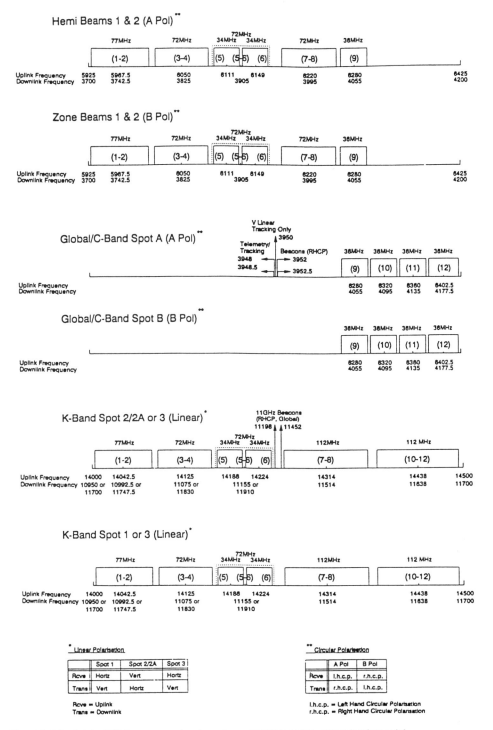

Fig. 16.15 Intelsat VII frequency plan (courtesy of British Telecommunications plc)

Table 16.12 Intelsat VII minimum G/T and EIRP within the specified coverages (courtesy British Telecommunications plc)

Beam	G/T (dB/K)	EIRP (dBW)
Hemi 1	−8.5	33.0
Hemi 2	−7.5	33.0(32.1)†
Zone 1	−5.5	33.0
Zone 2	−4.0	33.0
Zones 1 & 1A	−9.0	33.0
Zones 2 & 2A	−7.5	33.0
C-band spot A (9)*	−3.0	36.5
C-band spot B (9)*	−3.0	34.3
C-band spot A&B (10),(11)*	−3.0	33.3
C-band spot A&B (12)*	−3.0	36.3
Global A (9)*	−11.5	28.5
Global B (9)*	−11.5	26.0
Global A, Global B (10),(11)*	−11.5	26.0
Global A, Global B (12)*	−11.5	29
Spot 1 Inner (AOR/POR)	4.5	46.7/45.4
Outer (AOR/POR)	1.5	44.6/43.4
Spot 2 Inner (AOR/POR)	2.5	44.5/45.8
Outer (AOR/POR)	−1.0	41.4/42.6
Spot 2+A Inner	0.5	44.1
Outer	−3.0	41.2
Spot 3 Inner (1–6)(35W/50W)	3.8(3.5)‡	46.0/47.5(45.7/47.2)‡
Outer (1–6)(35W/50W)	0.8(0.5)‡	43.0/44.5(42.7/44.2)‡
Spot 3 Inner (7–12)(35W/50W)	3.8(3.5)‡	46.3/47.8(46.0/47.5)‡
Outer (7–12)(35w/50W)	0.8(0.5)‡	43.3/44.8(43.0/44.5)‡

* Channel numbers in brackets.
† Values with the beam broadened for IOR coverage in brackets.
‡ Values with the Spot 3 polarization switchability.

Changes have already occurred to the basic Intelsat VII design as the programme progresses. The final three satellites will have polarization switchability on the third Ku-band spot so that it can be operated co-polarized with either of the other two spot beams. Also, the final two satellites of the series have a broadened hemi beam coverage to allow optimization in the IOR.

An enhanced version of the Intelsat VII spacecraft (to be known as Intelsat VIIA) completed its preliminary design reviews in November 1991. An order for two spacecraft has already been placed with SS/Loral for delivery after the initial five VII satellites. An option exists for a further VIIA spacecraft.

The main differences are added Ku-band transponders and higher Ku-band and C-band EIRP.

For C-band the increase in output power is to be achieved using 30 W TWTAs in the global channels (10) and (11) giving an EIRP of 29 dBW for those transponders.

For Ku-band, the 35 and 50 W TWTAs of the VII series are to be replaced by 50 and 72 W devices to produce an increased EIRP of 47 dBW. Four additional wideband transponders are to be added with opposite polarization to the original transponders to give an additional 448 MHz of usable bandwidth. Also, any of the new cross-polar TWTAs can be individually coupled to the original wideband TWTA to give an EIRP increase of 2.5 dBW. Table 16.13 shows the modifications.

16.3 Ground network

Table 16.13 Summary of Ku-band modifications on Intelsat VIIA (courtesy British Telecommunications plc)

Parameter	INTELSAT VII	INTELSAT VIIA
No. 72 MHza transponders	6	6
No. 112 MHz transponders	4	8
TWTA powers	35 and 50W	50 and 72W
EIRP (dBW)		
Spot 1 inner coverage (AOR/POR)	46.7/45.4	47.2*
Spot 1 outer coverage (AOR/POR)	44.6/43.5	44.7
Spot 2 inner coverage (AOR/POR)	44.5/45.8	47.2*
Spot 2 outer coverage (AOR/POR)	41.4/42.6	43.7
Spot 2A inner coverage	44.1	45.4†
Spot 2A outer coverage	41.2	42.3
Spot 3 inner coverage LO/HI‡	46/47	43/44
Spot 3 outer coverage LO/HI‡	43/44	41/43
Spot 3 inner coverage size	$2.0 \times 2.0°$	$3.3 \times 3.3°$
Spot 3 outer coverage size	$2.76 \times 2.76°$	$4.4 \times 4.4°$

* 50 dBW is available in the wideband transponders and involves the use of two TWTAs. This renders another wideband transponder unusable.
† As above but 48 dBW only.
‡ LO/HI–LO is for lower power tube. HI for higher power tube selected.

Table 16.14 Current and future Intelsat satellites (courtesy Intelsat)

INTELSAT Designation	INTELSAT V	INTELSAT V-A	INTELSAT VI	INTELSAT K	INTELSAT VII	INTELSAT VII-A
Year of First Launch	1980	1985	1989	1992	1993	1995
Prime Contractor	Ford Aerospace	Ford Aerospace	Hughes	GE Astro Space	SS/Loral*	SS/Loral
Launch Vehicles	Atlas Centaur Ariane 1, 2	Atlas Centaur Ariane 1, 2	Ariane 4 Titan	Atlas IIA	Ariane 4, Atlas IIAS	Ariane 44L
Lifetime (Years)	7	7	13	10	10–15	10–15
Capacity	12,000 circuits and 2 TV	15,000 and 2 TV	24,000 and 3 TV (up to 120,000 with digital circuit multiplication equipment, DCME)	16 54 MHz Ku-band transponders; can be configured to provide up to 32 high quality TV channels	18,000 and 3 TV (up to 90,000 with DCME)	22,500 and 3 TV (up to 112,500 with DCME)

* Formerly Ford Aerospace

A list of current and projected Intelsat satellites is shown in Table 16.14.

16.3 Ground network

Telemetry, tracking, command and monitoring (TTC&M)

In 1987 Intelsat reviewed the existing TTC&M system and devised a new network to coincide with the introduction of Intelsat VI. The original network comprised eight

356   *The Intelsat Organization*

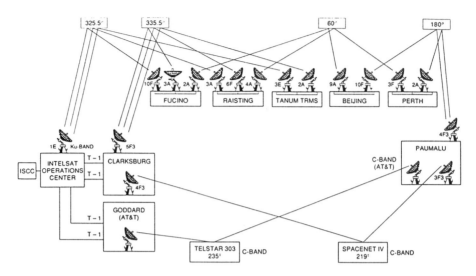

Fig. 16.16   Intelsat CCC network (reproduced by permission of *COMSAT Technical Review*, CTR, Vol.21, No.2, Fall 1991, p 400)

TTC&M stations and two communication system monitoring (CSM) stations, while the new system would consist of just six TTC&M stations. The new network has stations at Perth, Australia; Raisting, Germany; Fucino, Italy; Beijing, People's Republic of China; Clarksburg and Paumalu, United States. The hub of the network is the Satellite Control Centre (SCC) situated at Washington DC and each TTC&M station is linked to headquarters via two independent paths for reliability. The Intelsat Control Coordination (CCC) network is shown in Fig. 16.16.

The network consists of International Business Service (IBS) carriers linking all TTC&M stations with headquarters and it is over this CCC network that all telemetry, command, ranging, monitoring and station remote monitoring and control (M&C) signals are transmitted. Each station can process telemetry from four Intelsat V/VA satellites and four Intelsat VI satellites using information transmitted via the satellite beacon antennae. The received data is PSK modulated and after demodulation is sent to Intelsat headquarters via the CCC network. The telemetry data is used to verify system performance and monitor network response to the Intelsat remote commands.

Commands are essential to induce changes in satellite attitude and orbital position and may be achieved automatically using the command coordination system (CCS). Alternatively, the commands may be issued manually under the direction of the SCC. The TTC&M stations use command units and several different units are needed to cater for the different satellite series. A tone burst is used to frequency-modulate (FM) the station up-link under the control of the CCS.

Communications system monitoring (CSM) is a sub-system at the earth station used to evaluate emissions from the satellite communications sub-system. Frequency plan data is stored in a computer and each carrier emission is checked for EIRP, frequency offset and deviation. Also, the CSM measures the gain of each satellite transponder to check for degradation with time. Any earth station wishing to join the Intelsat network must pass antenna tests which evaluate EIRP stability, transmit gain, receive G/T, antenna sidelobe pattern etc. The CSM is involved in these tests via the TTC&M earth station personnel.

Ranging is also carried out by the TTC&M stations using a range processor which produces a stable sequential tone pattern using sine waves at 35 Hz, 283 Hz, 3.96 kHz and 27.7777 kHz which modulate, using FM, a signal in the up-link. This ranging signal is transmitted to the satellite, returned to the range processor via the telemetry path and the phase delay of each tone is measured. Ranging data consisting of phase angles, azimuth and elevation coordinates, time of measurement and satellite identification, are stored in the station computer and can be evaluated by SCC staff via the CCC network.

Intelsat Earth Station Standards

Intelsat provide the space segment of the communications link and the user provides the earth station and, where applicable, links to a terrestrial network. The term 'User' is intended to refer to:

- a signatory,
- a duly authorized telecommunications entity,
- neither of the above but authorized by an appropriate regulatory body to access the Intelsat system.

The earth station must be an approved type, compatible with all earth stations within the system and capable of meeting the Intelsat criteria for interference between earth stations and the space segment. Intelsat provides documents which outline the performance characteristics needed to achieve the following:

- satisfactorily meet Intelsat technical requirements;
- qualify as a 'standard' earth station.

Intelsat publish what is known as the Intelsat Earth Station Standards (IESS) which give users a common source of reference for the performance characteristics needed for earth stations and associated equipment in order to access the Intelsat space segment. Brief details of the IESS modules are mentioned in the sections that follow.

IESS documents
The IESS comprises six groups of documents as follows.

- Group 1, titled INTRODUCTORY, contains general guidelines for the assistance of earth station users in acquiring earth station facilities for the provision of certain communication service(s). Documents from this group are numbered 101, 102 etc.
- Group 2, titled ANTENNA AND RF EQUIPMENT CHARACTERISTICS, contains the performance characteristics of the various earth station categories (Standard A, B, C etc) approved for access to the Intelsat space segment. Documents from this group are numbered 201, 202 etc.
- Group 3, titled MODULATION AND ACCESS CHARACTERISTICS, contains the performance characteristics of the various types of modulation and access techniques (e.g. FDM/FM, SCPC/QPSK, TV/FM etc) approved for access to the Intelsat space segment. Documents from this group are numbered 301, 302 etc.
- Group 4, titled SUPPLEMENTARY, contains additional performance characteristics or technical information on specialized areas which may be required by users to construct their earth station facilities. Documents from this group are numbered 401, 402 etc
- Group 5, titled BASEBAND PROCESSING, contains the System Specification for Digital Circuit Multiplication Equipment (DCME) which may have an application to more than one modulation technique such as TDMA and IDR and may be located at an International Switching Centre (ISC) or an earth station dependent on the user's requirements. Documents from this group are numbered 501, 502 etc.

- Group 6, titled GENERIC EARTH STATION STANDARDS, contains boundary RF characteristics for earth stations accessing the Intelsat space segment for international services not covered by other earth station standards and for domestic earth stations accessing the Intelsat leased space segment. Documents in this group are numbered 601, 602 etc.

The complete list of IESS modules is given in Table 16.15.

Table 16.15 Intelsat earth station modules (IESS) (courtesy Intelsat)

Module number	Revision	Title
IESS-101	17	Introduction
IESS-201	1	Standard A Earth Station
IESS-202	–	Standard B Earth Station
IESS-203	2	Standard C Earth Station
IESS-204	1	Standard D Earth Station
IESS-205	1	Standard E Earth Station
IESS-206	–	Standard F Earth Station
IESS-301	1	FDM/FM
IESS-302	3	CFDM/FM
IESS-303	2	SCPC/QPSK
IESS-304	1A	SPADE
IESS-305	–	SCPC/CFM (VISTA)
IESS-306	1	TV/FM
IESS-307	A	TDMA
IESS-308	5	IDR
IESS-309	2	IBS
IESS-401	–	Intermodulation Criteria
IESS-402	3	EIRP Adjustment Factors
IESS-403	1A,1B	ESCs
IESS-404	–	Deleted
IESS-405	4	INTELSAT V Satellites
IESS-406	4	INTELSAT VA Satellites
IESS-407	4	INTELSAT VA(IBS) Satellites
IESS-408	5	INTELSAT VI Satellites
IESS-409	1	INTELSAT VII Satellites
IESS-410	1	Leased Transponder Definitions
IESS-411	2	Inclined Orbit Module
IESS-412	1	Earth Station Pointing Data
IESS-413	–	TDRS Module
IESS-414	1	INTELSAT K Satellites
IESS-501	2	DCME Specification
IESS-601	2	Standard G Earth Station
IESS-602	2	Standard Z Earth Station

INTELSAT CARRIER TYPES AND THEIR RELATIONSHIPS BETWEEN EARTH STATION STANDARDS

Table 16.16 Intelsat earth station standards (courtesy Intelsat)

Earth Station Standard	Type of service	Antenna diameter (m)	Frequency band (GHz)
A	International voice, data and TV including IBS and IDR	15.0 – 18.0	6/4
B	International voice, data and TV including IBS and IDR	10.0 – 13.0	6/4
C	International voice, data and TV including IBS and IDR	11.0 – 14.0	14/11 and/or 14/12
D1	Vista	4.5 – 6.0	6/4
D2	Vista	11.0	6/4
E1	IBS	3.5 – 4.5	14/11 and 14/12
E2	IBS and IDR	5.5 – 7.0	14/11 and 14/12
E3	IBS and IDR	8.0 – 10.0	14/11 and 14/12
F1	IBS	4.5 – 5.0	6/4
F2	IBS and IDR	7.0 – 8.0	6/4
F3	International voice and data including IBS and IDR	9.0 – 10.0	6/4
G	International Lease Services	all sizes	6/4 and 14/11 and 14/12
Z	Domestic Leased Services	all sizes	6/4 and 14/11 and 14/12

Earth Station Standards
Intelsat has specified guidelines for earth stations as follows.

- An earth station will normally comprise: one or more antennae with steering and/or tracking equipment as required, one or more RF transmitters, one or more low-noise RF receiving amplifiers, RF transmission lines, ground communications equipment for conversion from baseband-to-RF and vice versa, multiplex and terrestrial interface equipment.

360 The Intelsat Organization

- The number of antennae is dependent on the number of satellites through which communication is required and any facilities for system diversity and redundancy that may be deemed necessary.
- The minimum steering and tracking capabilities of antennae operating with the various satellite series are given in the Group 2 IESS modules. It is recommended that users allow variation of main beam pointing so that operation with satellites at different longitudes is possible. This would allow an on-site demonstration of compliance with the mandatory sidelobe envelope specification.
- Users working normally with one series of satellites may be required to switch to another series under certain contingency conditions in order to maintain continuity of service.
- The RF subsystem shall be capable of covering the minimum RF transmit and receive bands specified in the modules describing the earth station performance characteristics. For the RF transmit subsystem, this can be accomplished by means of a single power amplifier or by means of narrower band units properly combined, which can provide coverage of the required transmit band.
- Station design should accommodate changes of transmitted and received RF carrier frequencies that can be accomplished easily and without unacceptable interruption of service.
- Earth station reliability should be such that the space segment cannot be jeopardized by emissions that are in error due to carrier level, frequency, deviation, synchronization or polarization state.
- The required characteristics of a standard earth station have been determined on the basis of its ability to provide a channel performance consistent with CCIR recommended standards. The actual performance may vary according to climatic conditions and, if this is the case, enhanced standards of performance may be required.

The types of Intelsat standard earth stations and their performance characteristics are shown in Table 16.16.

At the time of writing, the standard modulation and access methods approved and specified in detail in the IESS modules for the provision of the various Intelsat services are as indicated in Table 16.17.

16.3 Ground network 361

Table 16.17 Intelsat carrier types to be used between standard earth stations (courtesy British Telecommunications)

Carrier Type	Standard A Transmitting to Std Std Std Std Std Std A B C D E F	Standard B Transmitting to Std Std Std Std Std Std A B C D E F	Standard C Transmitting to Std Std Std Std Std Std A B C D E F	Standard D Transmitting to Std Std Std Std Std Std A B C D E F
FDM/FM	yes – yes – – –	– – – – – –	yes – yes – – –	– – – – – –
CFDM/FM*	yes yes yes – – yes	yes yes yes – – yes	yes yes yes – – yes	– – – – – –
VISTA	– – – – – –	– – – – – –	– – – – – –	– – yes – –
TV**	yes yes – – – –	yes yes – – – –	– – – – – –	– – – – – –
SCPC/PSK	yes yes – – – –	yes yes – – – –	– – – – – –	– – – – – –
SPADE (CSC)	yes – – – – –	– – – – – –	– – – – – –	– – – – – –
TDMA	yes – – – – –	– – – – – –	– – – – – –	– – – – – –
IBS	yes yes yes – yes yes	yes yes yes – yes yes	yes yes yes – yes yes	– – – – – –
MCS	– – – – – –	– – – – – –	– – – – – –	– – – – – –
IDR***	yes yes yes – yes yes	yes yes yes – yes yes	yes yes yes – yes yes	– – – – – –

Carrier Type	Standard E Transmitting to Std Std Std Std Std Std A B C D E F	Standard F Transmitting to Std Std Std Std Std Std A B C D E F	MCS Coast (Ship) Station to Ship (Coast) Station
FDM/FM	– – – – – –	– – – – – –	–
CFDM/FM*	– – – – – –	yes yes yes – – yes	–
VISTA	– – – – – –	– – – – – –	–
TV**	– – – – – –	– – – – – –	–
SCPC/PSK	– – – – – –	– – – – – –	–
SPADE (CSC)	– – – – – –	– – – – – –	–
TDMA	– – – – – –	– – – – – –	–
IBS	yes yes yes – yes yes	yes yes yes – yes yes	–
MCS	– – – – – –	– – – – – –	yes (yes)
IDR***	yes yes yes – yes yes	yes yes yes – yes yes	–

* F1 and F2 earth stations are not allowed to transmit or receive CFDM/FM carriers.
** Occasional use Global Beam
*** E1 and F1 earth stations are not allowed to transmit or receive IDR carriers.

17
Intelsat services

17.1 Introduction

Because of its global coverage Intelsat is able to provide a variety of national and international services ranging from telephony to television. Almost 40 nations currently use the Intelsat system for domestic communications services. Intelsat has attempted to improve utilization of its system by offering transponders for unrestricted use (TUU) which allows leasing for bulk capacity for a mix of domestic and international traffic. The leasing may be on a long-term commitment basis in increments of 18 MHz. The use of TUU is likely to increase further the demand for Intelsat services and this, in turn, requires the provision of increased system capacity. The provision of the Intelsat K satellite, launched in 1992, was as a direct result of the upsurge in demand for Intelsat services in the Atlantic Ocean region. It is likely that demand for traffic capacity will continue to increase and the newer generation of satellites will have to cope with this higher density of traffic as well as an increasing range of traffic types. The way in which Intelsat is attempting to provide the service for the various types of traffic demands is outlined in the sections which follow.

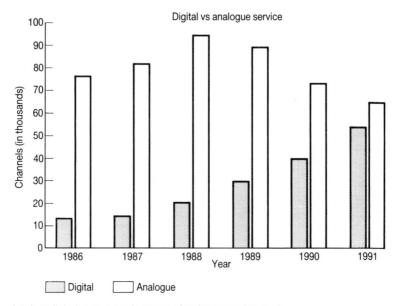

Fig. 17.1 Intelsat digital versus analogue service (courtesy Intelsat)

17.2 Public switched services

Intelsat is responsible for handling a large proportion of the billions of telephone calls that are made each year. Such links can be established via large national gateway earth stations or smaller urban and rural antennae. Processing the communications traffic using digital technology has become the aim of Intelsat and it is anticipated that the system will be completely digital by the end of the decade. The change from analogue to digital is proceeding well, as the graph of Fig. 17.1 shows.

The digital services provided by Intelsat are the integrated digital communications service IDR and TDMA.

IDR

The service is compatible with the evolving standards for the integrated services digital network (ISDN). It can also be used to establish dedicated private digital networks. For voice applications, the service allows use of low rate encoding (LRE) and digital circuit multiplication equipment (DCME). Information bit-rates can be as high as 45 Mbit/s and can be charged on a carrier or channel basis. A feature of IDR is that it can be operated using an earth station antenna as small as 5.5 m diameter and that it can be operated on a bearer channel basis. Using bearer channels means that multiple channels can be derived when using Digital Circuit Multiplication Equipment (DCME).

Earth stations for IDR are shown in Table 17.1.

The growth in the use of IDR bearer channels has shown a dramatic upsurge since

Table 17.1 Earth stations for IDR (courtesy Intelsat)

Earth Station Standard	Frequency band	G/T (dB/K)	Typical size (m)
A	C	35.0	15.0 – 18.0
B	(6/4 GHz)	31.7	10.0 – 13.0
F3		29.0	9.0 – 10.0
F2		27.0	7.0 – 8.0
F1		22.7	4.5 – 6.0

Earth Station Standard	Frequency band	G/T (dB/K)	Typical size (m)
C	Ku	37.0	11.0 – 14.0
E3	(14/11/12 GHz)	34.0	8.0 – 10.0
E2		29.0	5.5 – 7.0

1988, being in excess of 30,000 bearer channels in 1991 and it is likely to maintain an increased demand in the years to come.

TDMA

TDMA is another ISDN service used primarily for international public switched telephony traffic of reasonably large volume. TDMA is also offered on a bearer channel basis so that DCME may be used to derive multiple channels without an additional space segment charge. The growth in TDMA bearer channels has shown an annual increase since 1987, and in 1991 exceeded 9000 bearer channels.

17.3 Private network services

Intelsat has provided custom designed services for business use. These services are the Intelsat Business Service (IBS) and Intelnet.

IBS

Designed to meet the private network telecommunications needs of multinational corporations and organizations, IBS is a totally integrated, digital service providing ISDN quality communications on a global basis. IBS supports all private network business communications applications including voice, data, video-conferencing,

Table 17.2 Earth stations for IBS (courtesy Intelsat)

Earth Station Standard	Frequency band	G/T (dB/K)	Typical size (m)
A	C	35.0	15.0 – 18.0
B	(6/4 GHz)	31.7	10.0 – 13.0
F3		29.0	9.0 – 10.0
F2		27.0	7.0 – 8.0
F1		22.7	4.5 – 6.0

Earth Station Standard	Frequency band	G/T (dB/K)	Typical size (m)
C	Ku	37.0	11.0 – 14.0
E3	(14/11/12 GHz)	34.0	8.0 – 10.0
E2		29.0	5.5 – 7.0
E1		25.0	3.5 – 4.5

facsimile and telex. Carrier information data-rates range from 64 kbit/s to 8448 kbit/s and higher, in 64 kbit/s increments. Transponders may be leased on a full or fractional basis with allocations from 9 MHz. Earth stations used with IBS may be small, sited at the customer's premises, through to large country gateways. As shown in Table 17.2 below, IBS can be operated using antennae as small as 3.5 m, is located directly on the customer's premises and is offered over the three ocean regions through more than 200 earth stations.

Earth stations for IBS are shown in Table 17.2.

In 1991 the number of IBS earth stations in use was approaching 250.

Examples of the use of IBS include:

- an oil company in Africa communicating between its offshore drilling and coastal production facilities and its offices in Zaire and its facilities in the United Kingdom and the United States;
- in Latin America, a credit card company provides voice and data links between its offices in Chile and the United States;
- Intelsat's Japanese Signatory, Kokusai Denshin Denwa (KDD) offers HI-BITLINK, a high speed digital leased service for voice, electronic funds transfer and image data transfer for remote printing.

Occasional-use IBS, used predominantly for video-conferencing, has also made spectacular growth and, by the end of 1991, 30 countries were using occasional-use IBS for their video-conferencing requirements. As an example, Singapore Telecom is operating hundreds of hours of service via special video-conferencing links to Australia, Finland, France, Gemany, Hong Kong, Italy, Japan, the Netherlands, Sweden, the United Kingdom and the United States.

Intelnet

Intelnet is a digital data service designed for use with very small aperture terminals (VSATs) operating with a larger central hub station. News agencies in particular have used the service to provide world-wide networks; as an example ITAR-TASS, the main news agency of Russia and the other members of the Commonwealth of Independent States, has established a satellite network using Intelsat which provides interactive connections among its bureaux and subscribers via more than 200 VSATs in all parts of the world.

Intelnet is also being used for environmental monitoring, disaster relief and other critical areas requiring remote terminal operation. The European Space Agency (ESA) is operating an Intelnet network for an environmental project in Africa. ESA's Direct Information Access Network for Africa (project DIANA) uses Intelnet to exchange data on droughts, crop failures and desert locust movements between Ghana, Kenya, Zimbabwe and the Food and Agriculture Organization (FAO) in Italy.

There are no restrictions with respect to antenna size, G/T or modulation technique. Earth station parameters are defined by a transmission plan based on the transponder resources described by the Intelsat publication, IESS Module 410. Standards G (international service) and Z (domestic service) provide a set of earth station performance characteristics for leased services. Receive-only antennae do not require individual Intelsat earth station application or approval while transmit antennae may be type-approved.

The growth of Standard G and Standard Z earth stations has shown a steady annual increase since 1986.

Intelsat services

Table 17.3 Earth stations for Vista (courtesy Intelsat)

Earth Station Standard	Frequency band	G/T (dB/K)	Typical size (m)
D1	C	22.7	4.5 – 5.5
D2	C	31.7	11.0

17.4 Other services

Intelsat provides a thin-route service for communications to rural and remote communities with relatively low traffic requirements. There are two services, namely Vista and SCPC, which meet the situation in a cost effective way.

Vista

In June 1989, Intelsat approved for use with Vista, digital modulation techniques, such as SCPC/QPSK (single channel per carrier/quadrature phase shift keying) with voice activation. Before this date, the service had been restricted to the analogue SCPC/CFM (companded frequency modulation) technique. By 1990 there were 295 Vista channels operating via the analogue or digital service for voice or low-speed data transmissions. A Super Vista service, which utilizes the space segment more efficiently through the use of Demand Assigned Multiple Access (DAMA), had 82 channels as of 1990. Available at C-band, the Vista service can be used with a variety of earth stations, including the 4.5 m D1 antennae. Transmission parameters include:

- allocated satellite bandwidth of 30 kHz per SCPC/CFM carrier using frequency modulation (FM) with 2:1 syllabic companding;
- PSK carriers may use up to 30 kHz of allocated bandwidth at a down-link EIRP no greater than that required for an SCPC/CFM link to the same size station.

Earth station and networking options include:

- being provided through Standard A, B, F3, F2, F1, D2 and D1 earth stations;
- the D2, which is similar to the Intelsat Standard B, is designed to fulfil a versatile role as a central hub station for a star network;
- the D1, developed as a small, low cost earth station, can be used with simple access and signalling methods to operate a star, mesh or combined configuration;
- D1 earth stations may be type-accepted, minimizing on-site installation testing.

Earth stations for Vista are shown in Table 17.3.

SCPC

SCPC is a long standing Intelsat service designed primarily for international public switched service in low-density traffic streams. It is used for thin-route traffic at C-band between standard A and B earth stations.

Applications include:

- public-switched two-way telephony
- voice band data communications (up to 9.6 kbit/s);

17.4 Other services

Table 17.4 Earth stations for SCPC (courtesy Intelsat)

Earth Station Standard	Frequency band	G/T (dB/K)	Typical size (m)
A	C	35.0	15.0 – 18.0
B	C	31.7	10.0 – 13.0

- high-speed data communications (48, 50 and 56 kbit/s);
- two-way audio-conferencing.

Transmission parameters include:

- available in bandwidth units of 45 kHz;
- transmission rate of 64 kbit/s.

The bit error rate (BER) quoted for the service is 10^{-6} for voice and voice-band data and 10^{-9} for high-speed data.

Earth station and networking options are shown in Table 17.4.

Television services

International leases
Intelsat provides facilities for international transmission of television on a lease basis. Various lease periods exist ranging from 1 month to several years with the tariffs for the latter being lower due to the value of the commitment.

Occasional-use
Occasional-use television is used for rapid transmission of news and other material such as Olympic Games, World Cup Rugby, the Persian Gulf conflict, political summit meetings etc.

The use of small transportable antennae to provide satellite news gathering (SNG) has grown, with over 300 SNG terminals authorized for use by 28 countries on the Intelsat system in 1991.

In 1992, Intelsat introduced its computerized integrated booking and information service (IBIS-TV) so that signatories and authorized users can schedule and reserve services directly. On-line users, from anywhere in the world, can access real-time information on space segment availability directly, request new services, amend or delete existing services and obtain information on earth station capabilities or other reference information.

Domestic and regional services

Several Intelsat signatories, aware of the benefits of international links, sought to use Intelsat capacity for domestic services, particularly for telephony and television. As a result, the first domestic services were introduced in 1973 and their use has continued to grow. During 1991, for example, ASETA (the Association of Telecommunications' State Enterprises of the Sub-Regional Andean Agreement) decided to use a combination of Intelsat domestic leases and TUUs as a means of implementing a regional Andean satellite communications system. The five countries involved—Bolivia, Colombia, Ecuador, Peru and Venezuela—will use leased capacity from Intelsat while building regional traffic utilization and their terrestrial infrastructures.

Domestic services were first offered as leases of pre-emptible capacity. Today, Intelsat domestic users may also lease non-pre-emptible capacity as well as purchase transponders under the Planned Domestic Service (PDS) programme. New tariffs were established for domestic non-pre-emptible leases in 1989 as part of a policy of resource-based pricing and the overall effort for tariff simplification. Pre-emptible and non-pre-emptible leases for domestic services are currently carried on six Intelsat satellites in three ocean regions. As of April 1990, Intelsat provided domestic leases to the following countries:

Algeria, Chile, the People's Republic of China, Colombia, Côte D'Ivoire, Denmark, France, India, Libya, Malaysia, Mozambique, New Zealand, Nigeria, Pakistan, Peru, South Africa, Spain, Sudan, Thailand, the United Kingdom, Venezuela and Zaïre.

In addition, Intelsat provides service to the United Nations for peace-keeping purposes.

As of 1 April 1990, transponders on the Intelsat satellites had been sold to 20 different countries since the inception of the Planned Domestic Service (PDS) programme in 1985. The following countries had purchased capacity under the PDS programme:

Argentina, Bolivia, Central African Republic, Chad, Chile, People's Republic of China, Ethiopia, Gabon, Iran, Israel, Italy, Japan, Niger, Norway, Portugal, Sweden, Turkey, United States and Venezuela.

Cable restoration

Intelsat has been restoring cables throughout its history and is actively involved in establishing international technical standards compatible with satellite links. Despite the type of cable link (analogue or digital), Intelsat provides immediate restoration of service and for any length of time that may be necessary. During 1991, Intelsat restored four major fibre optic cables: TAT-8 in the Atlantic, once; TCS-1 in the Caribbean, once; and HAW4/TPC3 and NPC in the Pacific, once and three times respectively. These cable failures accounted for 447,120 digital bearer channel days of restoration.

Maritime Communications Sub-system (MCS)

As discussed in earlier sections, some Intelsat satellites were equipped with an additional payload known as the Maritime Communications Sub-system (MCS) which operates in the L-band (1.5/1.6 GHz) for the provision of mobile communications, primarily to ships. The shore-based communications to and from the satellite are in the C-band (6/4 GHz). The MCS facilities are leased to Inmarsat and are now used for spare capacity (see Section 2).

Sources of revenue

The various Intelsat services which produced revenues for 1991 are as indicated in Table 17.5.

Table 17.5 Intelsat revenue for 1991 (courtesy Intelsat)

Source	Revenue (%)
Analogue international switched service	43.1
Digital international switched service	19.8
International TV leases	8.2
Private network (business) services	7.7
Occasional-use TV	5.0
Domestic/regional telecommunications service	6.0
Transponders for unrestricted use (TUUs)	4.4
Cable restoration service	2.2
Other (includes Intelnet and MCS)	3.6

As of December 1991 about three-quarters of Intelsat traffic in service was covered by long-term commitments of 15 years and almost 80% of the traffic was covered by long-term commitments of five years or more.

18
The Eutelsat Organization

18.1 Introduction

The history of European satellite communications began in the period 1962 to 1964 when the European Space Research Organization (ESRO) and the European Launcher Development Organization (ELDO) were established. Studies concerning a European telecommunications satellite continued until 1969 under the auspices of the European Conference of Postal and Telecommunications Administrations (CEPT) and the European Broadcasting Union (EBU), together with ESRO and ELDO. In 1972 nine European states embarked on a telecommunications satellite programme ECS (European Communications Satellite) within the framework of ESRO. The initial programme concentrated on the development of an experimental satellite OTS (orbital test satellite). By 1975 ESRO and ELDO were replaced by the European Space Agency (ESA) and by 1976 the development of the ECS satellites was approved by CEPT and a study undertaken to define the organization necessary to operate the satellites. In 1977, 17 members of CEPT signed an agreement creating a provisional organization known as Interim Eutelsat. By 1978 the first successfully launched OTS satellite was deployed with Interim Eutelsat co-ordinating the utilization of the satellite. The satellite was OTS-2, with OTS-1 having been destroyed during an unsuccessful launch in 1977. By 1979 Interim Eutelsat had signed an agreement with ESA for the in-orbit procurement of five ECS satellites. In 1982 an Intergovernmental Conference attended by 26 European countries, whose administrations were members of CEPT, finalized the Convention establishing the definitive Eutelsat Organization. By 1985 the Eutelsat Convention had come into force providing Eutelsat with an organizational structure as follows:

- Assembly of Parties
- Board of Signatories
- Executive Organ.

The Assembly of Parties has overall responsibility for Eutelsat policy which includes supervision of the general approach and long-term objectives of the organization. The Assembly is composed of parties from each of the member countries, with each member state having one vote, who meet to consider resolutions and recommendations on policy and objectives.

The Board of Signatories comprises representatives of the member governments (Parties), i.e. designated telecommunications entities who have signed an Operating Agreement and attend the Board by invitation. The Board meets at least four times a year and is responsible for all operational matters such as design, development, procurement, establishment, operation and maintenance of the organization's space segment. Each of the signatories contributes to the organization's capital requirements and has voting rights on the Board in proportion to the investment share held.

The Executive Organ is responsible for the day-to-day activities of the organization

18.1 Introduction

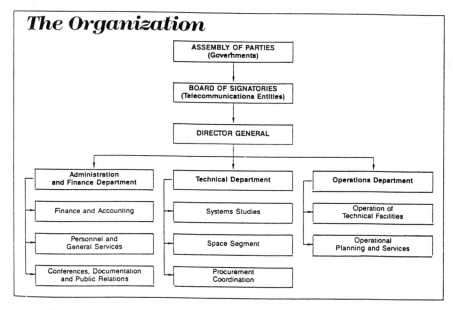

Fig. 18.1 Eutelsat organization (courtesy Eutelsat)

and is headed by a Director General. It is the function of the Director General to submit recommendations to the Board of Signatories and to implement decisions taken by the Board.

The organizational structure is shown in Fig. 18.1.

Eutelsat connections with other organizations include bodies such as the following.

- International Telecommunication Union (ITU), where discussions have taken place regarding such matters as the role of satellites in the ISDN and future access to frequency and orbital resources.
- European Conference of Postal and Telecommunications Administrations (CEPT), where a contribution is made to technical aspects such as the use of on-board processing etc.
- European Space Agency (ESA), where matters relating to the in-orbit management of the four first-generation Eutelsat satellites are discussed, together with suggestions for future payload applications.
- Commission of the European Communities, where satellite communication matters are discussed, including the development of High Definition Television standards.

Eutelsat signatories use the system as part of their domestic and international networks and, in addition, may lease capacity to other users. The shareholding of signatories is determined annually according to their use of the system capacity. The minimum share is 0.05%. As of early 1993 there were 34 signatories with other countries wishing to join and simply waiting for their governments to ratify the Eutelsat Convention. Signatories with major investment shareholding include France, at over 19%, Spain at over 17%, the United Kingdom at almost 16% and Germany at approximately 14.5%.

Eutelsat space segment charges are designed to cover:

- operating, maintenance and administrative costs,
- provision of operating funds, as determined by the Board,

- capital repayment and, as available, compensation to signatories for the use of their capital.

Ground equipment comprises the monitoring and control equipment sited at various European locations, and equipment at the Paris headquarters.

The ground segment provision is the responsibility of the signatories, and, in the main, the installation must be approved by the Board of Signatories for use with the Eutelsat space segment. In the case of receive-only earth stations, the persons receiving the transmissions may, in certain cases, be authorized to establish and operate their own installations. This would be the case, for example, for a TV receive-only (TVRO) installation.

18.2 Eutelsat space segment

Eutelsat I satellites

The first ECS satellite was launched in 1983, becoming Eutelsat I-F1 when operational; this was followed by the second ECS satellite in 1984 which became Eutelsat I-F2. A third ECS satellite was lost in 1985 following a launch failure. Eutelsat I-F4 and F5 were successfully launched in 1987 and 1988 respectively. The Eutelsat I satellites are maintained in orbit by ESA from its earth station at Redu in Belgium. Because of design improvements Eutelsat I-F1 is different from subsequent Eutelsat I satellites in many ways:

- it has only 12 transponders compared with 14 transponders on subsequent satellites;
- it does not have the two 14/12 GHz transponders fitted to later versions;
- from Eutelsat I-F2 onwards, three of the six telecommunications antennae were modified;
- of the ten transponders that can be used simultaneously by each spacecraft, only six can be used when in eclipse. Eutelsat I-F2 onwards are fully eclipse protected.

Table 18.1 Eutelsat satellites (courtesy Eutelsat)

	EUTELSAT I-F1	EUTELSAT I-F2 and following	EUTELSAT II
Stabilization	3 axis	3 axis	3 axis
Mass at launch	1 045 kg	1 160 kg	1 700 kg (7 years)
			1 800 kg (10 years)
Mass in orbit [1]	510 kg	550 kg	866 kg
Span (with solar panel deployed)	13.80 m	13.80 m	22.40 m
Electrical power [1]	900 W	900 W	3 000 W
Lifetime	7 years	7 years	7 to 10 years
Frequency bands	14/11 GHz	14/11 and 14/12 GHz	14/11 and 14/12 GHz
Number of transponders	12	14	16
Number of transponder for simultaneous use	10[2]	10	16
Transmit power of each transponder	20 W	20 W	50 W
Antennae:			
– receive/transmit	–	1	1
– receive only	2[3]	1	–
– transmit only	4	4	1

(1) at end of life – (2) 6 only in eclipse – (3) one antenna as a back-up for the other.

18.2 Eutelsat space segment

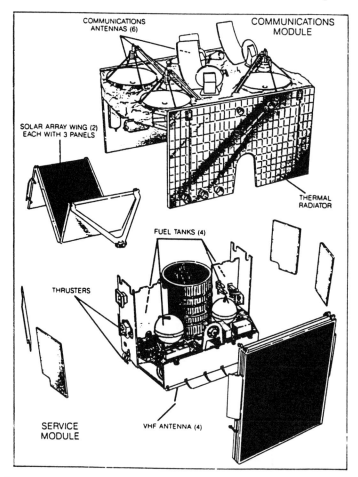

Fig. 18.2 Cut-away view of a Eutelsat I satellite (courtesy Eutelsat)

The frequencies used for the down-link are 10.95–11.20 GHz, 11.45–11.70 GHz on F1–F5 and also 12.50–12.75 GHz on F2–F5. High-power amplifiers are 20 W TWTAs and the transponder bandwidth is 72 MHz. The design life is seven years. The configuration is three-axis body-stabilized using the ECS/OTS platform with British Aerospace as the prime contractor. The power delivered by the bus sub-system is 1000 W falling to 900 W by the end of operational life. Details of the Eutelsat I satellites are shown in Table 18.1 which also includes details of the Eutelsat II satellites.

A cut-away view of Eutelsat I-F2 is shown in Fig. 18.2.

The power delivered by the Eutelsat I series is:

- Atlantic, West and East spots, EIRP 40.8–46 dBW.
- Eurobeam, EIRP 34.8–41 dBW.
- Satellite multiservice system (SMS), EIRP 39.8–43.5 dBW.

Eutelsat II satellites

The initial contract for Eutelsat II satellites was signed in 1986 and called for three satellites with an option for five more. A fourth satellite was ordered in 1987, a fifth in

374 The Eutelsat Organization

1989 and a sixth in 1990. To date Eutelsat II has four satellites operational, F1 launched in 1990, F2 and F3 in 1991 and F4 in 1992. The fourth and fifth Eutelsat II satellites have been modified to extend their widebeam coverage as far as Moscow and its surrounding area. The sixth Eutelsat II satellite is being modified for co-location with Eutelsat II-F1 at its orbital position of 13°E. Each Eutelsat II satellite has 16 transponders for instantaneous use with eight more providing back-up. Transponder reliability is provided for by dividing 24 TWTAs into two groups of 12 for eight active transponders. The power rating of the TWTA is 50 W on Eutelsat II-F1–F5 and 70 W on Eutelsat II-F6. Higher power means a reduction in the diameter of receive antennae on the ground for a large part of the coverage area. There are two multifeed antennae, each fitted with a 1.6 m dual reflector. One antenna operates in receive and transmit mode and the other in transmit mode only. The antennae give shaped beams (one receive beam and two transmit beams) adapted to suit the area being served. Switching the feed networks for each of the two antennae allows a choice to be made in orbit between a wide beam coverage or a more concentrated, higher gain beam (superbeam). The values of EIRP for these arrangements is as follows:

- Eutelsat II-F1–F5, Superbeam, EIRP 44–52 dBW
- Eutelsat II-F1–F5, Widebeam, EIRP 39–47 dBW
- Eutelsat II-F6, Widebeam, EIRP 40–49 dBW.

Bandwidth for Eutelsat II-F1–F5 is 72 MHz for seven transponders and 36 MHz for nine transponders. For Eutelsat II-F6, the bandwidth will be 36 MHz for all 16 transponders. Frequencies used for the down-link are 10.95–11.20 GHz, 11.45–11.70 GHz and 12.50–12.75 GHz for Eutelsat II-F1–F5 and for Eutelsat II-F6, 11.20–11.55 GHz.

The satellite configuration is the spacebus platform three-axis body-stabilized with Aerospatiale as the prime contractor. With contributions from Alcatel Espace, Alenia, CASA, Deutsche Aerospace, Ericsson Radio Systems, ETCA, Marconi Space Systems and Sextant Avionique, the second series of Eutelsat satellites is truly European. A Eutelsat II satellite in deployed state is shown in Fig. 18.3.

Fig. 18.3 Eutelsat II satellite in deployed state (courtesy Aerospatiale)

18.2 Eutelsat space segment 375

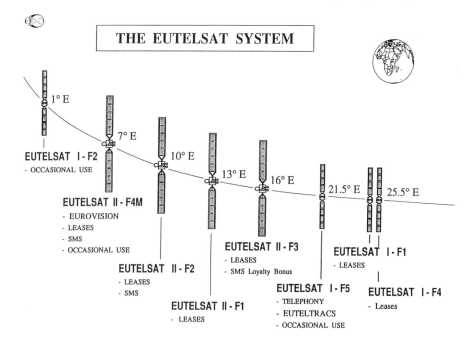

Fig. 18.4 Eutelsat satellite system (courtesy Eutelsat)

The power delivered by the power sub-system is 3000 W up to the end of operational life, which is expected to be about nine years.

The orbital position of the Eutelsat II satellites can be controlled within $\pm 0.08°$ E/W and $\pm 0.05°$ N/S over a period of at least seven years.

The geostationary positions of the eight Eutelsat satellites operational as of mid-1992 is shown in Fig. 18.4.

Some services, such as telephony with its large tracking gateway earth stations, can stand a change in orbital position. Other services, such as SMS which uses small antennae, could only receive successfully if the antennae are manually repointed. Eutelsat monitors traffic requirements and, as traffic demands it or because of the deployment of a new satellite, satellites can be moved to a new orbital position. Figure 18.4 was correct as of mid-1992. Eutelsat I-F2 was removed from service in mid-1993 and was due to be replaced by a fifth Eutelsat II satellite in January 1994. This satellite failed to reach its correct altitude due to a launch failure. Eutelsat II-F6 is likely to be launched late in 1994. It is anticipated that by early 1995 there will be eight satellites in seven orbital positions offering 100 transponders for communications throughout Europe and the Mediterranean Basin.

A new generation of Eutelsat satellites, to be known as Eutelsat III, is under development with a view to their introduction starting in 1998. Eutelsat III is likely to make use of the new 13.75 – 14.0 GHz band on the up-link allocated by WARC 92.

Frequencies and polarization
Frequency and polarization arrangements for Eutelsat II satellites are shown in Fig. 18.5.

In Fig. 18.5 channel numbers are ascribed to transponders and the bandwidth is either 36 MHz or 72 MHz. As an example, transponders 45 and 46 are 36 MHz transponders while transponders 25 and 26 are 72 MHz transponders. Some transponders are shown paired, i.e. 37/47 and 34/44; these are paired because only one transponder can be

376 The Eutelsat Organization

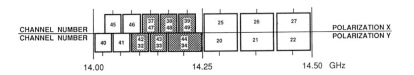

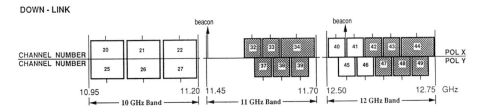

Fig. 18.5 Eutelsat II channel arrangement (courtesy Eutelsat)

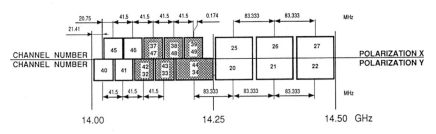

Fig. 18.6 Channel centre frequencies (up-link) (courtesy Eutelsat)

operational at any one time since they are identical in terms of frequency and polarization for reception. The only difference is the value of frequency translation and down-link frequency, 11 GHz or 12 GHz. It follows that only 16 out of a possible 22 transponders shown can operate at any particular time.

The transponder centre frequencies in the up-link frequency band are shown in Fig. 18.6.

Corresponding centre frequencies in the down-link bands are specified by frequency conversion as follows:

- 3.30 GHz for transponders with down-link in the 10.95–11.20 GHz band;
- 2.55 GHz for transponders with down-link in the 11.45–11.70 GHz band;
- 1.50 GHz for transponders with down-link in the 12.50–12.75 GHz band.

Polarization

The antennae on Eutelsat II satellites can transmit and receive simultaneously using orthogonal linear polarization at the same frequency. The two orthogonal polarizations are indicated X and Y on Fig. 18.5. Signals received on one polarization are transmitted on the other polarization, i.e. if reception is on polarization X, transmission is on polarization Y and vice versa.

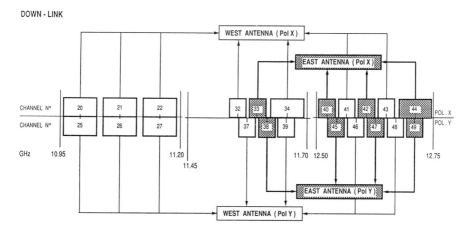

Fig. 18.7 Transmit antenna used for each channel (courtesy Eutelsat)

The reference X polarization is defined as that polarization whose plane contains the satellite antenna boresight and makes an angle of 93.535°, in an anticlockwise direction, looking towards the earth, with the plane defined by the pitch axis and the satellite antenna boresight direction. The reference Y polarization is defined as the polarization whose plane is orthogonal to the plane of X polarization as defined above. Polarization discrimination of the antennae is at least 33 dB for the receive antenna within the receive coverage area and at least 34 dB for the transmit antenna within the transmit coverage area.

Coverage patterns for receive and transmit have been shown in Chapter 9 for a particular Eutelsat II satellite at orbital location 10°E. Coverage diagrams for the other satellites are also available. The antennae are identified as 'EAST' or 'WEST' according to their position on the satellite. The 'EAST' antenna is used for reception and can generate a wide beam with X or Y polarization. The 'EAST' antenna or the 'WEST' antenna can be used for transmission. Figure 18.7 shows which antenna is used for each transponder.

The feed network for each of the two polarizations for each antenna is reconfigurable, by means of ground commands, to provide two different beams, i.e. for medium-gain or high-gain. The reconfiguration can be achieved for each antenna independently and for each of the two polarizations

Also, by ground command it is possible to select one of the two transponders in each of the six pairs 32/42, 33/43, 34/44, 37/47, 38/48 and 39/49 independently of the transponder selected in each of the other five pairs.

Reference to Fig. 18.7 shows that transponders are connected to the antennae in groups. Medium-gain or high-gain can be selected for any transponder, regardless of the choice of gain for any other transponder, provided the two transponders do not belong to one of the six, independent groups. The groups are:

- Group 1: Transponders 46 and 48
- Group 2: Transponders 41 and 43
- Group 3: Transponders 25,26,27,37 and 39
- Group 4: Transponders 20,21,22,32 and 34
- Group 5: Transponders 38,45,47 and 49
- Group 6: Transponders 33,40,42 and 44

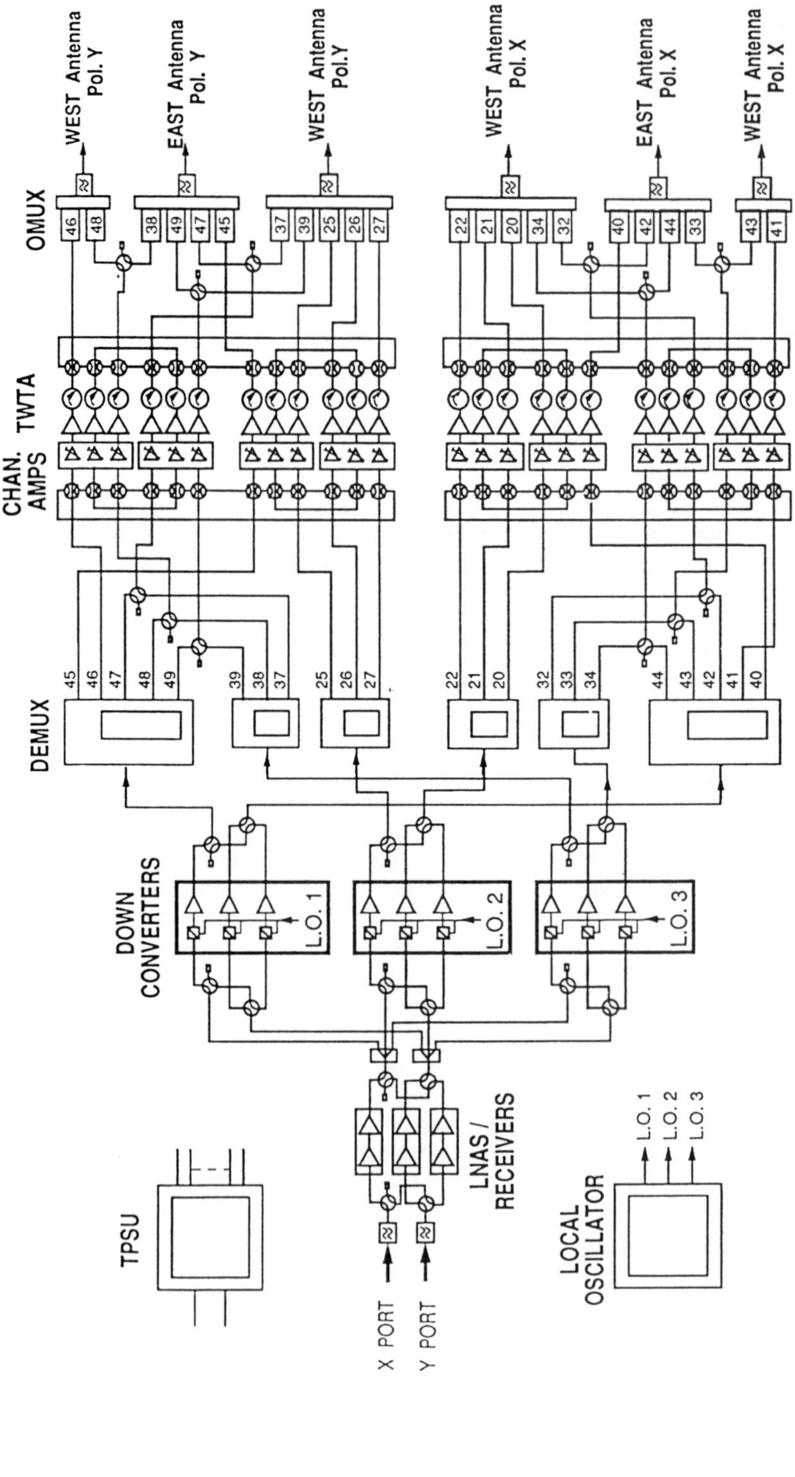

Fig. 18.8 Repeater block diagram (courtesy Eutelsat)

18.2 *Eutelsat space segment* 379

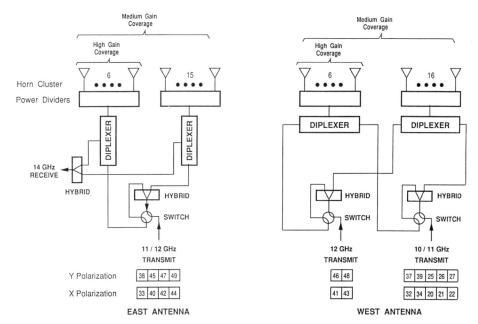

Fig. 18.9 Block diagram of feed assemblies and channel association with East/West antennae (courtesy Eutelsat)

Communications sub-system
As mentioned earlier there are 24 TWTAs of 50 W output power connected in two groups of 12 such that any of the eight transponders connected to a group can use any of the 12 TWTAs. This redundancy allows up to four failed TWTAs in either group without affecting the satellite capacity.

The repeater block diagram for Eutelsat II is shown in Fig. 18.8.

The two groups of 12 TWTAs can be seen in Fig. 18.8 together with the redundancy rings and the switches that allow selection of either the 11.45–11.70 GHz or the 12.50–12.75 GHz down-link frequency bands for the six pairs of transponders 32/42, 33/43, 34/44, 37/47, 38/48 and 39/49. Figure 18.9 shows a block diagram of the antenna feed assemblies from the output multiplexers, with the switches for the 'EAST' and 'WEST' antennae that allow selection of either the medium-gain or high-gain coverage configurations for each of the two antennae.

As an example, it can be seen from Fig. 18.9 that for the 'WEST' antenna, transponders operating in the 12.50–12.75 GHz band (transponders 41 and 43 for X polarization and transponders 46 and 48 for Y polarization) can be switched independently of those transponders operating in the 10.95–11.20 GHz and 11.45–11.70 GHz frequency bands.

Each of the Eutelsat II satellites has two beacons, one transmitting in the 11 GHz band and the other in the 12 GHz band. There are two 11 GHz beacon frequencies but only one is used at any time; the choice of 11 GHz frequency is made by ground command. The 11 GHz beacon is for telemetry or telemetry and ranging and has a value depending on the satellite flight number:

- Eutelsat II-F1: 11 451.091 MHz/11 451.830 MHz
- Eutelsat II-F2: 11 451.830 MHz/11 450.350 MHz
- Eutelsat II-F3: 11 452.570 MHz/11 451.830 MHz

- Eutelsat II-F4: 11 451.091 MHz/11 452.570 MHz
- Eutelsat II-F5: 11 450.350 MHz/11 451.091 MHz

The frequency of the 12 GHz beacon is 12 541.667 MHz.

The frequency stability of each beacon is better than $\pm 2 \times 10^{-5}$ over the operational lifetime of the satellite, and $\pm 2 \times 10^{-6}$ over any 24 hour period.

The EIRP of each of the two beacons is at least 9.0 dBW within the transmit coverage areas. Where the beacon is used as a telemetry or telemetry and ranging carrier (11 GHz beacon), the EIRP value is the residual carrier EIRP after modulation by the telemetry and/or the ranging signal is applied. EIRP stability is better than 0.2 dB over any two minutes for each beacon.

The beacons are transmitted on the X polarization and the level of the signal in the Y polarization is at least 33 dB below the X polarization level.

Any transponder can have its gain adjusted such that the input power flux density (PFD) for saturation, from locations where the satellite G/T is -0.5 dB/K, lies in the minimum range -92 dBW/m^2 to -77 dBW/m^2. The adjustments are carried out by ground commands in steps of 0.5 dB for any transponder independently of other transponders. Operationally, three gain settings are used corresponding to an input PFD for saturation, from locations where the satellite G/T is -0.5 dB/K:

- high, at input PFD of -83.0 dBW/m^2
- nominal, at input PFD of -80.0 dBW/m^2
- low, at input PFD of -77.0 dBW/m^2

18.3 Ground segment

The four Eutelsat I satellites are controlled in orbit by the European Space Agency (ESA) under an agreement signed with ESA. The generation of Eutelsat II satellites is controlled by Eutelsat itself using a Satellite Control Centre (SCC) in Paris, France. The SCC has the necessary data processing installations and incorporates a control room capable of handling as many as six satellites in orbit. The four second-generation satellites currently in orbit are dealt with by the SCC with functions that include the overall control of the satellite, ranging measurements, orbital manoeuvres, switching of the satellite payload according to customer requirements, control of the earth stations and storage of relevant data. The Satellite Control System (SCS) also includes two Telemetry, Command and Ranging (TCR) sites which are connected to the SCC. Telemetry, command and ranging can be carried out via a telemetry/telecommand antenna fitted to each orbiting satellite. The TCR sites are at Rambouillet in France, which also has back-up SCC facilities, and Sintra in Portugal, which has a UHF antenna used for satellite control purposes in an emergency. The wide separation of the TCR sites allows precise orbit determination and control system security. Nominal operational configuration has one TCR antenna per satellite at each of the TCR stations.

For network control a number of earth stations are used to monitor transmissions to both generations of satellites. These earth stations are connected to a Communications System Control (CSC) centre which co-ordinates the access to the satellite transponder (frequency, transmit power, time slots etc) over a 24-hour period and co-ordinates, monitors and manages so that any difficulties with a satellite or transponder can be dealt with. The CSC facilities are also situated in Paris. Eutelsat also has equipment at two earth stations connected to the public switched telephone network (PSTN). These are Time Division Multiple Access (TDMA) Reference and Monitoring Stations (TRMS) and are situated at Guadalajara in Spain and Fucino in Italy. The TRMS equipment permits synchronization of the TDMA transmissions of each earth station and fully automatic monitoring of earth station equipment operation under service contracts with the Spanish signatory (Telefonica) and the Italian signatory (Telespazio). EBU television transmissions are monitored from an earth station owned by Deutsche Bundespost Telekom at Usingen in Germany. Monitoring for the Satellite Multiservice System (SMS), occasional-use traffic and leased transponder services are provided at Rambouillet. Most of the monitoring centres on observation of the RF spectrum on all eight Eutelsat satellites. Earth stations used mainly for control and monitoring are also used for reception of the RF spectrum from each satellite and the results are processed and transmitted to the CSC.

Earth stations

The nature of Eutelsat traffic is such that it can be accessed and received by large and small earth stations for a variety of applications. The earth stations should, in principle, be provided and operated by the signatories (or by an authorized telecommunications entity if the installation is to be used from a territory that does not fall within the jurisdiction of a Member State). The earth stations must have the installation approved by the Board of Signatories before access to Eutelsat satellites can be authorized. The main characteristics of the earth stations used with the Eutelsat satellites must comply with the standards established by the Board of Signatories for the different types of utilization. These standards apply to three main categories of installation.

- Connection to the telephony network. These are transmit/receive stations, with antennae between 14 and 20 m diameter, and equipped for TDMA operation. The bit-rate used is 120 Mbit/s. These stations are also equipped for television with transmissions using Eurovision and on a demand-assignment basis.
- Transmit installations for the distribution of television programmes with antennae of between 8 and 12 m in diameter.
- Transmit/receive installations used for data communications and video-conferencing via a 14/12 GHz transponder using a 2.4 m diameter antenna.

At the end of 1992, 18 large earth stations serving 19 European countries had been approved to access Eutelsat satellites for telephony and for the European Broadcasting Union's EUROVISION and EURORADIO transmissions. Also, four new TDMA earth stations had been inaugurated during the year, namely at Lario in Italy, Seville in Spain, Elfordstown in Ireland and Cheia in Romania. There are further stations planned within this network and these should be operational shortly.

Another area of expansion has been in the use of transportable receive equipment with nearly 200 transportable terminals approved by the end of 1991 to access Eutelsat satellites for services such as Satellite News Gathering (SNG). For the Satellite Multiservice System (SMS) there are well over 1000 VSATs in operational use at the start of 1992 and this area also seems set for a big increase with time. There are thousands of unregistered Eutelsat earth stations in addition to the large number of registered receive-only units for direct-to-home services such as television and radio. In terms of dishes installed, Europe is the world's biggest market with the Eutelsat earth segment extending to a potential audience of over 600 million people.

Eutelsat publish Earth Station Standards and Specifications as follows.

Standard T and related specifications (TDMA, Telephony and Television).

- Performance characteristics of earth stations operating for telephony and television in the ECS system.

- Performance characteristics of standard earth stations for 120 Mbit/s TDMA transmission and high quality TV transmissions or for 120 Mbit/s TDMA transmissions only via a Eutelsat II satellite.
- TDMA/DSI system specification.
- DCME specification.

Standard V (Television).

- Standard for high quality TV transmission via Eutelsat II.

Standard S (SMS).

- SCPC satellite multiservice system specification.
- Eutelsat Telecom 1/SMS system specification.

Standard L (Access to leased transponders).

- Minimum technical and operational requirements for earth stations accessing leased capacity in the ECS space segment.

Standard M (access to SMS transponders for non-standard structured applications).

- Minimum technical and operational requirements for earth stations accessing a Eutelsat SMS transponder for non-standard structured types of SMS transmissions.
- Background information on the technical and operational requirements for SMS closed network earth stations.

All of these documents are available from Eutelsat or one of the signatories.

18.4 Eutelsat services

Eutelsat satellites carry international and domestic public fixed and mobile telecommunications (telephony, telegraphy, telex, fax, data, videotex, television and radio transmissions). The organization can also offer specialized telecommunications services such as radio-navigation, space research, meteorological and remote-sensing of earth resources.

The traffic can be broadly sub-divided as follows.

- TV and radio for cable, community and direct-to-home (DTH) markets.
- EBU Eurovision and Euroradio networks.
- Satellite news gathering (SNG).
- Domestic and international telephony.
- Business communications.
- Land mobile communications (EUTELTRACS).

TV and radio for cable, community and DTH markets

Television and radio channels generally access capacity on a full-time lease basis. Almost 40 television and over 40 radio channels currently use Eutelsat for reception by cable and satellite master antenna (SMATV) networks and domestic satellite reception systems. SMATV is used for community reception via a single antenna; the received programmes are then distributed to several homes in an apartment block (or several rooms in a hotel) via a small in-house cable network. A typical arrangement for TV reception is shown in Fig. 18.10.

With its more powerful satellite transmitters, the Eutelsat II generation allows direct reception using Superbeam coverage with 0.8 m diameter antennae in central

18.4 *Eutelsat services* 383

Fig. 18.10 TV reception methods used with Eutelsat II satellites (courtesy Eutelsat)

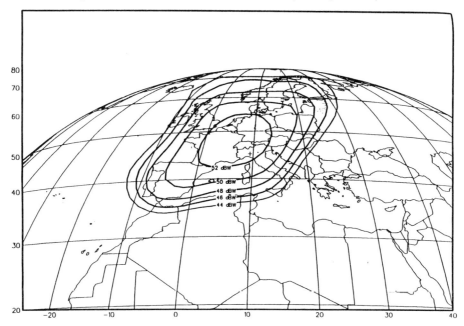

Fig. 18.11 Eutelsat II superbeam typical measured coverage (courtesy Eutelsat)

and western Europe. The Eutelsat II superbeam typical coverage is shown in Fig. 18.11.

At the edge of the superbeam coverage area, because of the reduced value of the EIRP, larger antennae may be necessary, see Table 18.2.

The wide beam coverage can also be used for direct reception by all member countries and parts of North Africa and the Middle East. Typical coverage for widebeam reception is shown in Fig. 18.12.

Again, the size of antenna required to receive the transmitted programmes is 0.8 m at the centre of the coverage with a requirement for increased antenna sizes at the edge of coverage.

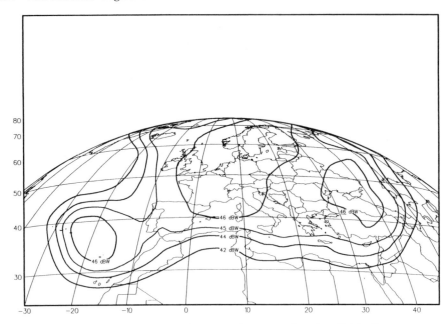

Fig. 18.12 Eutelsat II widebeam typical measured coverage (courtesy Eutelsat)

Table 18.2 Antenna size required within the Eutelsat II converage contours* (courtesy Eutelsat)

Coverage contour	Direct to home reception	TV Distribution (community reception and cable)
52 dBW	0.8 m**	0.8 m (15.5 dB/K)
50 dBW	0.8 m**	0.9 m (17.5 dB/K)
49 dBW	0.8 m**	1 m (18.5 dB/K)
48 dBW	0.8 m	1.1 m (19.5 dB/K)
47 dBW	0.8 m	1.2 m (20.5 dB/K)
46 dBW	0.9 m	1.3 m (21.5 dB/K)
45 dBW	1 m	1.5 m (22.5 dB/K)
43 dBW	1.2 m	1.8 m (24.5 dB/K)
40 dBW	1.8 m	2.8 m (27.5 dB/K)

* The precise contour may vary slightly from channel to channel. The contour maps provided are typical and indicative.
** Even with an antenna smaller than 80 cm within this contour you can still receive the programmes carried on the *Superbeam*. However, the smaller antennae are not recommended due to the possibility of interference from programmes carried on other satellites. Reception of channels sharing the same transponder may require the use of a slightly larger antenna. Please check the technical specifications for each channel.

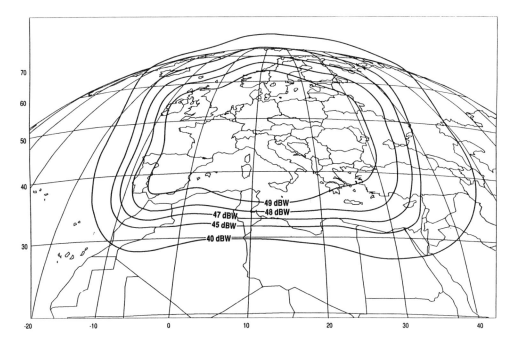

Fig. 18.13 Eutelsat II-F6 predicted coverage (courtesy Eutelsat)

The 'hot bird' coverage using Eutelsat II-F6, due for launch in late 1994, will add 16 new TV channels at 13° East. The 'hot bird' system evolves from the co-location of Eutelsat II-F6 with Eutelsat II-F1 at 13° East to provide increased capacity. The predicted coverage area is shown in Fig. 18.13.

The coverage of Fig. 18.13 suggests an increased reception area for a given value of EIRP, allowing the use of small antennae throughout that area. This is made possible by the increase in output power of the Eutelsat II-F6 satellite (70 W compared with 50 W of the other Eutelsat II satellites).

EBU Eurovision and Euroradio networks

The European Broadcasting Union (EBU) has been associated with the projected European telecommunications satellites from the very start. The EBU has used Eutelsat since 1984 for its Eurovision exchanges, under an agreement signed between the two organizations in 1982 for communications capacity on Eutelsat I. There are 25 earth stations at present registered for EBU exchanges of television programmes. In 1991, the EBU signed an agreement with Eutelsat to lease four Eutelsat II transponders for Eurovision purposes. In 1993, the EBU programme exchange network was moved from a Eutelsat I satellite to four wideband transponders on Eutelsat II-F4, and it has options on two additional transponders on the same satellite. The EBU operates six studio-quality digital video channels, with associated sound, and radio in this capacity. The extended coverage given by the change of satellite enables the entire European Broadcasting area in Europe, consisting of 39 active member broadcasters, as well as North Africa and the Middle East, to be reached.

Thirty-five radio stations broadcast via Eutelsat, which is the main carrier of satellite radio programming for the European continent. Radio programmes can be transmitted in analogue or digital mode, with a large number of radio programmes being transmitted as additional sound sub-carriers to television programmes. The EBU launched a Euroradio service in 1989 and broadcast to full compact disc quality is the eventual aim.

Satellite news gathering (SNG)

Capacity on Eutelsat transponders is used by TV broadcasters for satellite news gathering via transportable stations. Eutelsat operates a booking office which co-ordinates use of SNG capacity as well as use by regular part-time users. Time can be reserved on the satellites for periods ranging from as little as 10 minutes to a series of pre-booked slots totalling 180 hours or more. Transponders can also be booked on a full-time basis for short periods of one, three or six months. The Gulf War of 1991 saw a heavy increase in demand for occasional use capacity, and this was met by Eutelsat by tilting one of the satellites in orbit. Major news agencies, such as Reuters for example, have used Eutelsat, and in 1991 over 2000 hours were used for this application. The total number of SNG hours booked on Eutelsat (including short-term leases, reservation by subscription and occasional-use) reached 16,350 hours in 1992. Reservation by subscription (RBS) allows the use of capacity by news agencies on a regular basis.

Domestic and international telephony

Public switched digital telephony is carried on six transponders via a network of 19 earth stations across Europe. Further stations are planned to go into operation in 1993/4. By July 1993 over 15,000 telephony half-circuits were in use. The Eutelsat I-F5 satellite was assigned to TDMA in 1991 and, in that year, there were 220 traffic relations operational among 19 European countries carrying TDMA/DSI telephony. Full-time leases are also used for public telephony. For example, Tele Danmark uses one transponder for domestic telephony with the Faroe Islands while Spain uses one transponder for data transmissions.

Eutelsat see the opening up of eastern Europe as an area for expansion, and extra transponders could be made available for such an increase in traffic. Close co-operation exists between Eutelsat, signatories and earth station operators to ensure successful routeing of TDMA traffic. The organization is examining the introduction of an expanded TDMA network with a new TRMS to ensure that the current generation of existing traffic terminals continues to operate. The new expanded network is expected to be available by 1997/8 at the latest, allowing for a substantial increase in traffic.

Business communications

The Eutelsat Satellite Multiservice System (SMS) is a fully digital system serving the whole of Europe and operates with data rates from 9.6 kbit/s to 2 Mbit/s. The satellites are accessed directly by small earth stations or by VSATs (Very Small Aperture Terminals). Typical applications include video-conferencing, real-time distribution of financial and stock exchange information, remote printing of newspapers and computer transfers of data for business use. At the end of 1992 over 200 full-time SMS networks were in operation with nearly 400 registered SMS earth stations and over 1500 VSATs, in addition to a large number of receive-only VSAT terminals. Eutelsat regards the SMS system as a priority telecommunications service and guarantees its users optimum security in the unlikely event of a satellite or transponder failure. A total of 14 transponders (five on Eutelsat II-F2, eight on Eutelsat II-F4 and one on Eutelsat II-F3) carry traffic for SMS. Users connect to SMS either through a shared antenna (for example at a teleport) or through individual on-site antennae. Additional SMS transponders are expected to be activated in due course. Eutelsat is able to provide two types of SMS network for open and closed services respectively. The open system is compatible with the public telecommunications network and allows the user to establish digital links with any other business user accessing the system within the SMS coverage area. The closed network offers a customer-specific network either for individual users or for closed user groups. Small antennae (typically 1–2 m) are perfectly adequate and are installed at the customers' premises. The closed network can be established between as few as two separate points in eastern or western Europe or extended to include many thousands.

Land mobile communications (EUTELTRACS)

Eutelsat introduced its land mobile satellite service for Europe in 1991. This service is called EUTELTRACS and uses technology developed by the American company Qualcomm. EUTELTRACS is a two-way message-exchange and position reporting service for mobiles and operates via Eutelsat capacity in Europe, North Africa and the Middle East. Up to 45,000 mobiles can be accommodated on the capacity currently allocated on two Eutelsat satellites, and double that number with a third transponder. The system operational centre is contained in hub equipment in an earth station at Rambouillet near Paris under the control of Eutelsat and France Télécom. Tariffs have been approved for EUTELTRACS and authorization given for ten-year service contracts to be signed with a number of operators for the fixed-site Mobile Communications Terminals (MCTs) needed for the position-indication and messaging facility. Software and terminals are distributed by the joint venture company Alcatel Qualcomm. The EUTELTRACS service operates in the 12 GHz band. The service offered connects a transport operator's administrative centre to the individual vehicles in a fleet. Two separate satellites are used to provide accurate position-finding. MCTs installed in the vehicle cabs can transmit and receive messages while the administrative centre can transmit messages instantly to individual vehicles, groups of vehicles or even an entire fleet of vehicles. The value of EUTELTRACS is particularly applicable to heavy goods vehicle operators. The vehicle location can be seen on a screen at headquarters and any request to change the itinerary or diversion to an additional pick-up point can be transmitted quickly to the vehicle.

Other users that could have an interest in EUTELTRACs include emergency service operators and repair and maintenance services operators. Although EUTELTRACS has initially focused on the road transport industry there are other applications. The average positioning accuracy of vehicle reporting at 300 m and the large number of vehicles that can be accommodated are two factors in its favour.

Appendix

The decibel (dB)

Extremely large ranges of signal parameters are used in electronic communications work. For instance power may be quoted in values as small as microwatts or as large as megawatts. Ranges as large as this also apply to voltage, current, frequency and to a lesser extent temperature.

A BEL (whole unit) is a ratio of two different levels. The BEL unit was named in honour of Alexander Graham Bell famous for his research work leading to the invention of the telephone. However, for most electronic work the BEL as a unit is much too large and consequently a one tenth unit, the decibel (dB) is used.

As an example, an amplifier with signal power output level compared to the power input signal level may have the power gain figure as a ratio. This figure is often quoted in dB. Although initially used as a power ratio the decibel has come to be used for other ratios as well. In terms of power the ratio in decibels is given by:

$$\text{power ratio} = 10 \log_{10} \left(\frac{P_o}{P_i}\right) \text{ dB}$$

Power can be expressed in terms of voltage and current i.e. $P = V^2/R$ or $P = I^2 R$ so that the power ratio can be written as:

$$P_o/P_i = (V_o^2/R_o)/(V_i^2/R_i)$$

It follows that if $R_o = R_i$:

$$P_o/P_i = (V_o/V_i)^2$$

which expressed in decibels gives:

$$\text{power ratio} = 10\log_{10} \left(\frac{V_o}{V_i}\right)^2 = 20 \log_{10} \left(\frac{V_o}{V_i}\right)$$

A similar expression can be deduced for current. It is now common usage to express a voltage, or current, gain in decibels using the expression:

gain in decibels = $20\log_{10}(V_o/V_i)$ or $20\log_{10}(I_o/I_i)$.

The decibel is not a linear unit of measurement. This is primarily because it would be virtually impossible to find a linear unit which was readily available to show very large changes of signal parameters.

It should be noted that the decibel is based on ratio measurement and can have no absolute value unless it is referenced to a specific level. As an example the figure +20 dB simply means that one level is 20 dB greater than another; it does not give an absolute value. However +20 dB above 1 mWatt possesses an absolute value since a reference level is quoted. From the table, at the end of this appendix, it can be seen that +20 dB possesses a power ratio of 100. With respect to 0 dB there will be a positive increase of 100. A negative value for decibels indicates that the value measured is less than the reference i.e. −20 dB indicates a decrease in value by 100 compared to the reference i.e. to 0.01.

A figure quoted to indicate power levels is 3 dB. This corresponds to a doubling of the gain power ratio from 1.0 to 2.0 or a halving of the loss power ratio from 1.0 to 0.5. Thus the −3 dB points of a bandwidth graph are often called the 'half power points'. The gain power ratio is doubled for each +3 dB change and halved for each −3 dB change.

Decibel figures are calculated using logarithmic tables or electronic calculators. It is not always convenient to do this and if an approximate ratio is required reference can be made to a table of decibels.

Once the various signal parameters have been converted to the common ratio − decibels − the total figure for gain or loss in a system may be found by simply adding and/or subtracting the dB ratios. Examples of this technique can be found in the text.

Common Reference Levels Used in Communications Systems

The dBm

The reference level used is the milliwatt (mW). The suffix m signifies that the dB figure is quoted to a reference level of 1 mWatt. Therefore dB values will represent an absolute power level.

The dBW

Reference level; the Watt. The suffix W signifies that the dB figure is quoted as a power ratio to a reference level of 1 Watt.

The dBμV

Reference level; the microvolt. The suffix μV signifies that the dB figure is quoted as a voltage ratio to a reference level of 1 microVolt.

The dBi

Reference level; the mAmp. The suffix i signifies that the dB figure is quoted as a current ratio to a reference level of 1 milliamp.

The dBHz

Reference level; the Hertz. The suffix Hz signifies that the dB figure is quoted as a frequency ratio to a reference level of 1 Hertz.

The dBK

Reference level; the Kelvin. The suffix K signifies that the dB figure is quoted as a temperature ratio to a reference level of 1 degree Kelvin.

Table A

	Decibel ratio			
	Gain (+)		Loss (−)	
dB	Power ratio	Voltage ratio	Power ratio	Voltage ratio
0	1.00	1.00	1.00	1.00
1	1.26	1.12	0.80	0.89
2	1.59	1.26	0.63	0.79
3	2.00	1.41	0.50	0.71
4	2.51	1.59	0.40	0.63
5	3.16	1.78	0.32	0.56
6	3.98	2.00	0.25	0.50
7	5.00	2.24	0.20	0.45
8	6.31	2.51	0.16	0.40
9	7.96	2.82	0.12	3.36
10	10.00	3.16	0.10	0.32
15	31.60	5.62	0.03	0.18
20	100.00	10.00	0.01	0.10
50	10^5	316.20	10^{-5}	3.16×10^{-3}
80	10^8	10^4	10^{-8}	1.00×10^{-4}
100	10^{10}	10^5	10^{-10}	1.00×10^{-5}

Index

Above deck equipment (ADE) 222, 251, 253, 260, 267, 272, 288
Absorption 14
Access control and signalling 228, 289, 298, 317
Adaptive Predictive Coding (APC) 60, 87, 110, 114
Aeronautical fixed telecommunications network (AFTN) 282
ALOHA 15, 55, 81
  Slotted ALOHA 83, 301
  Slot reservation ALOHA 84
AM (amplitude modulation) 89
  double sideband, suppressed carrier (DSBSC) 93
  frequency spectrum 90
  noise 91
  single sideband, suppressed carrier (SSBSC) 92
  threshold 103
AM-PM conversion 30, 169, 195
Amplifiers
  back-off 30, 41, 168
  gallium arsenide GaAs FET 21, 169, 265, 338
  high power (HPA) 36, 165, 192, 214, 252, 265, 344, 348, 373
  klystron 164, 217
  low noise (LNA) 164, 165, 169, 192, 217, 252, 265, 348
  parametric 21, 41, 169
  solid state power (SSPAs) 32, 56, 167, 195, 334, 351
  travelling wave tube (TWT) 41, 56, 164, 167, 195, 212, 214, 217, 334, 348, 351, 354, 373, 379
Amplitude phase keying (APK) 123
Amplitude shift keying (ASK) 123
Analogue modulation
  amplitude modulation, see also AM 89
  frequency modulation, see also FM 93
Analogue transmission 49
Antennae 4, 169–77, 197
  alignment 278, 282

aperture 18
aperture efficiency 18, 175
array 198, 212, 316, 335
beam, see under Global, Hemi, Spot or Zone
beam waveguide 174
boresight 17, 377
calculations 42
Cassegrain 18, 170, 217
cup dipole array 197, 212
earth station 169–77
efficiency 18, 172
gain 7, 34, 174, 197, 202, 253, 272, 288, 309, 326, 334
horn 18, 170, 197, 338, 344
monitoring and control 177
noise 26, 41
offset fed parabolic 170, 327, 328, 335, 337
polarization 175, 253, 327, 335, 338, 341, 376
prime focus fed parabolic 170
reflector 18, 171, 197
rewind 260
satellite 197, 326, 327, 328, 333, 335–8
sidelobe level 34, 172, 253
tracking 162, 177, 188, 254, 274, 277
Apogee 10
Assignment channel 112, 289
Assignment message 112, 289, 236, 248, 299
Attenuation 12, 14, 16, 21, 27, 190
Attitude control 179, 181, 189, 211
Autocorrelation function 159
Automatic frequency control (AFC) 56, 165, 217
Automatic gain control (AGC) 57
Automatic Retransmission Request (ARQ) 81, 158, 298
Azimuth 172, 174, 189, 217, 254, 255, 260, 272, 291, 310, 315, 357

Back-off 30, 41, 167, 202

Index

Bandwidth   15, 16, 19, 29, 34, 40, 46, 49, 51, 53, 57, 76, 88, 97, 108, 121, 123, 134, 136, 164, 189, 196, 267, 286, 288, 295, 315, 325, 338, 344, 351, 354, 366, 373
Baseband   4, 15, 17, 49, 53, 61, 67, 73, 80, 86, 90, 100, 106, 109, 119, 127, 163, 217, 229, 252, 267, 274, 286, 357
Batteries   6, 186, 212, 278
Bearer channel   111
Below deck equipment (BDE)   236, 251, 265, 270
Bessel function   97
Bipolar phase shift keying (BPSK)   64, 81, 125, 134, 137, 143, 228, 285, 298
Bipropellant   181
Bit Error Rate (BER)   39, 61, 87, 125, 196, 367
  probability   131, 137, 143, 196
British Telecommunications (BT)   216

Carrier-to-noise ratio (C/N)   17, 36, 41, 57, 61, 165, 168
Carrier and bit timing recovery (CBT)   75
Channel   4
Channel amplifier   194
Chips   77
Coast earth station (CES)   28, 36, 64, 76, 87, 170, 205, 217, 237
Codec   58
Code
  block   138, 153
  Bose-Chadhuri-Hocquenghem (BCH)   143
  convolutional   81, 87, 145, 146, 153
  cyclic   141
  distance   138, 144
  error correction   5, 58, 136, 138, 143, 149, 151, 154, 157, 165, 286, 298, 306, 315
  error detection   5, 136, 138, 149, 237, 286
  extended Golay   139
  Golay   112, 139
  Gray   260
  Hamming   138
  orthogonal   77
  Reed Solomon   144
Code division multiple access (CDMA), see Multiple Access
Coding   136–61, 285
  gain   140
Common Signalling Channel (CSC)   217, 289, 317
Communications link   3
Communications Satellite Corporation (COMSAT)   211, 215, 222, 323
Companding   56, 103, 106, 230
Compression   56
Constraint length   81, 145

Convolutional code   81, 87, 145, 146, 153
  tree diagram   147
  trellis diagram   147
  polynomial representation   146

Data
  rate   6, 50, 87, 161, 221, 228, 236, 365, 386
  signal   50, 87, 110, 136, 162, 270
  transmission   87, 113, 124, 297, 366, 386
Decoder   138, 140, 149, 165
  Viterbi   87, 157
Decoding
  block codes   149
  convolutional codes   149
  sequential   151
  Viterbi   87, 150
De-emphasis   103, 164
Demand-Assigned Multiple Access (DAMA) see Multiple Access
Demodulation   4, 86, 91, 98, 101, 106, 252, 265, 277
Depolarization   176
Differential phase shift keying (DPSK)   127
Digital circuit multiplication equipment (DCME)   63, 110, 140, 161, 363
Digital modulation   15, 74, 123, 230, 285
Digital speech interpolation (DSI)   63, 110
Digital transmission   49, 53, 59, 65, 86, 108, 163
  errors   58, 138
Digitally non-interpolated (DNI)   110
Diplexer   192, 272
Doppler shift   217
Down-converter   24, 164, 192, 194, 265, 338, 348

Early Bird   323
Earth station   8, 162
  antennae   169–77, 251, 272
  CES, see Coast earth station
  GES, see Ground earth station
  G/T   27, 169, 170, 253, 285, 309, 316
  LES, see Land earth station
  low noise amplifier (LNA)   164, 165, 169, 217, 252, 265
  MES, see Mobile earth station
  operating FDM/FM/FDMA   162, 168
  operating TDM/QPSK/TDMA   165
Echo effect   5, 270
Eclipse   6, 185, 212
ECS   370, 372, 381
Effective isotropic radiated power (EIRP)   17, 34, 40, 48, 167, 170, 202, 228, 253, 274, 285, 288, 309, 316, 354, 380

Elevation angle   11, 14, 16, 26, 28, 34, 40, 48, 59, 170, 174, 189, 217, 254, 255, 260, 272, 291, 309, 315, 357
Encoder   108, 137, 141, 154, 286, 303
  Adaptive differential pulse code modulation (ADPCM)   113
  convolutional   81, 145, 157
  differential   64
  forward error correction (FEC)   66
  Quantizer   109
  Waveform   109
Encoding
  adaptive   109
  delta modulation   109
  pulse code modulation (PCM)   109
    differential PCM   109
  vocoder   109
Energy dispersal   41, 164
European Conference of Postal and Telecommunications Administrations (CEPT)   50, 63, 370
European Space Agency (ESA)   211, 370
Eutelsat   69, 125, 197, 321
  ground segment   380
  organization   370
  space segment   372
Eutelsat services   382
  business   386
  Eurovision and Euroradio   385
  EUTELTRACS   387
  Satellite news gathering (SNG)   386
  telephony   386
  TV and radio   382
Eutelsat I   372
Eutelsat II   373

Facsimile   4, 38, 60, 87, 110, 220, 230, 251, 291, 314, 365
FDM, see Multiplexing
FDMA, see Multiple Access
Federal Communication commission (FCC)   35
Fixed satellite service (FSS)   32, 321
FM (frequency modulation)   93, 229, 274
  Carson's rule for bandwidth   98
  deviation ratio   94
  frequency deviation   94, 229
  frequency spectrum   97
  frequency swing   94
  modulation index   94
  noise   99
  over-deviation   68–72
  phasor representation   95
  threshold   103
FM/TV   67, 89
  applications   69, 367
  interference   70
Forward error correction (FEC)   63, 74, 157, 285, 298, 315
Frequency
  division multiplexing (FDM) see Multiplexing
  modulation, see FM
  re-use   79, 197, 327, 333, 338, 349
  shift keying (FSK)   123

Global beam   14, 197, 202, 292, 326, 338
Global Maritime Distress and Safety System (GMDSS)   6, 228, 285, 308, 315
Global network services (GNS)   220
Ground earth station (GES)   205, 265
Ground segment   6, 214, 288, 298

High power amplifier (HPA)   36, 165, 192, 214, 252, 265
Hemi beam   43–6, 196–202, 327, 333, 338, 344, 350
Hydrazine   181

Improved multi-band excitation (IMBE)   119
Inmarsat   6, 125, 170, 186
  access control and signalling   228, 289, 298, 317
  ground segment   214, 288, 298, 315
  organization   205
  satellite status   210
  space segment   208, 286, 297, 315
Inmarsat-A   56, 75, 77, 87, 143, 205, 216
Inmarsat-B   28, 59, 64, 65, 75, 87, 110, 114, 157, 160, 205, 285–94
Inmarsat-C   9, 46, 81, 84, 155, 205, 215, 295–313
Inmarsat-M   59, 110, 119, 160, 205, 314–18
Intelsat   28, 50, 56, 59, 71, 125, 140, 144, 161, 170, 321, 323
  earth station standards (IESS)   357
  ground network   355–61
  organization   323
  space segment   325–55
Intelsat services
  Intelnet   76, 365
  Intelsat Business Service (IBS)   87, 157, 333, 351, 356, 364
  Intermediate data rate (IDR)   63, 74, 87, 109, 157, 161, 363
  TV/FM   70, 367
  Vista   56, 366
Intelsat 1   5, 15, 323, 325
Intelsat II   325
Intelsat III   326
Intelsat IV   326
Intelsat IV-A   327

Index

Intelsat V 32, 69, 327
Intelsat V-A 333
Intelsat VI 32, 69, 80, 180, 195, 197, 333
Intelsat VII 186, 348
Intelsat K 344, 362
Interference 32, 46
  intersymbol 130
Integrated services digital network
  (ISDN) 288, 295, 324
Interleaving 81, 140, 153
Intermodulation 15, 46
  distortion 15, 28, 30, 168, 214
International Maritime Organization
  (IMO) 285, 315
International Radio Consultative Committee
  (CCIR) 35, 40, 57, 61, 253, 324
International Telecommunication Union
  (ITU) 32, 181, 324, 333, 371
International Telegraph and Telephone
  Consultative Committee (CCITT) 50,
  74, 86, 87, 109, 161, 288, 324

Land earth station (LES) 36, 205, 215, 285,
  295, 298
  LES services 219
    facsimile 220
    high speed data 221
    telephony 219
    telex 220
Leased transponder 67, 365, 381
Link budget 36
  Eutelsat-II 48
  Inmarsat-B 39
  Inmarsat-C 46, 81, 84
  Intelsat-V 43
Link parameters 16–52
  system power levels 17
  system noise 19
Low noise amplifier (LNA) 164, 165, 169,
  192, 252
Low rate encoding (LRE) 63, 108, 109, 363

Marecs 211, 297
Marisat 211, 222
Maritime communications sub-system
  (MCS) 211, 297, 331, 368
Microwave integrated circuit (MIC) 338
Mobile earth station (MES) 8, 36, 81, 170,
  205, 285, 295, 308, 314
Mobile satellite service (MSS) 32
Modem 86, 127
Modulation 4, 14, 86, 252, 265
  amplitude (AM) 49, 89
  frequency (FM) 15, 49, 55, 93, 229, 274
  pulse code (PCM) 49, 108
Monopropellant 181

Monopulse tracking 177, 189
Multipath 153
Multiple access 15, 17, 53
  ALOHA 55, 81, 82
  CDMA 54, 76
  DAMA 58
  Direct sequence CDMA 77
  Frequency division multiple access
    (FDMA) 41, 53, 55, 161, 167, 189, 334
  Frequency-hopping CDMA 79
  Packet access 55, 81
  Random access 55, 82
  SDMA 79
  TDMA 54, 64, 74, 161, 165, 167, 189, 228,
    237, 248, 333, 338, 340, 364, 380
Multiple channels per carrier (MCPC)
  MCPC/FM/FDMA 61, 164, 325
  MCPC/PSK/FDMA 63
Multiplexed analogue components
  (MAC) 67, 89
Multiplexer 252
  input 193–4, 338
  output 192–5, 344, 379
Multiplexing 49, 64, 66
  Frequency division multiplex (FDM) 49,
    61, 101, 105, 162
  Time division multiplex (TDM) 36, 50, 55,
    63, 75, 81, 143, 162, 228, 248, 267, 270,
    285, 299

NASA 323
Network Co-ordination Centre (NCS) 36,
  75, 228, 237, 246, 267, 288, 293, 315, 317
Noise 3, 19–27
  antenna 26, 41
  bandwidth 59, 63, 134
  figure 22, 25, 334, 338
  power 20, 32, 57, 63, 91, 100, 105, 134
  temperature 20, 22, 25, 28, 169, 199
  white 20, 39, 77, 104, 131, 136
NTSC 67, 89

Offset QPSK (O-QPSK) 15, 66, 124, 129,
  285, 315
Orbit
  elliptically inclined 8
  geostationary 6, 8, 255
  geosynchronous 5, 8, 208
  polar 8
Orbital parameters 6
Orbital Test Satellite (OTS) 370, 373

Packet access 55, 81
Packet switched public data network
  (PSPDN) 87
PAL 67, 74, 89

Phase locked loop (PLL)   99, 127, 270, 274
Phase shift keying (PSK)   15, 30, 41, 51, 55, 57, 63, 123, 125, 270, 274, 356
Polarization   16, 34, 48, 79, 175, 191, 197, 253
  circular   34, 79, 175, 253, 272, 309, 327, 335, 338, 341
  cross polarization (XPD)   34, 172, 175, 197
  linear   34, 175, 328, 338, 343, 345, 376
  orthogonal   34, 79, 191, 327, 376
Polynomial   31, 142, 146, 157, 160
Power flux density (PFD)   17, 34, 40, 326, 380
Pre-emphasis   56, 103, 164
Pseudo-random noise (PN)   77, 158
Psophometric weighting   57, 61
Public switched telephone network (PSTN)   87, 108, 162, 296, 314, 380
Pulse amplitude modulation (PAM)   109
Pulse code modulation (PCM)   49, 67, 77

Quadrature phase shift keying (QPSK)   58, 61, 74, 124, 129, 134, 137, 161, 165
Quantization   109, 115, 149, 286
  noise   109, 122
Quasi-elliptic filters   338

Random access   55, 82
Reaction Control System (RCS)   180
Receive gain/system noise temperature ratio (G/T)   27, 48, 52, 169, 191, 253, 285, 288, 309, 334, 380
Receiver   4, 17, 21, 41, 63, 68, 77, 80, 104, 109, 124, 131, 153, 155, 158, 162, 169, 192, 252, 265, 338, 348
Redundancy   108, 136, 158, 165, 177, 192, 195, 202, 214, 286, 338, 344, 348, 360, 379
Repeater   6, 212, 260, 333, 339, 349, 351, 379
  regenerative   191, 196
  transparent   189, 191
  channel amplifiers   194
  down-converter   194
  High-power amplifier (HPA)   194
  input filter   193
  input multiplexer   194
  low-noise amplifier (LNA)   193
  output multiplexer   195

Safety of Life at Sea Convention (SOLAS)   288, 308
Satellite
  antennae   197, 326-8, 333, 335-8
  attitude control   179, 181, 189, 211
  body-stabilized   181, 327, 348, 374
  cup dipole array   197, 212
  EIRP   17, 40, 202, 354, 380

footprint   14, 34, 210
gain   197
G/T   191, 199, 334, 380
link budget   36
link parameters   16-52
mobile frequency bands   319
primary power   185
propulsion   180
sensors   184, 188, 211, 326
spin-stabilized   181, 326, 333
structure   179
support sub-systems   179
switching   196
thermal control   187
three-axis stabilized, see body-stabilized
transmission path loss   14
transmission path range variation   12
transponder   5, 53, 191, 213, 286, 325, 338, 372
TTC&R   187
Satellite Control Centre (SCC)   215, 315, 356, 380
Satellite master antenna TV (SMATV)   382
Satellite news gathering (SNG)   367, 381, 386
Scrambler   66, 160
SECAM   67, 74, 89
Sensors   184, 188, 211, 274, 308, 326
Ship earth station (SES)   28, 36, 64, 75, 87, 170, 191, 217, 222, 237, 250-82, 285, 288
Signal-to-noise ratio (S/N)   4, 17, 57, 61, 70, 101, 103, 124, 136
Single channel per carrier (SCPC)   55, 87, 103, 108, 167, 214, 285, 315, 339, 366
  SCPC/FM/FDMA   56, 103, 164, 214
  SCPC/PSK/FDMA   57, 165, 214
Single channel per carrier (SCPT)   74
Societé Europeanne des Satellites (SES)   108
Solar cell   6, 185
Solid state power amplifiers (SSPAs)   32, 56, 167, 195, 334, 351
Space domain multiple access (SDMA), *see also* Multiple Access   79
  SDMA/CDMA   80
  SDMA/FDMA   79
  SDMA/SS/FDMA   80
  SDMA/TDMA   80
  SDMA/SS/TDMA   80, 196, 333, 340
Space segment   6, 208, 286, 297
SPADE   58
Spot beam   43, 45, 196-202, 286, 315, 326, 337, 350, 373
Spread-spectrum multiple access (SSMA), *see* Code division multiple access
Step tracking   177
Switching matrix   330
  dynamic (MSM)   339

static (SSM)   339
Syllabic compandor   106, 230

Telemetry, tracking and control (TT&C)   6, 187, 189, 215, 321
Telemetry, tracking, command and monitoring (TTC&M)   355
Telemetry, command and ranging (TCR)   380
Telephony   4, 15, 38, 50, 53, 59, 61, 92, 102, 105, 110, 114, 123, 216, 219, 222, 230, 236, 246, 267, 271, 274, 282, 289, 291, 293, 318, 326, 339, 362, 364, 366, 375, 381, 386
Television   14, 15, 67, 362, 367, 381
  audio sub-carrier   72
  channels   325, 326, 327, 330, 333, 344, 382
  chrominance   67
  digital   74, 87, 108
  digital SIS   68
  emphasis/weighting   68
  frequency modulation   67
  full transponder   69
  half transponder   69
  interference   70
  luminance   67
  Multiplexed analogue components (MAC)   67, 89
  NTSC   67, 89
  occasional use   367
  over-deviation   71
  PAL   67, 74, 89
  pre-emphasis   67
  receive-only (TVRO)   170, 372
  SECAM   67, 74, 89
  signal   4, 67, 71, 87, 89
  standards   371, 381, 382
  sound   68
Telex   4, 38, 60, 63, 75, 108, 110, 136, 157, 216, 218, 220, 222, 228, 230, 236, 248, 253, 267, 271, 282, 286, 289, 293, 296, 365, 382
Telstar   5
Threshold   41
  FM   41
  PSK   41
Time Assignment Speech Interpolation (TASI)   110
Time division multiple access (TDMA), see Multiple Access
  TDMA/FDMA   64, 76
Time division multiplex (TDM), see Multiplexing

TDM/FDMA   63
TDM/TDMA   76, 81, 165, 248, 288, 317
Tracking
  conical scan   177
  monopulse   177
  programmed   177
  step   177
Transducer   3
Transmission delay   5
Transmitter   4, 17, 41, 77, 155, 158, 162, 170, 202, 217, 251, 265
Transponder   5, 53, 67, 191, 213, 286, 325, 338, 372
  bandwidth   29, 40, 56–59, 61, 64, 67, 325, 338, 351, 354, 366, 373
Transponders for unrestricted use (TUU)   362, 368
Travelling wave tube amplifier (TWTA)   41, 56, 164, 167, 195, 212, 214, 217, 334, 348, 351, 354, 373, 379
TV receive-only (TVRO)   170, 372

Unique word (UW)   75

Very small aperture terminal (VSAT)   6, 76, 170, 321, 365, 381, 386
Video   1, 53, 67, 70, 76, 86, 108, 162, 339, 385
  bandwidth   88
  conferencing   364, 381, 386
  display unit (VDU)   312
Videotel   220
Viterbi code   87, 150, 286
Vocoder   109, 119
Voice   36, 55, 63, 76, 86, 136, 162, 170, 186, 218, 295, 314, 359, 363, 365
  activation   55, 57, 60, 164, 285, 366
  circuits   15, 50, 53, 57, 59, 66, 110, 157, 165, 199, 270, 325
  coding   114
  signals   49, 106, 108
  telephony   38, 274
Voltage controlled oscillator (VCO)   99, 270, 277

World Administrative Radio Conference (WARC92)   319
White noise   20, 39, 77, 104, 131, 136

XPD, see Polarization

Zone beam   43–46, 196–202, 327, 333, 338, 344, 350